Vol. 32. **Determination of Organic Compounds: Methods and Procedures.** By Frederick T. Weiss

Vol. 33. **Masking and Demasking of Chemical Reactions.** By D. D. Perrin

Vol. 34. **Neutron Activation Analysis.** By D. De Soete, R. Gijbels, and J. Hoste

Vol. 35. **Laser Raman Spectroscopy.** By Marvin C. Tobin

Vol. 36. **Emission Spectrochemical Analysis.** By Morris Slavin

Vol. 37. **Analytical Chemistry of Phosphorus Compounds.** Edited by M. Halmann

Vol. 38. **Luminescence Spectrometry in Analytical Chemistry.** By J. D. Winefordner, S. G. Schulman and T. C. O'Haver

Vol. 39. **Activation Analysis with Neutron Generators.** By Sam S. Nargolwalla and Edwin P. Przybylowicz

Vol. 40. **Determination of Gaseous Elements in Metals.** Edited by Lynn L. Lewis, Laben M. Melnick, and Ben D. Holt

Vol. 41. **Analysis of Silocones,** Edited by A. Lee Smith

Vol. 42. **Foundations of Ultracentrifugal Analysis.** By H. Fujita

Vol. 43. **Chemical Infrared Fourier Transform Spectroscopy.** By Peter R. Griffiths

Vol. 44. **Microscale Manipulations in Chemistry.** By T. S. Ma and V. Horak

Vol. 45. **Thermometric Titrations.** By J. Barthel

Vol. 46. **Trace Analysis: Spectroscopic Methods for Elements.** Edited by J. D. Winefordner

Vol. 47. **Contamination Control in Trace Element Analysis.** By Morris Zief and James W. Mitchell

Vol. 48. **Analytical Applications of NMR.** By D. E. Leyden and R. H. Cox

Vol. 49. **Measurement of Dissolved Oxygen.** By Michael L. Hitchman

Vol. 50. **Analytical Laser Spectroscopy.** Edited by Nicolo Omenetto

Vol. 51. **Trace Element Analysis of Geological Materials.** By Roger D. Reeves and Robert R. Brooks

Vol. 52. **Chemical Analysis by Microwave Rotational Spectroscopy.** By Ravi Varma and Lawrence W. Hrubesh

Vol. 53. **Information Theory As Applied to Chemical Analysis.** By Karel Eckschlager and Vladimir Štěpánek

Vol. 54. **Applied Infrared Spectroscopy: Fundamentals, Techniques, and Analytical Problem-solving.** By A. Lee Smith

Vol. 55. **Archaeological Chemistry.** By Zvi Goffer

Vol. 56. **Immobilized Enzymes in Analytical and Clinical Chemistry.** By P. W. Carr and L. D. Bowers

Vol. 57. **Photoacoustics and Photoacoustic Spectroscopy.** By Allan Rosencwaig

Vol. 58. **Analysis of Pesticide Residues.** Edited by H. Anson Moye

Vol. 59. **Affinity Chromatography.** By William H. Scouten

Vol. 60. **Quality Control in Analytical Chemistry.** By G. Kateman and F. W. Pijpers

Vol. 61. **Direct Characterization of Fineparticles.** By Brian H. Kaye

Vol. 62. **Flow Injection Analysis.** By J. Ruzicka and E. H. Hansen

Vol. 63. **Applied Electron Spectroscopy for Chemical Analysis.** Edited by Hassan Windawi and Floyd Ho

Vol. 64. **Analytical Aspects of Environmental Chemistry.** Edited by David F. S. Natusch and Philip K. Hopke

Vol. 65. **The Interpretation of Analytical Chemical Data by the Use of Cluster Analysis.** By D. Luc Massart and Leonard Kaufman

Vol. 66. **Solid Phase Biochemistry: Analytical and Synthetic Aspects.** Edited by William H. Scouten

(*continued on back*)

Chemometrics

CHEMICAL ANALYSIS

A SERIES OF MONOGRAPHS ON ANALYTICAL CHEMISTRY AND ITS APPLICATIONS

VOLUME 82

A WILEY-INTERSCIENCE PUBLICATION

JOHN WILEY & SONS

New York / Chichester / Brisbane / Toronto / Singapore

Chemometrics

MUHAMMAD A. SHARAF
Department of Chemistry
University of Petroleum and Minerals
Dhahran, Saudi Arabia

DEBORAH L. ILLMAN
Center for Process Analytical Chemistry
University of Washington
Seattle, Washington

BRUCE R. KOWALSKI
Laboratory for Chemometrics
Department of Chemistry
University of Washington
Seattle, Washington

A WILEY-INTERSCIENCE PUBLICATION

JOHN WILEY & SONS

New York / Chichester / Brisbane / Toronto / Singapore

Library of Congress Cataloging in Publication Data:

Sharaf, Muhammad A.
Chemometrics.

(Chemical analysis, ISSN 0069-2883; v. 82)

"A Wiley-Interscience publication."
Includes bibliographies and index.
1. Chemistry—Mathematics. 2. Chemistry—Statistical methods. I. Illman, Deborah L. II. Kowalski, Bruce R., 1942– . III. Title. IV. Series.

QD39.3.M3S47 1986 543′.001′5195 85-22619
ISBN 0-471-83106-9

Printed in the United States of America

10 9 8 7 6 5 4 3 2 1

PREFACE

The role of the analytical chemist has changed profoundly over the past few decades. The advent of intelligent instruments and laboratory automation has catalyzed the transformation of analytical chemistry into an information science. In addition to mastering chemistry, analytical chemists must now acquire skills in mathematics, statistics, and computer science in order to function efficiently. During this time, chemometrics has evolved as the field in which mathematical and statistical techniques are used to extract valuable, but often hidden, information from measurements.

The chapters in this book represent a logical and systematic introduction to chemometrics as incorporated into analytical chemistry as well as other areas of experimental chemistry. The development and application of mathematical methods to aid in moving efficiently from experimental data to useful information to an advanced level of understanding of chemical processes or systems is a reoccurring theme throughout this book.

A single volume of average size cannot give a complete description of the seven major concepts treated in this book. Our aim is to highlight each area in order to introduce the field of chemometrics to advanced students of chemistry. The treatments are focused on chemistry with mathematical derivations avoided as much as possible. The material in this book has evolved over the years from a course in chemometrics taught at the University of Washington since 1974.

MAS wishes to acknowledge the support of The University of Petroleum and Minerals, Dhahran, Saudi Arabia, during the writing of this book. DLI and BRK acknowledge the Center for Process Analytical Chemistry and its staff who provided invaluable aid in its preparation. Finally, BRK and the members of the Laboratory for Chemometrics wish to thank the following organizations who fueled the fires as a new science was born: Office of Naval Research, Department of Energy, National Science Foundation, The 3M Company, Lilly Research Laboratories, and the Murdock Trust Fund.

MUHAMMAD A. SHARAF
DEBORAH L. ILLMAN
BRUCE R. KOWALSKI

Dhahran, Saudi Arabia
Seattle, Washington
January 1986

CONTENTS

1. SAMPLING THEORY **1**

Probability Distributions 1
Standard Normal Variate 6
Populations and Samples 7
Student's t-Distribution 10
Binomial Distribution 12
Bias 13
Estimation of μ 14
Estimation of σ^2 14
Analytical Applications 15
Minimizing the Variances of Sampling and Analysis 17
References 21
Suggested Readings 21

2. FUNDAMENTALS OF EXPERIMENTAL DESIGN AND OPTIMIZATION **23**

Assessment of Performance 24
Comparative Experiments 27
Paired Observations 34
Randomized Blocks 36
Latin Squares 41
Factorial Designs 44
ANOVA for Linear Models 50
Response Surfaces 54
References 63
Suggested Readings 63

3. SIGNAL DETECTION AND MANIPULATION **65**

Signal Detection 68
Point Estimation of the Detection Limit 76

Point Estimation of the Detection Limit by t-Tests 79
Signal-to-noise ratio 80
t-test based on the difference between individual measurements $(x_A - x_B)$ 82
The Wilcoxon Test 84
Precision at the Detection Limit 85
Increasing the S/N Ratio 86
Optimization 86
Signal averaging 86
Boxcar integration 87
Signal filtering and modulation 88
Multiplex spectroscopy 89
Hadamard transform spectroscopy 91
Fourier transform spectroscopy 92
Decoding $f(t)$ [computing $F(\nu)$] 94
Fast Fourier transform 95
Signal Manipulation 96
Curve fitting 97
Curve fitting of nonlinear functions 97
Estimating peak parameters 98
Estimating the area under the peak 100
Smoothing of data 102
Boxcar averaging 102
Moving window averaging 103
Least-squares polynomial smoothing 104
Fourier transform smoothing 108
Differentiation of the signal 109
References 112
Suggested Readings 115

4. CALIBRATION AND CHEMICAL ANALYSIS 119

Comparison with Standards 120
Constructing a Calibration Curve 120
Response function 121
Linear calibration 122
Examination of the Residuals 125

Utilizing the calibration curve for chemical analysis 126
Constructing a calibration curve with heteroscedastic data 127
Estimating the detection limit from a linear calibration curve 128
Intersection of two regression lines 129
Linear model when both variables are subject to error 131
Nonlinear calibration 132
Effects of the Sample's Matrix 132
Standard addition method (SAM) 133
Error Propagation in Calibration and Analysis 134
Multicomponent Analysis 135
Generalized Standard Addition Method (GSAM) 139
Experimental Designs in GSAM 142
Total difference calculations 142
Incremental difference calculations 144
References 145
Suggested Readings 147

5. RESOLUTION OF ANALYTICAL SIGNALS 149

Determining the Complexity of Signals 150
Visual inspection 150
Differentiation of signals 150
Factor analysis 155
Geometric approach to factor analysis 155
Algebraic approach to factor analysis 158
Resolving Composite Signals 159
Deconvolution of overlapping signals 159
Resolving signals by mathematical modeling (curve fitting) 159
Taylor series linearization 160
Grid search 163
Method of steepest descent 164
Newton method 164
Resolving signals using multiple regression and optimization techniques 166

Method of rank annihilation 168
Biller–Biemann technique 169
Resolution using eigenvectors scores space 170
Other resolution techniques 171
References 172
Suggested Readings 175

6. EXPLORATORY DATA ANALYSIS **179**

Multivariate Leverage 180
Category Versus Continuous Property Data 182
Pattern Recognition: The Approach 183
Preprocessing Techniques 188
Missing data 188
Redundant/constant variables 189
Translation 189
Normalization 190
Scaling 191
Autoscaling 193
Feature weighting 195
Rotation 197
Eigenvector rotation 198
Varimax rotation 206
Factor analysis 214
Nonlinear factor analysis 214
Display Techniques 216
Introduction 216
Linear methods 216
Nonlinear methods 217
Unsupervised Learning 219
Supervised Learning 228
Introduction 228
Linear learning machine 229
K-nearest neighbor method 234
Feature selection in classification 239
SIMCA 242
Cross-validation 254
Bayes classification rule 256

Pattern Recognition Analysis in Practice: Classification of Archeological Artifacts on the Basis of Trace Element Data 257
Partial Least-Squares Path Modeling 281
References 292
Suggested Readings 294

7. AN INTRODUCTION TO CONTROL AND OPTIMIZATION **297**

Single Input/Single Output: Optimization 298
Single Input/Single Output: Control 301
Multiple Input/Single Output: Optimization 305
Multiple Input/Multiple Output Systems 310
References 311

LIST OF TABLES AND THEIR ORIGINS **313**

INDEX **327**

Chemometrics

CHAPTER

1

SAMPLING THEORY

"Very well," said Henchard quickly, "please yourself. But I tell you, young man, if this holds good for the bulk, as it has done for the sample, you have saved my credit, stranger though you be. What shall I pay you for this knowledge."

THOMAS HARDY—*The Mayor of Casterbridge*

PROBABILITY DISTRIBUTIONS

Probability is a character associated with an event indicating its tendency to take place. To speak of probabilities one needs to define an *event-space* that contains all possible (or known) outcomes of a certain process. If there are n possible outcomes associated with a certain process and an event, A, occurs in m of these outcomes, then the probability of obtaining event A, $P(A)$, is defined as

$$P(A) = \frac{m}{n}. \tag{1.1}$$

This definition has the convenience of expressing probabilities as real numbers between 0 and 1, inclusive. Probabilities calculated as defined above are also known as *relative* or *objective probabilities*.

In many situations it is difficult, if not impossible, to enumerate all possible or known events. For example, it is hard to define an event-space to adequately calculate the probability that oil prices will change or that the University of Washington Huskies will win the Rose Bowl. The probabilities of these and similar events can be estimated from experience and trends. Such probabilities are known as *subjective probabilities*. Our discussion will be restricted to objective probabilities.

A plot of events versus their probabilities constitutes a probability

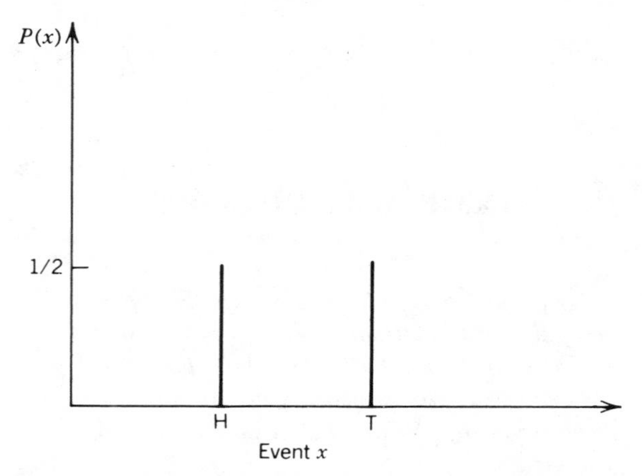

Figure 1.1 Probability distribution of tossing a fair coin.

distribution. The probability distributions of tossing a fair coin and rolling a fair die are shown in Figures 1.1 and 1.2, respectively. In each of these examples all events have the same probability. This is not generally the case. Consider the probability distribution of rolling a pair of dice with the outcome being taken as the sum of the digits on each die. The probability distribution of such a process is shown in Figure 1.3. For

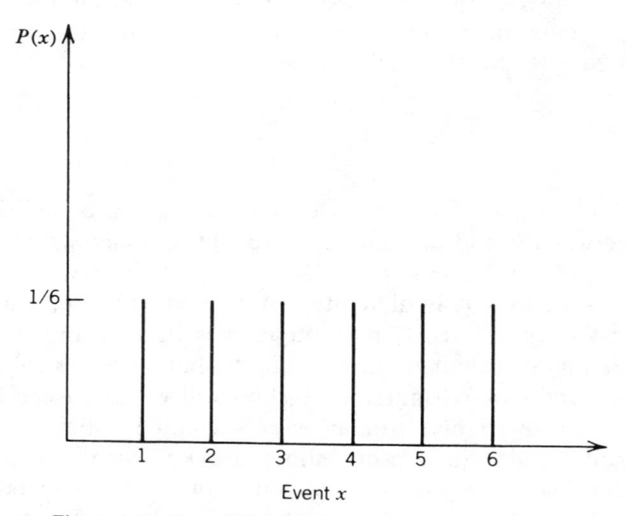

Figure 1.2 Probability distribution of rolling a fair die.

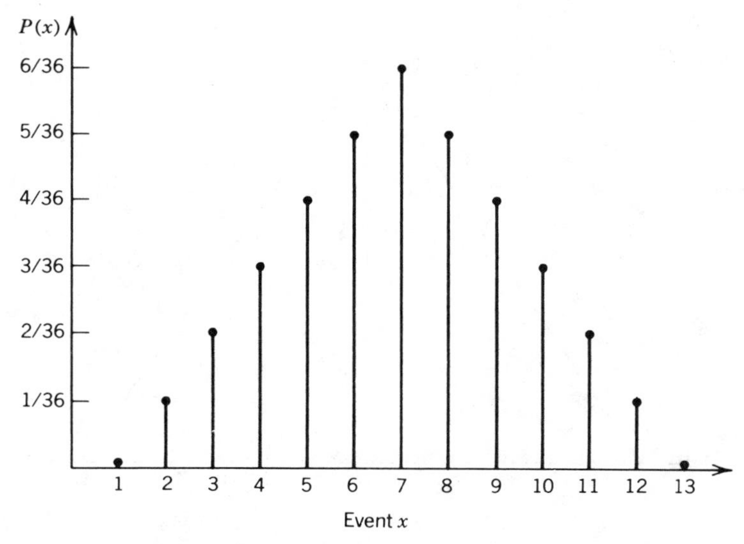

Figure 1.3 Probability distribution of rolling a pair of dice.

convenience, the event-space may be considered continuous. If it is assumed that all values between 1 and 13 are possible, and that extrapolation is valid, Figure 1.3 can be redrawn as shown in Figure 1.4.

Assume that the heights of the students at a certain school are being measured. The measurements may range from 150 to 200 cm. Even though our measurement may be precise only to 0.5 or 1.0 cm, we can assume that our event-space (the height) is continuous with any value between 150 and 200 cm possible. The probability distribution of such an event may look like that shown in Figure 1.5.

Probability distributions can assume any shape depending on the event-space under consideration. Distributions that show variations in probability (usually maximizing at a particular event) are of great importance in chemical analysis. Analysts rarely deal with probability distributions of the type shown in Figures 1.1 and 1.2. For convenience, a probability distribution is expressed in terms of a *probability density function*. Generally, if

$$P(x_a < x < x_b) = \frac{\int_{x_a}^{x_b} f(x)\,dx}{\int_{-\infty}^{+\infty} f(x)\,dx}, \tag{1.2}$$

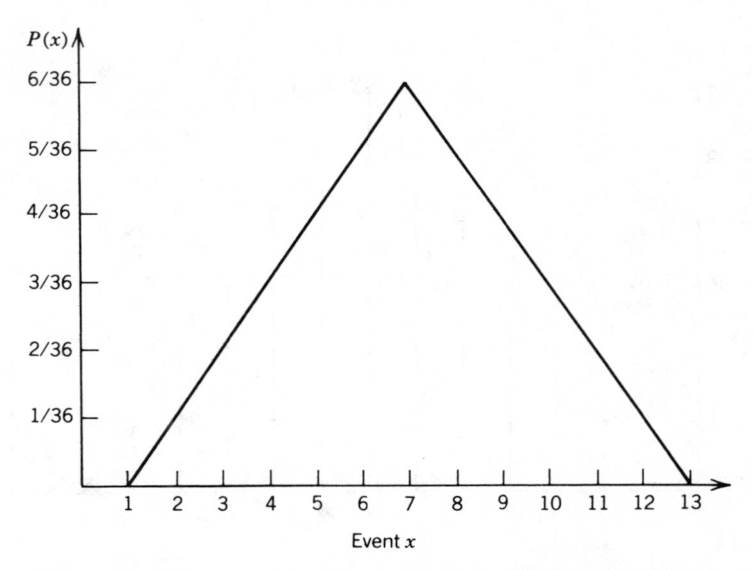

Figure 1.4 Figure 1.3 if the event-space is considered continuous.

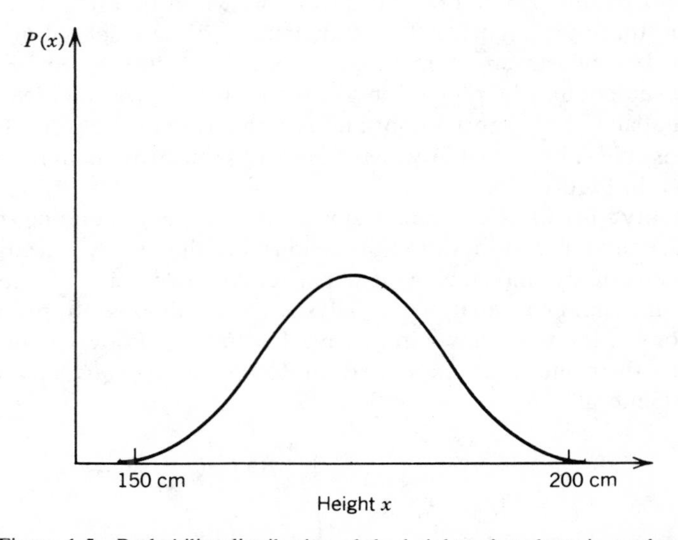

Figure 1.5 Probability distribution of the heights of students in a school.

then $f(x)$ is called the probability density function for the variable x. If the probability distribution shown in Figure 1.5 is expressed in terms of a probability density function, $f(x)$, then the probability of finding a student whose height is between 185 and 190 cm is the ratio of the shaded area to the total area under the curve in Figure 1.6.

The most commonly studied probability distribution is the normal (also called the Gaussian) distribution (1). For any normally distributed variable, x, the probability density function, $f(x)$, is given by

$$f(x) = \frac{1}{\sqrt{2\pi\sigma^2}} \exp\left[-\frac{(x-\mu)^2}{2\sigma^2}\right], \qquad -\infty < x < \infty. \tag{1.3}$$

Thus the normal distribution is defined only by two parameters. The first is the mean, μ, and the second is the variance, σ^2. These parameters are given by

$$\mu = \int_{-\infty}^{\infty} x f(x)\, dx \quad \text{or} \quad \mu = \frac{\sum_{i=1}^{N} x_i}{N} \tag{1.4}$$

and

$$\sigma^2 = \int_{-\infty}^{\infty} (x-\mu)^2 f(x)\, dx \quad \text{or} \quad \sigma^2 = \frac{1}{N} \sum_{i=1}^{N} (x_i - \mu)^2, \tag{1.5}$$

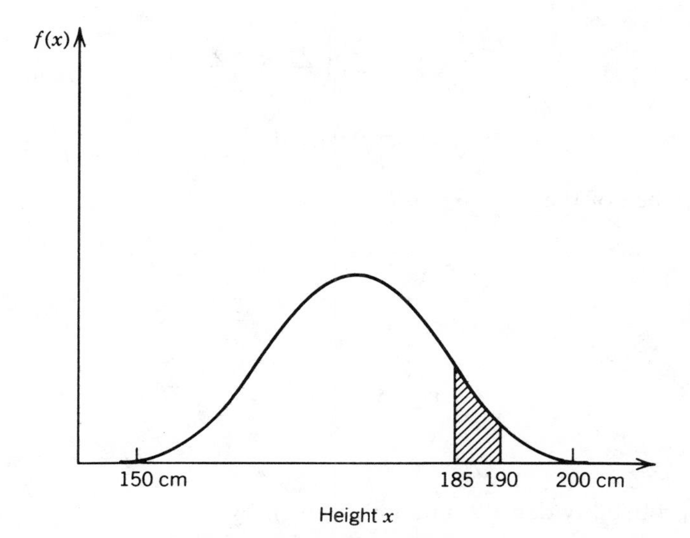

Figure 1.6 Probability density function associated with Figure 1.5. Probability of finding a student whose height is between 185 and 190 cm is the ratio of the shaded area to the total area.

where N is the total number of elements under consideration. A normal distribution of a variate x is expressed as $x = N(\mu, \sigma^2)$ and has the following properties:

1. It has a maximum at $x = \mu$.
2. It is symmetric with respect to $x = \mu$.
3. It has two points of inflection at $x = \mu \pm \sigma$.
4. A change in μ causes a translation of the curve without changing its shape.
5. A change in σ^2 will widen or narrow the curve without a change in μ.

STANDARD NORMAL VARIATE

Consider a variate $x = N(\mu, \sigma^2)$. Define Z such that

$$Z_i = \frac{x_i - \mu}{\sigma}. \tag{1.6}$$

The mean of the Z_i's, $\bar{Z}$, is

$$\bar{Z} = \frac{\sum_{i=1}^{N} Z_i}{N} = \frac{1}{N\sigma}\left[\sum_{i=1}^{N} x_i - \sum^{N} \mu\right]$$

$$= \frac{1}{N}[N\mu - N\mu] = 0.$$

The variance of the Z_i's, σ_Z^2, is

$$\sigma_Z^2 = \frac{\sum^{N} (Z_i - \bar{Z})^2}{N} = \frac{\sum^{N} (Z_i)^2}{N}$$

$$= \frac{1}{N} \frac{\sum^{N} (x_i - \mu)^2}{\sigma^2} = \frac{\sigma^2}{\sigma^2} = 1.$$

Therefore $Z = N(0, 1)$.

The probability density function is given by

$$f(Z) = \frac{1}{\sqrt{2\pi}} e^{-Z^2/2}.$$

This function is also known as the *standard normal function*. It describes all normally distributed variates regardless of the different values of their parameters (i.e., μ and σ^2). Therefore, all normal distributions with different means and variances, after transformation to a standard form by Equation (1.6), can be represented by a single table of probabilities. The probabilities of the standard normal variable Z are given in Table A.1 in the appendix.

POPULATIONS AND SAMPLES

When a phenomenon like the heights of students is to be studied, two statistical approaches are usually used. The first is to measure the height of every student of the N students enrolled in the school. Conducted in this manner one can find the "true" average height (i.e., μ). The data set of such an experiment contains information about every single element under observation and is thus called *population data*. Parameters such as μ and σ^2 calculated from such data are referred to as *population parameters*. The values of μ and σ^2 are calculated according to Equations (1.7) and (1.8):

$$\mu = \sum_{i=1}^{N} \frac{x_i}{N} \tag{1.7}$$

and

$$\sigma^2 = \sum_{i=1}^{N} \frac{(x_i - \mu)^2}{N}. \tag{1.8}$$

The second approach is to measure the heights of only a group (*sample*) of n students. This approach may be advantageous in terms of economics and time. One can calculate an average, $\bar{x}$, and a variance, s^2 (*sample parameters*) from the sample data. The sample parameters are calculated according to Equations (1.9) and (1.10):

$$\bar{x} = \sum_{i=1}^{n} \frac{x_i}{n} \tag{1.9}$$

and

$$s^2 = \sum_{i=1}^{n} \frac{(x_i - \bar{x})^2}{n-1}. \tag{1.10}$$

The crucial question at this point is: How do $\bar{x}$ and s^2 differ from the true values μ and σ^2, respectively? Before an adequate answer to this question is considered, the sampling process must be examined.

The chosen sample must be *representative* of the population. This can be guaranteed if we have a *random sample*; that is, when every element in the population has an equal chance of being included in the sample. A random sample can be obtained, for example, by drawing names from a well mixed box that contains the names of all the students. Our sample will not be random if, for instance, we choose the students from The Dean's Honor List (indeed, the authors would not be chosen). Sample size is also of great importance. The larger the sample size the closer the agreements between σ^2 and s^2 and between μ and $\bar{x}$. In practice, sample size is usually determined by the economics of the experiment being conducted.

As depicted in Figure 1.7, the sample distribution can be anywhere within the parent distribution. The calculated sample mean can be a good approximation of μ (e.g., $\bar{x}_B$) or a bad one (e.g., $\bar{x}_A$ or $\bar{x}_C$). Samples A and C may be the product of nonrandom sampling or a small sample size.

The mean, $\bar{x}$, and the variance, s^2, describe the distribution of one sample ($x \pm rs$, where r is a real number). Also of interest is the *distribution of the means*. If m different samples are obtained and m different averages are calculated, one can treat the m means as a separate population. An average, $\bar{m}$ or $E(\bar{x})$, and a variance, σ_m^2 or $\sigma_{\bar{x}}^2$, can be obtained for this distribution. These important parameters can be

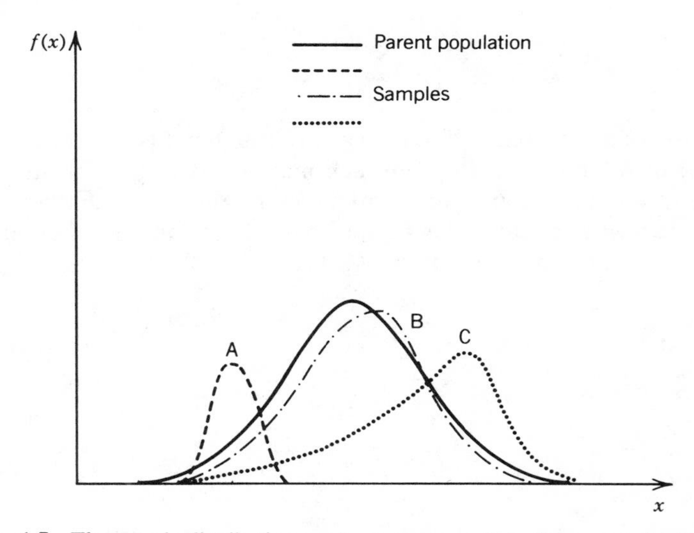

Figure 1.7 The sample distribution can be anywhere within the parent distribution.

estimated from one normally distributed random sample. The sample average, $\bar{x}$, is an unbiased estimate of $\bar{m}$:

$$\bar{m} = E(\bar{x}) = \frac{\sum_{i=1}^{n} x_i}{n}. \tag{1.11}$$

The variance of the mean, $\sigma_{\bar{x}}^2$, can be derived as follows. From Equation (1.9),

$$\bar{x} = \frac{x_1}{n} + \frac{x_2}{n} + \cdots + \frac{x_n}{n}.$$

Let Var(x) denote the variance of variate x. By using the propagation of error principle (2) and assuming that the x_i's are uncorrelated:

$$\text{Var}(\bar{x}) = \left(\frac{1}{n}\right)^2 [\text{Var}(x_1) + \text{Var}(x_2) + \cdots + \text{Var}(x_n)].$$

Assuming that the x_i's have equal variances, s^2, the above equation can be rewritten:

$$\text{Var}(\bar{x}) = \left(\frac{1}{n}\right)^2 ns^2 \quad \text{or} \quad s_{\bar{x}}^2 = \frac{s^2}{n}. \tag{1.12}$$

Thus the average and the variance of the normally distributed means can be estimated from Equations (1.11) and (1.12). When these parameters are known, the interval in which $\bar{x}$, the mean of any random sample drawn from the same population, exists can be estimated. Table A.1 shows that the value of a normally distributed variate lies in the interval $\bar{x} \pm 1.96s$ 95% of the time. By the same token, the mean of the population, μ, as calculated from any random sample taken from the population, lies in the interval $\bar{x} \pm 1.96s_{\bar{x}}$ 95% of the time. Generally, for a random sample of size n, the range

$$\bar{x} - Z\frac{s}{\sqrt{n}} \leq \mu \leq \bar{x} + Z\frac{s}{\sqrt{n}} \tag{1.13}$$

is an estimate of μ at a certain confidence level defined by Z. Some popular values of Z are given in Table 1.1. Thus a random sample of size n, drawn from a population of size N, has an average $\bar{x}$ and a variance s^2. The probability that the true mean, μ, is in the interval $\bar{x} \pm 1.96s/\sqrt{n}$ is 0.95 (95% of the time). The variance of the mean, $s_{\bar{x}}^2$, may be corrected for the finite sample size and expressed as

$$s_{\bar{x}}^2 = \left(1 - \frac{n}{N}\right)\frac{s^2}{n}. \tag{1.14}$$

Table 1.1 Values of Z at Various Confidence Levels

Z	Confidence Level
1.64	90%
1.96	95%
2.58	99%

The sample variance is an unbiased estimator of the variance of the population,

$$\sigma^2 = s^2. \tag{1.15}$$

If the population is not normally distributed, large random samples ($n > 30$) must be taken to estimate μ and σ^2 (3). The means of the random samples, however, are considered normally distributed (3, 4). The reliability of the estimates of μ and σ^2 ($\bar{x}$ and s^2) requires that s^2 be a good approximation of σ^2. Although large samples are needed, this sampling technique has the advantage of tolerating non-normally distributed parent populations. If the parent population is known to be normal, only a small random sample is sufficient to estimate μ and σ^2. However, different statistics are needed to treat small samples taken from normally distributed parent populations.

STUDENT'S *t*-DISTRIBUTION

Take all possible small random samples of size n from a normally distributed population with a mean μ and variance σ^2. For each sample compute $\bar{x}$ and s^2 and define

$$t = \left(\frac{\bar{x} - \mu}{s}\right)\sqrt{n}. \tag{1.16}$$

The distribution of the t values is known as the Student's t-distribution (5). Every value of n gives rise to a characteristic t-distribution that will be associated with $n - 1$ degrees of freedom. The variable t exists within a certain range given a certain sample size and a particular confidence level. In general,

$$\bar{x} - t\frac{s}{\sqrt{n}} \leq \mu \leq \bar{x} + t\frac{s}{\sqrt{n}}, \tag{1.17}$$

where the value of t is defined by the confidence level and $(n-1)$. If $n=6$ and the desired confidence level equals 0.95, Table A.2 in the appendix shows t to be 2.57 for 5, or $(n-1)$ degrees of freedom. Therefore, the 0.95 confidence level interval for μ is given by

$$\bar{x} - 2.57\frac{s}{\sqrt{6}} \leq \mu \leq \bar{x} + 2.57\frac{s}{\sqrt{6}}. \tag{1.18}$$

Equation (1.18) implies that μ exists 95% of the time in the range $\bar{x} \pm 2.57s/\sqrt{6}$. To estimate σ^2 we refer to yet another distribution—the χ^2 distribution. We define

$$\chi^2_{n-1} = \frac{(n-1)s^2}{\sigma^2}. \tag{1.19}$$

The population variance, σ^2, can be estimated from the sample variance, s^2. An interval in which σ^2 exists is given by

$$\frac{(n-1)s^2}{\chi^2_{(1-\alpha/2,n-1)}} \leq \sigma^2 \leq \frac{(n-1)s^2}{\chi^2_{(\alpha/2,n-1)}}, \tag{1.20}$$

where $1-\alpha$ is the desired confidence level. Assume that ten determinations for lead in Lake Sammamish are reported with a standard deviation, s, of 4 mg/L. From Table A.3 in the appendix, the population variance, at the 98% ($\alpha = .02$) confidence level, is in the following interval:

$$\frac{9(16)}{\chi^2_{(.99,9)}} \leq \sigma^2 \leq \frac{9(16)}{\chi^2_{(.01,9)}}$$

$$6.65 \leq \sigma^2 \leq 68.90.$$

To summarize, one only needs small samples if the population is known to be normally distributed. Equation (1.17) calculates an *exact* confidence level interval for μ. If the population is not normally distributed, we have to obtain large samples ($n > 30$) and Equation (1.13) gives an *approximate* confidence interval for μ. Sample size and the allowable error (the degree of accuracy required) are important since their values determine the estimation interval. If we denote the maximum allowable error by ε and the variance in the population whose mean is being estimated by σ^2, the sample size needed, n, is given by

$$n = \frac{Z^2\sigma^2}{\varepsilon^2}, \tag{1.21}$$

where Z is the value of the standard normal variate associated with the desired confidence level. We can obtain an estimate of σ^2 from past experience or from a preliminary sample. Alternately, one can manipulate sample size, n, until

$$\varepsilon = Zs_{\bar{x}}. \tag{1.22}$$

Thus far, we have considered parameter estimation of population distribution from only one sample. Better estimations are possible if more than one random sample are drawn from the population. If all possible samples of size n are drawn from a population of N, one can calculate $\bar{x}$ for each sample. The $\bar{x}$'s will be normally distributed when $n > 30$ even if the parent population is not. If the parent population is normally distributed, the $\bar{x}$'s will be normally distributed regardless of n. The distribution of $\bar{x}$'s will have a mean, $E(\bar{x})$, and a variance, $s_{\bar{x}}^2$. Population parameters are given by

$$\mu = E(\bar{x}) \tag{1.23}$$

and

$$\sigma^2 = s_{\bar{x}}^2 \frac{n(N-1)}{(N-n)}. \tag{1.24}$$

If $n \ll N$,

$$\sigma^2 = ns_{\bar{x}}^2. \tag{1.25}$$

BINOMIAL DISTRIBUTION

Consider a particular outcome, O, of a single trial of a certain process and let

$$p = \text{probability that } O \text{ will take place}$$

and

$$q = \text{probability that } O \text{ will not take place}$$

such that $p + q = 1$.

The binomial distribution is a discrete distribution that predicts the probability of O taking place X times in N trials. This is given by

$$P(X, N) = \frac{N!}{X!(N-X)!} p^X q^{N-X}. \tag{1.26}$$

Consider the tossing of a fair coin. In a single trial the probability of getting a head, p, is $\frac{1}{2}$. The probability of *not* getting a head is equal to the probability of getting a tail, q, and is also $\frac{1}{2}$: $p+q=1$. The probability of getting three heads in four trials is

$$P(3, 4) = \frac{4!}{3!1!}(\tfrac{1}{2})^3(\tfrac{1}{2})^1 = 0.25.$$

Another example is when a die is rolled. The probability of getting a 4, p, is $\frac{1}{6}$. The probability of not getting a 4, q, is equal to the probability of getting a 1 or a 2 or a 3 or a 5 or a $6 = \frac{1}{6}+\frac{1}{6}+\frac{1}{6}+\frac{1}{6}+\frac{1}{6}=\frac{5}{6}$ $(q+p=1)$. The probability of getting two 4's in six rolls is

$$P(2, 6) = \frac{6!}{2!4!}(\tfrac{1}{6})^2(\tfrac{5}{6})^4 = 0.20.$$

The mean and the variance of the binomial distribution are

$$\mu = Np \tag{1.27}$$

and

$$\sigma^2 = Npq. \tag{1.28}$$

Thus the expected number of heads obtained when a fair coin is tossed 100 times is $100 \times \frac{1}{2} = 50$ with a variance of $100 \times \frac{1}{2} \times \frac{1}{2} = 25$ or a standard deviation of five heads.

BIAS

There are two levels of bias. The first is when some elements in the population have unequal probabilities of being sampled. An example of that is when only the small (or the large) aggregates of a soil sample are chosen for analysis. This kind of bias gives a distorted image of the population mean (samples A and C in Fig. 1.7). It should and can be eliminated by random sampling (random particle size selection, grinding, etc.) (6).

The second level of bias is the improper estimation of population parameters μ and σ^2. Only unbiased estimators should be calculated for μ and σ^2. A statistic is considered to be an unbiased estimator of a parameter if the mean of the sampling distribution of the statistic equals the parameter. If $\hat{\theta}$ is a statistic calculated from a single sample to estimate the parameters θ, then $\hat{\theta}$ is considered an unbiased estimation of θ if the distribution of $\hat{\theta}$'s (as calculated from several samples) has a mean equal to θ. That is, if

$$E(\hat{\theta}) = \theta. \tag{1.29}$$

ESTIMATION OF μ

Let $\bar{x}$ be an estimation of μ such that

$$\bar{x} = \frac{\sum_{i=1}^{n} x_i}{n}$$

$$E(\bar{x}) = E\left[\frac{\sum_{i=1}^{n} x_i}{n}\right]$$

$$= \frac{1}{n} E\left[\sum_{i=1}^{n} x_i\right]$$

$$= \frac{1}{n} \sum_{i=1}^{n} E(x_i)$$

$$= \frac{1}{n} \sum_{i=1}^{n} \mu$$

$$= \frac{1}{n} n\mu$$

$$E(\bar{x}) = \mu \tag{1.30}$$

$\therefore \quad \bar{x} = \dfrac{\sum_{i=1}^{n} x_i}{n}$ is an unbiased estimation of μ.

ESTIMATION OF σ^2

Let s'^2 be an estimation of σ^2 such that

$$s'^2 = \frac{\sum_{i=1}^{n} (x_i - \bar{x})^2}{n}. \tag{1.31}$$

This is a valid assumption since

$$\sigma^2 = \frac{\sum_{i=1}^{N} (x_i - \mu)^2}{N}.$$

However,

$$E(s'^2) = \frac{1}{n} E\left[\sum_{i=1}^{n} (x_i - \bar{x})^2\right]$$

$$E(s'^2) = \frac{n-1}{n}\left[\sum_{i=1}^{n} \frac{(x_i - \mu)^2}{n}\right] = \frac{n-1}{n}\sigma^2 \neq \sigma^2.$$

$$\therefore \quad s'^2 = \sum_{i=1}^{n} \frac{(x_i - \bar{x})^2}{n} \quad \text{is } \textit{not} \text{ an unbiased estimator of } \sigma^2.$$

If we multiply s'^2 by $n/(n-1)$ we obtain

$$E\left(\frac{n}{n-1} s'^2\right) = \sigma^2.$$

Therefore, $[n/(n-1)]s'^2$ is an unbiased estimator of σ^2 or

$$s^2 = \frac{n}{n-1} \sum_{i=1}^{n} \frac{(x_i - \bar{x})^2}{n} = \sum_{i=1}^{n} \frac{(x_i - \bar{x})^2}{n-1} \tag{1.32}$$

is an unbiased estimator of σ^2.

ANALYTICAL APPLICATIONS

Sampling represents an important source of variance that becomes embedded in the final results. For every unit in the sample

$$\underset{\text{variance in results}}{S_R^2} = \underset{\text{variance in analysis}}{S_A^2} + \underset{\text{variance in sampling}}{S_S^2} \quad . \tag{1.33}$$

Example 1. You are to examine fish in a river for tumors. Let A = fish with tumors present in n_A quantity and B = fish without tumors present in n_B quantity.

$$P(A) = \frac{n_A}{n_A + n_B};$$

$$P(B) = \frac{n_B}{n_A + n_B};$$

and

$$P(A) + P(B) = 1.$$

If n fish are sampled, the binomial distribution statistics give

$n_A = n \cdot P(A)$ = number of fish with tumors expected in the sample

and

$$s^2 = n \cdot P(A) \cdot P(B) = \text{variance of the sample.}$$

If 20% of the fish have tumors and 10,000 fish are sampled, then

$$n_A = 2000 \quad \text{and} \quad s = 40.$$

The sampling error $= s/n_A = 0.02$ or 2%;

$$\therefore \quad s = \sqrt{n \cdot P(A) \cdot P(B)}$$

$$\therefore \quad \text{sampling error} = \frac{s}{n_A} = \frac{s}{nP(A)} = \sqrt{\frac{P(B)}{P(A)}} \cdot \frac{1}{\sqrt{n}}. \tag{1.34}$$

To minimize the sampling error one needs large samples (large n) and should be looking for major components [high $P(A)$]. If 80% of the fish have tumors, the sampling error in 10,000 fish becomes

$$\text{sampling error} = \sqrt{\frac{0.2}{0.8}} \cdot \sqrt{\frac{1}{10{,}000}} = 0.005 \text{ or } 0.5\%.$$

Equation (1.34) can be rearranged to express n in terms of the sampling error:

$$n = \frac{P(B)}{P(A)} \cdot \frac{1}{(\text{sampling error})^2}. \tag{1.35}$$

It can be shown that to reduce the sampling error by half, n must be increased fourfold.

Example 2. Assume that you are analyzing a solid sample having a mass of 1 g for a certain metal and that you wish the sampling error not to exceed 1%. If the metal present is 50% of the total sample, the probability of finding a metal particle, $P(A)$, is 0.5. The number of particles needed $= (0.5/0.5)(1/0.01)^2 = 10{,}000$ particles.

If the metal is present in minor quantities (100 ppm), then

$$P(A) = \frac{100}{1{,}000{,}000} = 0.0001.$$

The number of particles needed $= (0.9999/0.0001)(1/0.01)^2 = 10^8$ particles.

If the metal is present in trace amounts (1 ppb), then

$$P(A) = 10^{-9}.$$

The number of particles needed $= [(1 - 10^{-9})/10^{-9}](1/0.01)^2 = 10^{13}$ particles.

Example 3. Homogeneous solutions (e.g., aqueous solution) offer a drastically different perspective. Consider 1 L of a 0.0010*M* NaCl solution. Define

NaCl = type *A* particle,

$$n_A = 10^{-3} \times 6.0 \times 10^{23} = 6 \times 10^{20} \text{ molecules;}$$

H_2O = type *B* particle,

$$n_B = 55.5 \times 6.0 \times 10^{23} = 3.3 \times 10^{25} \text{ molecules;}$$

$$P(A) = \frac{6.0 \times 10^{20}}{6.0 \times 10^{20} + 3.3 \times 10^{25}} = 1.8 \times 10^{-5}$$

and

$$P(B) = 1.0 - P(A) \simeq 1.0.$$

To keep the error in sampling the NaCl particles at a maximum of 0.1%, we need

$$n = \frac{1.0}{1.8 \times 10^{-5}} \left[\frac{1}{0.001}\right]^2 = 5.5 \times 10^{10} \text{ molecule}$$

or a sample size of

$$\frac{5.5 \times 10^{10}}{6.0 \times 10^{23} \times 55.5} = 1.6 \times 10^{-15} \text{ L}$$

Only a minute amount of solution is needed to make the sampling error negligibly small. This offers a great advantage. Another advantage of homogeneous solutions is that random sampling is assured by simple mixing.

MINIMIZING THE VARIANCES OF SAMPLING AND ANALYSIS

Let the sampling variance and the analysis variance per unit be s_S^2 and s_A^2, respectively. If we have *n* units to be analyzed, two schemes of treatment are available.

Scheme 1. Analyze *k* units of the *n* units and report the mean. The variance of the mean due to analysis, $(s_m^2)_A$, and the variance of the mean due to sampling, $(s_m^2)_S$, are

$$(s_m^2)_A = \frac{s_A^2}{k} \tag{1.36}$$

and

$$(s_m^2)_S = \frac{s_S^2}{k}. \tag{1.37}$$

Scheme 2. Mix the n units and choose k units of the mixture for analysis. The sampling variance per unit after mixing, s_s^2, and the analysis variance per unit after mixing, s_a^2, are

$$s_s^2 = \frac{s_S^2}{n} \tag{1.38}$$

and

$$s_a^2 = s_A^2. \tag{1.39}$$

While mixing reduces the sampling variance, it does not affect the analysis variance. Analysis variance is reduced only by replicate analysis. If k units are analyzed and their mean is reported, the variance of the mean due to analysis and that due to sampling are given by

$$(s_m^2)_A = \frac{s_a^2}{k} = \frac{s_A^2}{k} \tag{1.40}$$

and

$$(s_m^2)_S = \frac{s_s^2}{k} = \frac{s_S^2}{nk}. \tag{1.41}$$

Example 4. You are given ten samples of river water; each has 5 ppb DDT with a standard deviation of 0.1 and every single determination of DDT has a standard deviation of 0.05. Here we have $s_A^2 = (0.05)^2$ and $s_S^2 = (0.1)^2$. Consider the following plans of analysis.

PLAN 1. The ten samples are analyzed separately and the mean is reported. The variance in the mean, s_m^2, is given by

$$s_m^2 = (s_m^2)_A + (s_m^2)_S.$$

From Equations (1.36) and (1.37),

$$s_m^2 = \frac{(0.05)^2}{10} + \frac{(0.1)^2}{10} = 1.25 \times 10^{-3}.$$

PLAN 2. Mix all ten samples and analyze one sample that is 1/10 of the whole mixture. The variance of the mean is given by

$$s_m^2 = (s_m^2)_A + (s_m^2)_S$$

$$= s_A^2 \qquad + \qquad \frac{s_S^2}{10}$$

analysis variance per unit is unchanged after mixing | sampling variance per unit due to mixing

$$s_m^2 = \frac{(0.05)^2}{1} + \frac{(0.1)^2}{10} = 3.50 \times 10^{-3}.$$

PLAN 3. Mix all ten samples and analyze ten samples, each sample being 1/10 of the mixture. The variance of the mean is given by

$$s_m^2 = (s_A^2)/10 + \left[\frac{s_S^2}{10}\right] \Big/ 10$$

$$= \frac{(0.05)^2}{10} + \frac{(0.1)^2}{100} = 3.50 \times 10^{-4}.$$

PLAN 4. Pool all ten samples and analyze only three samples, each sample being 1/10 of the mixture. The variance of the mean is given by

$$s_m^2 = \frac{(0.05)^2}{3} + \frac{(0.1)^2}{30} = 1.17 \times 10^{-3}.$$

This is a comparable variance to Plan 1 with less analysis.

PLAN 5. If the analysis variance is independent of sample size, the ten samples can be pooled and one analysis is performed on the whole mixture. The standard deviation due to analysis of n mixed samples will be equal to s_A/n. The analysis variance becomes s_A^2/n^2. Then

$$s_m^2 = \frac{s_A^2}{n^2} + \frac{s_S^2}{n} = 1.02 \times 10^{-3}.$$

We get a smaller variance than all other plans except Plan 3 (where ten analyses are involved) by performing one analysis. This is the plan most recommended when economics is the deciding factor. Note, however, that s_A must be independent of sample size.

Example 5. The standard normal variate transformation can be used to test for "abnormalities" in experimental results (e.g., outliers). Consider six determinations of a pollutant,* x, in a water well to be

$$x_1 = 0.664 \text{ mg/L}$$
$$x_2 = 0.655 \text{ mg/L}$$
$$x_3 = 0.686 \text{ mg/L}$$
$$x_4 = 0.657 \text{ mg/L}$$
$$x_5 = 0.653 \text{ mg/L}$$
$$x_6 = 0.661 \text{ mg/L}$$

These data are assumed to be obtained from a normally distributed parent population. A sample average, $\bar{x}$, and a sample standard deviation, s, are calculated as

$$\bar{x} = \frac{1}{n} \sum x_i = 0.663 \text{ mg/L}$$

$$s = \left[\frac{1}{n-1} \sum (x_i - \bar{x})^2 \right]^{1/2} = 0.0121 \text{ mg/L}.$$

A close look at the data shows that the third determination, x_3, is approximately two units of standard deviation higher than the average. Consequently, we wish to determine if it is as reliable as the others. We first transform x_3 into standard variate form, x'_3:

$$x'_3 = \frac{x_3 - \bar{x}}{s} = 1.90.$$

From Table A.1 it is found that the probability for obtaining $z \geq 1.90$ is about 3%. In view of this, x_3 can be either kept or withdrawn from the data. This indicates that a value greater than or equal to x_3 should not be obtained very frequently (only three times in every 100 determinations). Therefore, the number of times x_3 or a similar result is obtained should help the analyst revise either the method of sampling or the method of analysis.

It is also possible to estimate the likelihood of obtaining abnormal values. Assume, for instance, that DDT is being assayed by a procedure whose relative standard deviation is 10%. If a sample with a DDT content of 130 ppb is being analyzed, the probability of obtaining a result that is <100 ppb is calculable as follows: Set

$$\mu = 130 \text{ ppb}, \quad \sigma = 13 \text{ ppb } (130 \times 0.1), \quad \text{and} \quad x = 100 \text{ ppb}.$$

*These data are taken from an example in *Statistical Methods in Research and Production*, by O. Davies and P. Goldsmith. Longmans, 1976.

Then

$$Z = \frac{100 - 130}{13} = -2.3.$$

From Table A.1, $P(Z \leq -2.3) = 1.1\%$.

REFERENCES

1. A. J. Hastings and J. B. Peacock, *Statistical Distributions*, Halsted Press, New York, 1975.

   A compilation of the parameters of and statistical facts about the leading 25 statistical distributions. Information covers parameter estimation, variate relationships, and random number generations.

2. G. E. P. Box, William G. Hunter, and J. Stuart Hunter, *Statistics for Experimenters*, Wiley, New York, 1978, pp. 87–89.

   This section covers the variance of a linear combination of observations and demonstrates the propagation of errors for correlated and uncorrelated variates.

3. Ernest Kurnow, J. Glasser, and Frederick R. Ottman, *Statistics for Business Decisions*, Irwin, Homewood, Ill., 1959, pp. 175–184.

   A demonstration of the distributions of the means obtained when random samples of different sizes are taken from normally and non-normally distributed populations.

4. William G. Cochran, *Sampling Techniques*, 3rd ed., Wiley, New York, 1977, pp. 39–44.

   A brief discussion and bibliography pertaining to the validity of the normal approximation. The problems of sampling from a skewed parent distribution are demonstrated.

5. Student, "The Probable Error of a Mean," *Biometrika* **6**, 1 (1908).

   A discussion of the development of the t-distribution. Student's t-distribution was originated by William S. Gosset, a research chemist with the Guinness brewery who used the pseudonym "student," and who devoted part of his time to the interplay between theoretical statistics and practical experiments.

6. Herbert A. Laitinen and Walter E. Harris, *Chemical Analysis*, 2nd ed., McGraw-Hill, New York, 1975, pp. 569–574.

   A discussion on sampling particulate matter.

SUGGESTED READINGS

William G. Cochran, *Sampling Techniques*, 3rd ed., Wiley, New York, 1977.

One of the most comprehensive accounts on the subject.

G. Kateman and Frans W. Pijpers, *Quality Control in Analytical Chemistry*, Wiley, New York, 1981, Chapter 2.

A discussion of sampling theory covering homogeneous and heterogeneous objects, correlated variates, and sampling from a system over a period of time.

Byron Kratochvil and John K. Taylor, "Sampling for chemical analysis," *Anal. Chem.* **53**, 924A (1981).

A well written report on the relevance of sampling to analytical results.

Herbert A. Laitinen and Walter E. Harris, *Chemical Analysis*, 2nd ed., McGraw-Hill, New York, 1975, Chapters 26 and 27.

Chapter 26 discusses basic statistics, propagation of error, and analysis of variance in addition to other important concepts and operations. Sampling theory is treated in Chapter 27. The discussion covers aspects like variance reduction, sampling particulate matter, and sampling procedures.

Murray R. Spiegel, *Statistics*, Schaum's Outline Series in Mathematics, McGraw-Hill, New York, 1961.

A good introduction for those with very little or no background on statistics. The theory of statistics is explained with several examples covering the use of the z, t, and χ^2 tables and tests. Parameters of sampling distributions and their errors are demonstrated.

CHAPTER

2

FUNDAMENTALS OF EXPERIMENTAL DESIGN AND OPTIMIZATION

All too often the analyst is asked to analyze samples without being told the reasons behind sample(s) and analyte(s) selection. If qualified statisticians were involved in the study, chances are that the data will be useful in drawing helpful conclusions. If, on the other hand, the experimental plans were not tailored to answer the important questions, the analyst's time may be wasted and the data may be more misleading than helpful.

The tools presented in this chapter can help in avoiding wasted scientific efforts. They are good but not perfect tools that have been around for a long time. A complete treatment of all aspects of experimental design would take several volumes and is beyond the intended scope of this book. This introduction should alert the analyst to methods leading to better designed experiments that ensure that the flow from measurements to information to knowledge can proceed in the most cost effective way.

An *experiment* is a process by which information is acquired by observing the reaction of a subject to certain *stimuli*. The basic elements of any experiment are the observer (the experimenter), the subject (the experimental unit), the stimuli, and the information that the experiment yields. The stimuli are also known as the *factors*. They constitute the environment which is created and controlled by the experimenter and in which the experimental unit is observed. Factors are divided into two categories—those completely controlled by the observer (experimental factors) and those characteristic of the experimental units themselves (*classification factors*). Consider a study of the addition of acids to carbon–carbon multiple bonds. The temperature, solvent, and kind of acid are entirely controlled by the experimenter and are considered experimental factors. However, the type of multiple bond (double or triple) is a characteristic of the organic compound involved in the study. It is therefore a classification factor. The extent to which an experimental factor is applied is called the *level* of the factor, or the *treatment*. If temperature is a factor, three levels or treatment may involve temperatures of 100, 200, and 300°C.

The information obtained from an experiment can either increase or correct our state of knowledge with respect to a particular system. Consequently, our improved knowledge enables us to make better or more substantiated decisions. The results of experiments are therefore expected to "contain" adequate answers to the questions that inspire the experiment in the first place. A well designed experiment is one that answers all the questions with the least experimental effort. The *plan* of the experiment entails selecting the subjects, deciding what the factors and their levels (the treatments) are, and choosing a schedule for application. It is referred to as the *experimental design*.

A successful experiment has several prerequisites. The experiment must have a *well defined objective*. The questions which the experimenter is trying to address must be clearly stated in advance. This aids the experimenter in choosing the relevant materials, equipment, the experimental environment, and the level(s) of the factor(s) involved in the experiment. *An estimate of experimental error* must also be provided by the experiment. This is usually obtained by repetitive application and is indicative of the reliability of the results. The effects of any experimental factor *must not* be overshadowed (not even in part) by the uncontrolled variables or other experimental factors. The true effect of each experimental factor can be estimated in the presence (or the absence) of other factors by *randomization* and *factorial design*. The experiment must be of *sufficient precision* to satisfy the main objective. The size (or number) of the experimental unit(s) can be chosen to increase the experimental precision. The *experimental pattern* (the plan to take measurements) can also be modified to accommodate the requirement of a more (or less) precise experiment. The experiment *must be unbiased*. Even if the economic circumstances are poor (insufficient funds and/or time allowed) one must not be tempted to bias the experiment. No useful or statistically valid results can be obtained from a biased experiment because the experimental error can not be estimated and the confidence intervals become incalculable.

ASSESSMENT OF PERFORMANCE

Evaluating the results of analytical systems is critical to making sound decisions. The *signals* we obtain from an analytical instrument or the numbers we receive from the laboratory must be reliable and, to the best of our knowledge and abilities, foolproof. Laboratories are routinely subjected to "quizzes" in which known samples are submitted to them and their results evaluated.

Suppose that you are called upon to test the claims of Truth, Inc. This commercial analytical organization advertised that it had developed a system to detect the presence of polychlorinated biphenyls (PCBs) in very minute amounts (1–5 ppb) in drinking water. On your final word (whether the system is sound or unreliable) hangs a multimillion dollar contract which the government is considering to grant Truth, Inc. to monitor all the drinking water sources in the country. What do you do?

The most logical method to test the claims of Truth, Inc. would be to send them several known samples and upon receiving their results decide whether or not their claims are substantiated. Let us say that you sent them a PCB-free sample of distilled water and asked them to either confirm or deny the presence of PCBs in it. If they report that PCBs are present, a wrong classification, you may decide that their detection tools are misleading. However, their conclusion may simply be a blunder. Their system may be reliable, though occasionally wrong. More samples have to be examined before a decision can be made. If they report that PCBs are not present (a correct classification), it would be impetuous to give them your approval simply because this answer, in light of what you have asked of them, could have been arrived at by random guessing. Indeed, and without any analytical effort, they have a 50% chance of guessing the correct answer. The sample either contains or does not contain PCBs. In this case, more samples have to be examined before a decision is made.

Sending only one sample to Truth, Inc. is a bad experimental design because no matter what answer you receive you need more evidence to make a statistically sound decision. The results of the experiment will not answer the question upon which the experiment is based. No statistical reliability can accompany a single experiment.

Since you want to test a positive result, you need to decide, before anything else, the significance level of your decision. In other words, what are the odds that Truth, Inc. should beat in order to convince you of their ability to detect the presence (or the absence) of PCBs in drinking water? If you require that the probability of randomly guessing the "correct answer" should not exceed 0.05 (1 chance in 20 or more chances), then any design in which this probability is smaller or larger is either unnecessary (costly and/or time consuming) or inadequate. The "correct answer" should therefore be defined in advance. You must decide whether you will allow Truth, Inc. to be (or not to be) occasionally in error. Your experiment should thus be sensitive enough to detect the required degree of their accuracy. The experiment must be free of bias. To minimize the sampling variance, the samples must be made as "identical" as possible in terms of concentration range, ionic strength, acidity or alkalinity, and so on. They should also be submitted for analysis in a randomized fashion to allow uncontrolled factors such as time of day and laboratory temperature and procedures to average out. This minimizes the analysis variance.

A simple design is to make up six samples, only three of which contain PCBs. The samples are then submitted in a randomized order to Truth, Inc., which is asked to identify the three samples containing PCBs. This guarantees that an error in one group is always balanced by an error in the other. Since there are 20 different ways [6!/(3! 3!)] to choose three objects from a group of six (order being unimportant) and only one of these combinations corresponds to the correct classification of the samples, Truth, Inc. has only 1 chance in 20 chances to randomly guess the correct answer. With the probability of random guessing the correct answer minimized to your satisfaction, their results can be judged objectively. If they classify each of the six samples correctly, you may conclude that their system yields adequate results.

The number of samples determines both the significance level and the sensitivity (in this case the ability of the experiment to detect lower degrees of accuracy) of the experiment. For example, given the above confidence level (probability of random guessing the correct answer <0.05), the above experiment (testing six samples) is intolerant of mistakes. If you allow Truth, Inc. to have a lower degree of accuracy by permitting them to make one error (classify only two of the three PCB-free samples correctly and one of the PCB-free samples incorrectly), then the "correct answer" becomes: *either all three or at least two of the PCB containing samples are classified correctly*. The probability of random guessing this answer is 0.5 (1/20 for classifying all three correctly plus 9/20 for classifying any two of them correctly). This is a much higher probability than the critical 0.05 upper limit.

A more sensitive experiment is one that is identical in principle to the above (testing six samples) except the number of samples is raised to 12. The probability of randomly classifying all six PCB-containing samples or at least five of them correctly can be shown to equal $1/924 + 36/924$, which is less than 0.05. This design enables you to make a substantiated decision even when one sample is misclassified. The price you pay for having this sensitive experiment is obviously the cost of analyzing twice as many samples. You can design a cheaper and yet sensitive experiment with only eight samples. However, the structure of the experiment has to be modified. Each of the eight samples is considered an independent case and submitted, again, randomly for analysis. The probability of randomly classifying all eight samples correctly is 1/256 which is less than 0.05. If you permit Truth, Inc. to make one error, then the probability of randomly guessing the correct answer (classifying all eight or at least seven of them correctly) becomes 9/256, which is also less than 0.05.

The probability of random success (random guessing of the correct answer) can be raised by reducing the sensitivity of the experiment. If six samples are divided such that four samples are contaminated with PCBs and two are not, the probability of randomly classifying the samples correctly increases to 1/15 compared to 1/20 when the samples are divided in two groups (contaminated and uncontaminated) of three samples each.

The foregoing designs are valid tests of the discriminatory ability of any detection system. They can be used to qualify (or disqualify) the claims of Truth, Inc. They can also be used to test any signal detection system under study or development. Similar tests can be designed to examine the performance of analytical systems such as ion selective electrodes or a gas chromatographic detector. A probability of 0.05 for random success is a common significance level for most experimenters. The experiment may be repeated several times to eliminate any misgivings. Since there is always a finite probability for random success, one may be, and in some instances is encouraged to be, suspicious when and if the experiment shows positive results. Consider, for example, that eight samples are to be classified into two groups (those with PCBs and those without) with four samples in each. The probability of randomly classifying all eight samples correctly is 1/70. If the classification is accomplished successfully in one trial, one may still argue that the results represent this 1-in-70 chance. Indeed, the experiment is incapable of disproving this argument. Repeating the experiment can aid us to substantiate our evaluation. The probability of randomly classifying all eight samples correctly twice in a row is 1/4900, a very small frequency indeed. The appearance of this event (correct classification in both groups) at a frequency higher than 1/4900 is considered a *nonrandom* process.

These designs are concerned mainly with *qualitative* decisions. However, the majority of analytical operations are conducted to make *quantitative* decisions. In industrial quality control or environmental studies, the main concerns may be that the presence of a certain metal does not exceed 10 ppb. In these cases the experiment must answer the question: Is the concentration of the metal >10 ppb? In other situations the analyst may be asked to decide which catalyst offers a larger yield. These quantitative (and semiquantitative) decisions will be discussed in the next section.

COMPARATIVE EXPERIMENTS

Estimating differences between two or more processes (as represented by two or more statistical distributions) is of the utmost importance in research and industry. Decisions such as whether a catalyst does indeed increase the yield or a process does improve the quality of a product are critical in industry. In choosing between two methods of analysis one has to decide which is more accurate and/or more precise. One often needs to evaluate the differences, if any, between various detection systems or between analyses from several laboratories of the same standard or sample. Comparative experiments and the analysis of variance (ANOVA) have become indispensable tools in these scientific endeavors.

Comparing the Means of Two Samples Whose Parent Populations Have Equal Variances

Two different catalysts are considered to be used in order to increase the yield of an industrial product. An experiment must be designed such that both catalysts are implemented at one level in a randomized order eight times each. The yields are shown in Table 2.1. Is the difference between the mean yields significant?

We shall assume that each observation obtained with catalyst A comes from a *normally distributed population* with mean μ_1 and variance σ^2. Similarly, each yield with catalyst B comes from a normally distributed population with mean μ_2 and an *equal* variance σ^2. All observations are assumed to be *uncorrelated*. These assumptions can be tested but are assumed to be correct here.

The assumption of equal population variances can be tested by using

Table 2.1 Percent Yield of an Industrial Product with Two Catalysts A and B

	A	B
	85	87
	86	85
	83	86
	82	93
	87	89
	90	88
	80	86
	81	89
Mean, $\bar{x}$	84	88
Variance, s^2	11	6.4

the F-test (1) as follows:

$$F = \frac{s_A^2}{s_B^2} = \frac{11}{6.4} = 1.7. \qquad (2.1)$$

This value of F is less than 3.79, which is the critical value for the 0.05 significance level and seven degrees of freedom in each sample. The assumption of equal population variances is thus verified, even though the sample variances are unequal, and we proceed to test the difference between the two means.

The test statistic is defined as

$$t_\nu = \frac{[(\bar{x}_1 - \bar{x}_2) - (\mu_1 - \mu_2)][n_1 + n_2 - 2]^{1/2}}{[1/n_1 + 1/n_2]^{1/2}[(n_1 - 1)s_1^2 + (n_2 - 1)s_2^2]^{1/2}}, \qquad (2.2)$$

where n_1 = the size (the number of trials) of the first sample (data with catalyst A)

n_2 = the size of the second sample

μ_1 = the mean of the population from which the first sample is drawn (the true average of the first sample)

μ_2 = the mean of the population from which the second sample is drawn

s_1^2 = the variance of the first sample (A)

s_2^2 = the variance of the second sample (B)

$\nu = n_1 + n_2 - 2$ and is referred to as the number of degrees of freedom

The quantity

$$\frac{(n_1 - 1)s_1^2 + (n_2 - 1)s_2^2}{n_1 + n_2 - 2}$$

is known as the pooled (combined) variance. It is the best unbiased estimator of the population variance σ^2.

The following hypotheses are to be tested:

1. Equality of the means:
   H_0: $\mu_1 = \mu_2$
   H_1: $\mu_1 \neq \mu_2$
2. Inequality of the means:
   H_0: $\mu_1 \leq \mu_2$
   H_1: $\mu_1 > \mu_2$

μ_1 is set equal to μ_2 in Equation (2.2) and the experimental t_{14} statistic for the data in Table 2.1 is calculated to be

$$t_{14} = 2.71 \tag{2.3}$$

with catalyst B data as sample 1 and catalyst A data as sample 2. For a confidence level of 95%, in the equality hypothesis, H_0 is rejected if the experimental value of t_ν is greater than the upper 0.025 value or less than the lower 0.025 value of the t-distribution with $\nu = 14$ degrees of freedom. For a confidence level of 95%, in the inequality hypothesis, H_0 is rejected if the experimental value of t_{14} is greater than the upper 0.05 value of the t-distribution with ν degrees of freedom. Comparing the value of t calculated for data in Table 2.1 with the value of the t-distribution in Table A.2 with 14 degrees of freedom at $\alpha = 0.05$ ($t_{14,0.05} = 1.76$) indicates that catalyst B has a significantly higher yield than catalyst A.

Comparing the Means of Two Samples Whose Parent Populations Have Unequal Variances

Frequently, the observations are either believed or suspected to come from populations that are normally distributed but have unequal variances. Flame atomic absorption (FAA) and graphite furnace atomic absorption (GAA), for example, are two analytical methods where the former offers a much better precision (a smaller variance) than the latter (precision of GAA can be improved by using automatic injectors). Suppose that you are to examine both methods in order to quantify a metal in drinking water in the neighborhood of 5 ppm in order to decide which technique is more suitable. An NBS standard (5 ppm) is tested by both methods and the data are summarized in Table 2.2. The variance of the GAA data is about nineteen times as high as the FAA sample. An F-test shows that the variances are unequal:

$$F = \frac{1.07 \times 10^{-3}}{5.67 \times 10^{-5}} = 18.9. \tag{2.4}$$

The F value is greater than the critical value of $F_{5,5,0.05}$. If precision is important, FAA is obviously the method of choice. On the other hand, if precision is not so important and replicate analyses are easy to obtain, you will have to decide whether or not the cost of GAA analysis is justifiable in view of the differences in results. *Are the two means significantly different*?

The t-statistic in this case is defined as

$$t_\nu = \frac{\bar{X}_1 - \bar{X}_2}{[S_1^2/n_1 + S_2^2/n_2]^{1/2}}, \tag{2.5}$$

Table 2.2 Analysis of a 5 ppm Standard by FAA and GAA

Sample	FAA	GAA
1	4.99	5.01
2	4.99	4.95
3	5.00	5.03
4	5.01	4.96
5	5.00	4.99
6	5.00	5.02
Mean	5.00	4.99
Variance	5.67×10^{-5}	1.07×10^{-3}

where ν is the number of degrees of freedom rounded to the nearest integer:

$$\nu = \frac{[S_1^2/n_1 + S_2^2/n_2]^2}{\left(\frac{S_1^2}{n_1}\right)^2\left(\frac{1}{n_1 - 1}\right) + \left(\frac{S_2^2}{n_2}\right)^2\left(\frac{1}{n_2 - 1}\right)} - 2. \quad (2.6)$$

For the data in Table 2.2,

$$\nu = 4 \quad \text{and} \quad t_\nu = 0.73.$$

The t value indicates that no significant difference is present between the means of the analyses obtained by FAA and GAA.

Equation (2.5) is only an approximate test. If no evidence is available to support the assumption of unequal population variances, the test given by Equation (2.2) must be used. The test statistic given in Equation (2.2) can tolerate, to some extent, cases of unequal population variances. This is especially true when $n_1 = n_2$. The above conclusion about FAA and GAA can be arrived at by considering the test statistic in Equation (2.2) and completely ignoring the fact that the population variances are unequal.

The experimental t_{10} value, by Equation (2.2), is

$$t_{10} = \frac{(5.00 - 4.99)(10)^{1/2}}{(\frac{1}{6} + \frac{1}{6})^{1/2}[5 \times 5.67 \times 10^{-5} + 5 \times 1.07 \times 10^{-3}]^{1/2}} = 0.73,$$

thus showing no significant difference.

COMPARING SEVERAL MEANS

Several means can be compared by one-way analysis of variance, ANOVA. If we are testing, for example, a third catalyst, C, or examining a third analytical technique, an ion selective electrode or polarography, Tables 2.1 and 2.2 would contain a third column each for the third factor being tested. Consider that a third catalyst had been incorporated in the earlier study. Table 2.3 summarizes the data for the experiment with three catalysts.

The observations are assumed to be independent and each sample is obtained from a normally distributed population with variance σ^2. However, each sample has a different mean. Define

n_1 = size of the first sample

n_2 = size of the second sample

n_3 = size of the third sample

k = total number of factors (catalysts) being tested

x_{ij} = data (yield) for the ith trial in the jth sample

$$T_j = \sum_{i=1}^{n_j} X_{ij}.$$

$$T = \sum_{j=1}^{k} T_j.$$

$$n = \sum_{j=1}^{k} n_j.$$

One-way ANOVA consists of performing the following calculations:

1. Sum of squares, $SS = \sum_i \sum_j x_{ij}^2$.
2. Total sum of squares, $TSS = SS - T^2/n$.
3. Between samples sum of square, $BSSS = \sum_{j=1}^{k} T_j^2/n_j - T^2/n$.
4. Residual (random error), $R = TSS - BSSS$.
5. Between samples mean square, $BSMS = BSSS/(k-1)$.
6. Residual mean square, $RMS = R/(n-k)$.
7. The test statistic F at α confidence level $= F_{(k-1,n-k,\alpha)} =$ BSMS/RMS; α is usually taken as 0.05.

A significant value for F indicates a significant variance due to the different factors, that is, the means of the three catalysts are *not the same*.

Table 2.3 Percent Yield of an Industrial Product Using Three Different Catalysts (A, B, and C)

	A	B	C
	85	87	89
	86	86	85
	83	85	90
	82	93	86
	87	89	83
	90	88	88
	80	86	87
	81	89	91
Mean	84	88	87
Variance	11	6.4	7.1

The one-way ANOVA results for the data in Table 2.3 are summarized in Table 2.4.

The F-test gives

$$F_{2,21} = \frac{30.88}{8.30} = 3.72 > F_{2,21,0.05}(3.47).$$

The F value is significant. This indicates that the mean yields are different. Note that the test does not show where the differences are. We have already shown that catalyst B has a significantly higher average than catalyst A. Let us compare catalysts B and C.

Table 2.4 ANOVA Table for Data in Table 2.3

Source of Variation	Sum of Squares	Degrees of Freedom	Mean Square
Between samples (due to catalyst)	61.75	2	30.88
Residual (random analytical error)	174.25	21	8.30
Total	236	23	

Table 2.5 ANOVA Results for Catalysts B and C of Table 2.3

Source of Variation	Sum of Squares	Degrees of Freedom	Mean Square
Between samples	1	1	1
Residual (random analytical error)	94.75	14	6.76
Total	95.75	15	

By Equation (2.2),

$$t_{14} = 0.77.$$

The value of t shows that there is no significant difference between the average yields of the two catalysts. One-way ANOVA can also be used to compare two means. Table 2.5 shows the ANOVA results for the data given in Table 2.3 with catalyst A data omitted. The F test shows an insignificant difference between the average yields, the same conclusion obtained by the t-test.

The t-test is the most common method used to compare two means. On the other hand, one-way ANOVA gives extra information, namely, the *magnitude and the sources of variations*. As shown in Table 2.5 the variance between replicas (sampling and analysis variance) is large compared to variance between the factors so that if there is any "real" difference between catalysts it is completely obscured. This is contrasted by the ANOVA results shown in Table 2.4 where the variance between samples is clearly larger than random variations and it can not be accounted for on that basis alone. Knowing the magnitude of random variations is helpful because it aids us to improve the sampling methods, the raw materials, and the analytical techniques involved in estimating the yields in order to keep them in an acceptable range. The ratio of BSMS to RMS and the magnitude of RMS can be of equal importance.

PAIRED OBSERVATIONS

To study the effect of a certain variable (a factor or a stimulus) on some system, several subjects are observed in the presence of the variable and "identical" subjects are observed in the absence of the same variable with all other factors held constant. This is a very popular design in research, industry, and quality control. One can, for example, test the effects of drugs on the rate of heart beat, catalysts on reaction rates, a particular stage of a process (a particular treatment) on the quality of a product, or the effects of temperature or ionic strength on the response of an ion selective electrode or a UV-Vis absorption band.

Suppose that a process by which mercury is removed from an industrial waste, before it is discharged, is being evaluated. Table 2.6 shows the concentration of mercury in ten homogeneous samples; half of each sample is analyzed before the treatment and the other half analyzed after the treatment. The sample of differences (last column in Table 2.6) is treated as a separate distribution and a mean, $\bar{d}$, and a variance, s_d^2, are calculated. The differences, d_i's, are also assumed to be independent and normally distributed around a true mean, μ, and have a variance σ^2. The estimated average, $\bar{d}$, can be compared to the true mean, μ, by the following statistic:

$$t_{n-1} = \frac{\bar{d} - \mu}{s_d} \sqrt{n}, \tag{2.7}$$

where n is the number of samples.

This statistic is similar to the Student's t-distribution with $n-1$ degrees of freedom. Since we are examining a reduction in mercury content of the samples, μ is set equal to zero in Equation (2.7) and the value of t_{n-1} will thus test if d is significantly different from zero. In the above example, if the t_{n-1} value is greater than the 0.05 point of the t-distribution

Table 2.6 Concentration of Mercury (ppm) in Ten Samples Before and After Treatment

Sample	Before	After	Difference, d_i (Before − After)
1	19.3	18.6	0.7
2	18.5	18.2	0.3
3	18.9	19.2	0.3
4	20.0	19.5	0.5
5	18.8	18.2	0.1
6	17.3	17.4	0.4
7	18.6	18.2	0.1
8	18.5	18.6	0.3
9	17.9	17.6	0.3
10	17.6	17.3	0.3
Mean	18.5	18.3	0.3

with $n - 1$ degrees of freedom, one concludes that $\bar{d}$ is significantly greater than zero with 95% confidence.

For the data shown in Table 2.6:

$$\bar{d} = 0.3,$$

$$s_d = 0.18,$$

and

$$t_9 = \frac{0.3}{0.18}(10)^{1/2} = 5.27 > t_{9,0.05}(1.833).$$

Therefore, the mercury removal process is effective; the untreated samples contain more mercury than the treated ones.

The preceding concept has crucial applications and implications in the process of signal detection—an extremely important step in chemical analysis. To determine whether or not a "significant signal" is present, one needs to compare the response of an instrument (i.e., the magnitudes of the current or the voltage outputs) in the presence of the analyte-containing sample to the response obtained in the absence of the sample (or in the presence of a suitable blank). If, for example, the results shown in the second and third columns of Table 2.6 are regarded as the analytical responses obtained for ten sample replicas and ten blanks, respectively, a t-test can show that the signal (the output or the response of the instrument) is indeed significant. Consequently, the presence of the analyte is confirmed. If the blanks are replaced by primary standards, one can conclude that the samples are more concentrated than the standards and thus a lower limit for the concentration of the samples is established. A more detailed discussion will be given in Chapter 3.

RANDOMIZED BLOCKS

A *block* is a subset of the total observations that is more homogeneous than all the observations combined. In evaluating data from several laboratories, the data obtained from one laboratory should be more "alike" than the pooled data. Similarly, the variations of the output of a plant during a single shift may be much smaller than those occurring over a period of 1 month. In randomized designs the subjects are randomly assigned to the various treatments with the assumption that all subjects are observed under identical conditions except for the factor(s) being tested. This, however, is not generally true. If the yields in Table 2.1 are obtained by experimenting in different laboratories or at different shifts or locations, the differences in *experimental conditions*, if present, will certainly bias our conclusions. The variations caused by experimental conditions have to be separated from those resulting from treatments (in this case the different catalysts). The samples from each group of the data that is believed to be more homogeneous than the whole (results obtained from one laboratory or analyses performed by one operator) are grouped together into a separate *block*. *Randomized blocks* thus amounts to subdividing the observations into blocks of relatively uniform conditions. A two-way ANOVA gives the variations due to factors and those due to blocks. Table 2.7 shows the general form of the data collected in a randomized block designed experiment.

Each observation in Table 2.7 is assumed to be the sum of four components:

$$x_{mk} = \mu + \alpha_m + \beta_k + e_{mk}, \qquad (2.8)$$

where μ = an overall mean
α_m = the effect of the mth block
β_k = the effect of the kth factor
e_{mk} = random error

All e_{mk}'s are uncorrelated and have a mean of zero and variance of σ^2.

Table 2.7 General Form of Randomized Blocks Data

Blocks		Factors				
		1	2	3 . . .	k	Block Total
	1	x_{11}	x_{12}	x_{13}	x_{1k}	TB_1
	2	x_{21}	x_{22}	x_{23}	x_{2k}	TB_2
	3	x_{31}	x_{32}	x_{33}	x_{3k}	TB_3
	⋮	⋮	⋮	⋮	⋮	⋮
	m	x_{m1}	x_{m2}	x_{m3}	x_{mk}	TB_m
Factor total		TF_1	TF_2	TF_3	TF_k	$T = \sum_{i=1}^{m} \sum_{j=1}^{k} x_{ij}$

Where

$$TF_j = \sum_{i=1}^{m} X_{ij} \quad \text{and} \quad TB_i = \sum_{j=1}^{k} X_{ij}$$

A two-way ANOVA table is constructed by calculating the following:

1. Sum of squares, $SS = \sum_{i=1}^{m} \sum_{j=1}^{k} X_{ij}^2$.
2. Total sum of squares, $TSS = SS - T^2/mk$.
3. Between blocks sum of squares, $BBSS = \sum_{i=1}^{m} (TB_i)^2/k - T^2/mk$.
4. Between factor sum of squares, $BFSS = \sum_{j=1}^{k} (TF_j)^2/m - T^2/mk$.
5. Residual (random error), R = TSS − BBSS − BFSS.
6. Between block mean square, BBMS = BBSS/$(m-1)$.
7. Between factor mean square, BFMS = BFSS/$(k-1)$.
8. Residual mean square, RMS = R/$(m-1)(k-1)$.
9. The effect of factors:
   H_0: $\beta_1 = \beta_2 = \beta_3 = \cdots = \beta_k$
   H_1: β_i's are not all equal
   The test statistic is F = BFMS/RMS.
10. The effect of blocks:
    H_0: $\alpha_1 = \alpha_2 = \alpha_3, \ldots, \alpha_m$
    H_1: α_i's are not all equal
    The test statistic is F = BBMS/RMS.

The F ratios are interpreted as has been discussed for one-way ANOVA. Table 2.8 represents a typical two-way ANOVA table.

Table 2.9 shows the concentration of chromium in a river at four different locations and three different depths at each location near an industrial waste disposal site. Table 2.10 summarizes the analysis of variance for the data in Table 2.9. The variations due to location are clearly significant. Also, the depths of the samples are shown to have a significant effect.

Table 2.8 Two-Way ANOVA Table

Sum of Variations	Sum of Squares	Degrees of Freedom	Mean Square
Between factors (column)	BFSS	$k-1$	BFSS/$(k-1)$
Between blocks (rows)	BBSS	$m-1$	BBSS/$(m-1)$
Residuals	R	$(m-1)(k-1)$	R/$(m-1)(k-1)$
Total	TSS	$mk-1$	

Table 2.9 Concentration of Chromium Near Waste Disposal Site

	Distance from Site (km)				
Depth, m	1	2	3	4	Total
0	50.0	30.5	20.2	10.3	111.0
0.5	46.0	30.4	18.0	8.0	102.4
1.0	45.0	27.5	15.0	6.0	93.5
Total	141.0	88.4	53.2	24.3	366.9

The rows in Table 2.9 could as well represent three different laboratories or three different instruments being used to analyze three "identical" samples from each location. If this were the case, one can conclude that the methods of analysis show significantly different results. The design can be used to test consistency between several laboratories, instruments or operators. Consider the following arrangement where each of six different samples of a certain metal pollutant is analyzed by five different methods (instruments, laboratories, procedures, or operators) and con-

Table 2.10 Two-Way ANOVA for Data in Table 2.9

Source of Variation	Sum of Square	Degrees of Freedom	Mean Square
Between locations	2523.13	3	841.04
Between depths	38.28	2	19.14
Residual	4.41	6	0.735
Total	2565.82	11	

Effect of location: $F = \dfrac{841.04}{0.735} = 1144.27 > F_{3,6,0.05}$

Effect of depth: $F = \dfrac{19.14}{0.735} = 26.04 > F_{2,6,0.05}$

centration reported in ppm:

	Samples					
Method	1	2	3	4	5	6
1	5.02	9.96	15.30	12.10	13.00	17.00
2	5.01	10.02	15.40	12.05	13.02	17.02
3	4.40	9.40	14.70	11.10	12.20	16.30
4	4.98	10.00	15.30	12.13	13.05	16.98
5	4.99	10.01	15.35	12.00	13.01	17.00

A two-way ANOVA model is adopted to examine the variations due to the methods. The variation due to the samples is expected to be significant. Indeed, the variations due to the methods are significant ($F = 112.74 > F_{4,20,0.05}$). This significant F value implies that at least one method gives significantly different results than the others. This serves to alert the analyst to the presence of certain elements, inherent to at least one method, with analytically serious consequences. These elements can be as simple as operational errors (errors due to inadequate sampling, detection, and/or calibration) or as complex as the presence of interferences (one or more interfering analytes) or matrix effects. Since operational errors are virtually absent under "optimal" analysis conditions, interferences and matrix effects are usually the most suspected causes of the inconsistency obtained when several methods of analysis are being compared. Note that the analyst does not have to know the true concentration of the samples in order to infer that the five different methods are *not* equivalent.

Unfortunately, two-way ANOVA *does not* indicate which methods are responsible for the variations. Naturally it is of utmost importance to know which method is most reliable for our analysis purpose. The analyst must determine whether or not a recommended method of analysis is subject to interferences. The direct, but not necessarily the most efficient, approach is to examine *each* method separately by testing primary standard samples. There are, however, alternative procedures that aid the analyst to determine not only the qualitative effects of interferences but also their quantitative impact on the analytical results. These procedures are usually more efficient than the direct approach mentioned above and can determine which methods are responsible for significant (systematic) variations. They rely on multivariate data analysis (e.g., principal component analysis and multiple regression) and they will be presented in later chapters. The impatient reader may consult the literature (2–4) for various applications on the use of multivariate data analysis to estimate the presence of interferences both qualitatively and quantitatively.

LATIN SQUARES

Blocking of observations aids in evaluating any *nonrandom* variations in experimental conditions. However, it does not detect nonrandom variations in the *implementation* of the various treatments. While grouping of data into homogeneous blocks separates day-to-day or laboratory-to-laboratory variations in results, it does not separate variations due to, for example, operator-to-operator *application* of the treatment.

Consider an oil refinery where four different kinds of gasoline (A, B, C, and D) are tested for octane numbers. A random sample is drawn from each stock every day and submitted to one of four groups (testers) in charge of testing its octane number. If day-to-day variations and/or tester-to-tester variations are either negligible or nonexistent, a randomized block design is sufficient to evaluate the variations in octane numbers. However, those variations may be critical. In addition, if each tester examines only one particular kind of gasoline, a nonrandom source of variation will be introduced.

It is thus imperative to eliminate day-to-day and tester-to-tester variations by ensuring that each kind of gasoline is tested with *the same daily frequency* and by every tester. The octane number of each kind of gasoline is, consequently, an average over all days and over all testers. The testing results are typically arranged as shown in Table 2.11. Such a table is

Table 2.11 An $m \times m$ Latin Square ($m = 4$)

	Tester				
Day	1	2	3	4	Total
1	A	B	C	D	TB_1
2	B	C	D	A	TB_2
3	C	D	A	B	$\vdots$
4	D	A	B	C	TB_m
Total	TF_1	TF_2	. . .	TF_m	$T = \sum_{i=1}^{m} TB_i = \sum_{i=1}^{m} TF_i$

known as an $m \times m$ Latin square. Each entry is expressed as x_{ijk} where i stands for the row, j stands for the column, and k stands for the treatment. The variation in the Latin square entries (results) is the sum of the *variations between rows* (day-to-day), the *variations between columns* (tester-to-tester), the *variation between treatments* (gasoline-to-gasoline), and the *random errors*. The basic assumption is that these four components act independently. In practice they may interact to some degree. The latin square design will still be valid if their interactions are very small compared to the effects being estimated. Consequently, Latin square designs are not generally recommended to study highly correlated factors such as temperature, solubility, and reaction rates.

Each observation, x_{ijk}, is assumed to be expressible as

$$x_{ijk} = \mu + \alpha_i + \beta_j + \gamma_k + \varepsilon_{ijk} \tag{2.9}$$

where μ = an overall mean
α_i = effect of the ith row (day)
β_j = effect of the jth column (tester)
γ_k = effect of the kth treatment (gasoline)
ε_{ijk} = random analytical error; ε_{ijk}'s are distributed as $N(0, \sigma^2)$.

The analysis of variance of an $m \times m$ Latin square involves the following calculations.

1. Sum of square = SS = $\sum_{i=1}^{m} \sum_{j=1}^{m} \sum_{k=1}^{m} X_{ijk}^2$. Add the squares of all observations.
2. Total sum of square = TSS = SS − T^2/m^2, $T = \sum_{i=1}^{m} \sum_{j=1}^{m} \sum_{k=1}^{m} X_{ijk}$.
3. Between column sum of square = BCSS = $\sum_{i=1}^{m} (TF_i)^2/m - T^2/m^2$.
4. Between rows sum of square = BRSS = $\sum_{i=1}^{m} (TB_i)^2/m - T^2/m^2$.
5. Between treatment sum of square = BTSS = $\sum_{i=1}^{m} (T_i)^2/m - T^2/m^2$, where T_i is the sum of all observations for the ith treatment.
6. Residual = R = TSS − BCSS − BRSS − BTSS.
7. Between columns mean square = BCMS = BCSS/$(m-1)$.
8. Between rows mean square = BRMS = BRSS/$(m-1)$.
9. Between treatment mean square = BTMS = BTSS/$(m-1)$.
10. Residual mean square = RMS = $R/[(m^2-1)-3(m-1)]$.

An ANOVA table is constructed and the effect of each source of variation is examined by an F-test as shown before.

Table 2.12 Latin Square Design for Testing Octane Numbers

Day	Tester 1	2	3	4	Total (TB_i)
1	72(B)	78(D)	88(A)	77(C)	315
2	75(D)	79(B)	75(C)	83(A)	312
3	90(A)	90(C)	74(B)	77(D)	331
4	76(C)	89(A)	74(D)	64(B)	303
Total (TF_i)	313	336	311	301	$T = 1261$

Table 2.12 shows the results of testing octane numbers of four kinds of gasoline (A, B, C, and D) over a period of 4 days by four different testers. From Table 2.12, we have the following:

$$\frac{T^2}{m^2} = 99{,}382.563$$

$$\text{TSS} = 812.44$$

$$\text{SS} = 100{,}195$$

$$T_A = T_1 = 90 + 89 + 88 + 83 = 350$$

$$T_B = T_2 = 72 + 79 + 74 + 64 = 289$$

$$T_C = T_3 = 76 + 90 + 75 + 77 = 318$$

$$T_D = T_4 = 75 + 78 + 74 + 77 = 304.$$

The analysis of variance is shown in Table 2.13. F-tests show that all

Table 2.13 ANOVA Table for Data in Table 2.12

Source of Variation	Sum of Squares	Degrees of Freedom	Mean Square
Between testers	164.18	3	54.73
Between days	102.18	3	34.06
Between gasoline	507.69	3	169.23
Residual	38.39	6	6.40
Total	812.44	15	

effects are significant. The design has thus separated day-to-day and test-to-test effects that could have obscured the actual gasoline-to-gasoline effect variations. The gasoline-to-gasoline effect is probably expected. On the other hand, significant variations from day-to-day and tester-to-tester are an indication of *heterogeneity* in equipment, performance, personal skills, and raw material. This information can serve as a feedback to adjust operation conditions in order to bring this heterogeneity to an acceptable level or preferably eliminate it.

FACTORIAL DESIGNS

There are numerous cases where the underlying factors do not display their effects independently of each other. Industrial catalysts often contain *promoters*. Assume that the activity of a certain catalyst is determined by two promotors, T and I. One can study the effect of T while the amount of I is held constant or vice versa. The results of such an experiment, however, are misleading. Since the effect of each factor is certain to depend on the level of the other, the experiment is bound to yield an untrue image of the effect of each factor and, most importantly, is incapable of detecting the effect of the interdependence of both factors. Phenomena that involve factors whose effects are not independent of the levels of other factors must be examined by varying all factors' levels simultaneously. Conducted in this manner the experiment yields the effects of each factor accurately and also evaluates the effects of the interactions between factors.

If n factors are examined at $\ell_1, \ell_2, \ldots, \ell_n$ levels, the array of $\ell_1 \times \ell_2 \times \ell_3 \times \cdots \times \ell_n$ observations is known as an $\ell_1 \ell_2 \ell_3 \ldots \ell_n$ factorial design. For simplicity, a 2×2 factorial design will be demonstrated. The activity of an industrial catalyst is examined as the amounts of the promotors T and I vary. The effects of both factors are suspected to be dependent on each other. The amount of I is varied from 0.2% to 0.5% and the amount of T is varied from 20% to 40%. Each observation is duplicated (two trials for each treatment). Table 2.14 shows the different activities. Each observation in Table 2.14, x_{ijk}, is assumed to be expressible as

$$x_{ijk} = \mu + \alpha_i + \beta_j + \gamma_{ij} + \varepsilon_{ijk}, \tag{2.10}$$

where x_{ijk} = kth trial of ith I and jth T
μ = an overall mean
α_i = effect of the ith I
β_j = effect of the jth T
γ_{ij} = effect of the interaction between the ith I and the jth T
ε_{ijk} = random error; ε_{ijk} are distributed as $N(0, \sigma^2)$

Table 2.14 Percent Activity of an Industrial Catalyst (max = 100%) at Different Compositions of Promotors *T* and *I*

I	T	
	20%	40%
0.2%	29,24	35,40
0.5%	76,72	45,47

The analysis of variance proceeds as follows:

1. Total $= T_{tot} = 29 + 24 + \cdots + 45 + 47 = 368$.
2. Sum of square $= SS = (29)^2 + (24)^2 + \cdots + (45)^2 + (47)^2 = 19{,}436$.
3. Total sum of squares $TSS = SS - T_{tot}^2/8 = 2508$.
4. Effect of T: Add the *four* observations in each column of Table 2.14.

	$T = 20\%$	$T = 40\%$
Sum:	201	167

   Sum of square due to $T = SS_T = \frac{1}{4}[(201)^2 + (167)^2] - T_{tot}^2/8$.
   $SS_T = 144.5$.
5. Effect of I: Add the *four* observations in each row of Table 2.14.

	$I = 0.2\%$	$I = 0.5\%$
Sum:	128	240

   Sum of square due to $I = SS_I = \frac{1}{4}[(128)^2 + (240)^2] - T_{tot}^2/8$.
   $SS_I = 1568$.
6. Effect of interaction between T and I (TI): Add the two replicas of each treatment.

	$I = 20\%$	$T = 40\%$
$I = 0.2\%$	53	75
$I = 0.5\%$	148	92

   Sum of square due to $TI = SS_{TI} = \frac{1}{2}[(53)^2 + (75)^2 + (92)^2 + (148)^2] - T_{tot}^2/8 - SS_I - SS_T$.
   $SS_{TI} = 760.5$.
7. Residual (analytical error) $SR = TSS - SS_T - SS_I - SS_{TI}$.
   $SR = 35$.

8. ANOVA table: for two factors at m and p level and n replicas of each F-tests show that the effects of T, I, and TI are significant. Note that the interaction of T and I is more significant than T. This implies that, due to the relatively large interaction, the effects of T are *meaningless* unless the level of I is specified.

Source of Variation	Sum of Squares	Degrees of Freedom	Mean Square
T (m levels)	144.5	$(m-1)=1$	144.5
I (p levels)	1568	$(p-1)=1$	1568
TI	766.5	$(m-1)(p-1)=1$	760.5
Residual	35	$pm(n-1)=4$	8.75
Total	2508	$mpn-1=7$	

Factorial designs have a wide range of applications in research and industrial laboratories. Their only serious disadvantage is that the analysis of variance becomes more laborious as the number of factors and/or levels increases. The "Suggested Readings" section at the end of this chapter has more to offer on this topic as well as on the important tool of Fractional Factorial Designs.

Modeling

The analysis of variance helps the analyst to decide *which* factors display significant effects on the results. However, it does not give any information about *how* the factors affect the results. For example, we have inferred from the data shown in Table 2.3 that the type of catalyst does affect the yield. We do not know, however, what the yield will be if we double the amount of each catalyst. The inferences that were made are thus *qualitative* in nature. In several situations the purpose of the experiment is fulfilled at this stage. Once catalyst B, for example, is established as a better catalyst (causing higher yields) the experiment ends. This is sufficient for most experiments involving only qualitative factors (e.g., type of catalyst, kind of fertilizer). When the factors are quantitative in nature (e.g., temperature, concentration) one can not only find out *which* factors are important but also estimate *how* the factors affect the results. The latter is accomplished by evaluating the *quantitative* relationship between the factors and the results.

It is often desirable to express the results of experiments as a mathematical function of the experimental conditions—the factors and their levels. This practice has two advantages. First, it replaces large tables of data with a single equation and second, it provides a means to *predict and estimate* the results at levels that were not directly studied. The mathematical equation that expresses the results (e.g., solubility of inorganic salts) in terms of the experimental factors (e.g., temperature and ionic strength) is referred to as the *model*. The experimental results are referred to as the *responses*. A model can be very crucial if these responses are to be optimized.

Models can be classified into two major categories. The first category includes what are known as *theoretical models*. These models apply when the phenomenon under investigation is well understood and the results are expected to follow a known (theoretical) pattern. An example is the well known and widely used calibration curve method. In this technique the response of an instrument (e.g., the measured absorption of a colored complex in a UV-Vis spectrophotometer) is linear with respect to concentration over a certain range of concentrations. The modeling of such data is best described as a linear function. The second category involves *empirical models*. These models are used when the responses show such a complex behavior that no simple theory or model can account for the complete set of results. The analyst, nevertheless, can assume that a simple (linear or quadratic) model is adequate over a limited range of the independent variables. For instance, the true quantitative relationship governing the dependence of the activity of the catalyst on the amount of promoters (the data shown in Table 2.14) may be a complex one. One can *assume* that the activity at the amount of $I = 0.2\%$ is linear (or quadratic) over the range $T = 20\text{–}40\%$. An empirical model can then be estimated, and tested, over this range of compositions. Such a model may be used to predict what the activity is at $I = 0.2\%$ and $T = 30\%$.

Modeling of experimental data is an indispensable tool for research and analysis endeavors. It aids in establishing the quantitative relationships between responses and experimental variables. Later in this chapter we shall show how models are used to optimize experimental conditions. In subsequent chapters the use of models to resolve analytical signals will be discussed. More detailed treatments of modeling are available and the reader is advised to consult them or other equivalent discussions.

Table 2.15 lists the values of the limiting diffusion currents (Y's) obtained during the polarographic reduction of a certain metal at various concentrations (X's). It is convenient to express the relationship between Y and X mathematically. The mathematical relationship aids us in *quickly* predicting the concentrations of unknown samples after measuring the limiting diffusion current of their polarograms. The data in Table 2.15

Table 2.15 The Response of a dc Polarographic Wave (in terms of i_d) of a Certain Metal at Different Concentrations

Concentration, m*M* (X)	i_d, μA (Y)
0.10	0.100
0.15	0.224
0.20	0.398
0.25	0.628
0.30	0.896
0.35	1.231
0.40	1.608
0.45	2.035

represent the *calibration* process. Estimating the concentration of unknown samples graphically can be time consuming.

The algebraic equation relating Y and X in Table 2.15 and any other set of variables in general (the model) is postulated, and later analyzed, in view of the theory and the practical aspects of the process. The magnitude of i_d, for example, can be shown to be linearly related to concentration. On this basis, the following linear model is postulated for the data in Table 2.15:

$$y = \beta_1 x. \tag{2.11}$$

This model is a simple linear model which can be graphically represented by a straight line passing through the origin. The y's and x's are known. The only unknown in the model is the parameter β_1. The model is completed when β_1 is estimated.

Estimation of β_1 (Linear Model). Let b_1 be an estimate of β_1. A *modeled* value for each observation, $\hat{y}_i$, is given by

$$\hat{y}_i = b_i x_i . \tag{2.12}$$

Assuming that the x_i's are precisely known (i.e., $\sigma^2_{x_i} = 0$), a value for b_1 is chosen such that the following quantity is a *minimum*:

$$\sum \frac{(y_i - \hat{y}_i)^2}{\sigma^2_{y_i}},$$

where y_i = experimental value for the ith observation
$\hat{y}_i$ = modeled value for the ith observation
$\sigma^2_{y_i}$ = variance of y_i

Let

$$\text{SE} = \sum \frac{(y_i - \hat{y}_i)^2}{\sigma^2_{y_i}} = \sum \frac{1}{\sigma^2_{y_i}} (y_i - b_1 x_i)^2$$

and assume that the y_i's are homoscedastic (i.e., $\sigma^2_{y_1} = \sigma^2_{y_2} = \cdots = \sigma^2_{y_n} = \sigma^2$). Then,

$$\text{SE} = \frac{1}{\sigma^2}\left[\sum y_i^2 - 2b_1 \sum x_i y_i + b_1^2 \sum x_i^2\right]. \tag{2.13}$$

To find the value of b_1 that minimizes SE, we set $d(\text{SE})/db_1 = 0$. Equation (2.13) thus yields

$$b_1 \sum x_i^2 = \sum x_i y_i. \tag{2.14}$$

Solving for b_1,

$$b_1 = \frac{\sum x_i y_i}{\sum x_i^2}. \tag{2.15}$$

This solution for b_1 (simplified by assuming that the y_i's are homoscedastic) is known as the *unweighted least squares solution*. When the y_i's are not homoscedastic, b_1 is given as

$$b_1 = \frac{\sum (x_i y_i / \sigma^2_{y_i})}{\sum (x_i^2 / \sigma^2_{y_i})}. \tag{2.15a}$$

The solution given by Equation (2.15a) is known as the *weighted least squares solution*. Each measurement is weighted by the reciprocal of its variance.

Assuming that the linear model is applicable for the data in Table 2.15 and using Equation (2.15), $b_1 = 3.576$. Therefore the linear model is

$$y = 3.576x \tag{2.16}$$

or

$$i_d = 3.576[\text{M}],$$

where [M] is in units of mM and i_d is in μA.

ANOVA FOR LINEAR MODELS

An ANOVA table is constructed as follows [n = number of observation, p = number of parameters being estimated, and $\bar{y}$ is the average value for y (i.e., $\bar{y} = \sum y_i/n$)].

Source of Variation	Sum of Squares	Degrees of Freedom	Mean Square
Model	$\sum (\hat{y}_i - \bar{y})^2$	p	$\frac{1}{p}\sum (\hat{y}_i - \bar{y})^2$
Residual	$\sum (y_i - \hat{y}_i)^2$	$n-p-1$	$\frac{1}{n-p-1}\sum (y_i - \hat{y}_i)^2$
Total	$\sum (y_i - \bar{y})^2$	$n-1$	

Table 2.16 summarizes the analysis of variance for the model in Equation (2.16).

An F-test shows a significant variance due to the model. This test by itself, however, is *not* evidence of a good model.

Let us see, graphically, how good our model fits the data. Figure 2.1 shows the data points and the model given in Equation (2.16). The model is inappropriate. Table 2.16 shows that the total variance is much different from the sum of the variance due to the model and the residual variance. Also, the residuals, $y_i - \hat{y}_i$, exhibit a suspicious (negative at low [M] and positive at high [M]) behavior (Table 2.17).

Examining the residuals will be disscussed in more detail in Chapter 4. At this point it suffices to say that the residuals suggest a systematic

Table 2.16 ANOVA for Linear Model for i_d and [M]

Source of Variation	Sum of Squares	Degrees of Freedom	Mean Square
Model	1.412	1	1.412
Residual	0.580	6	0.0967
Total	3.321	7	

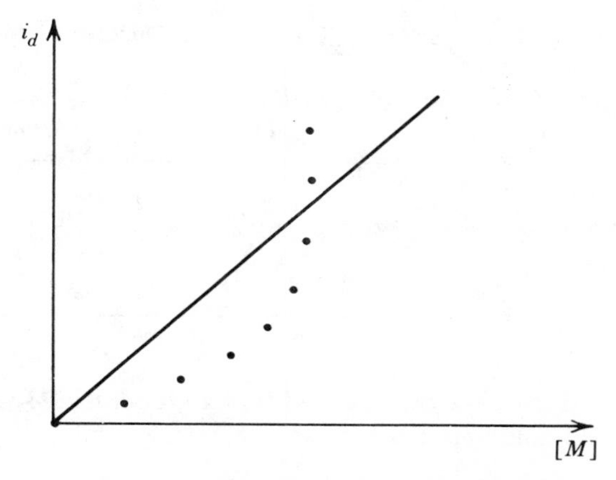

Figure 2.1 The data and model of Equation (2.16).

(nonrandom) variation. In addition, the model only accounts for $(1.412/3.321) \times 100 = 42.5\%$ of the total variance. These observations suggest that either the linear model is incorrect or that other factors are missing.

If no other factors need be included, another model must be postulated. Figure 2.1 indicates that a quadratic model may be appropriate. Consider the following model

$$y = \beta_1 x^2. \tag{2.17}$$

Table 2.17 List of Residuals

n	$y_i - \hat{y}_i$
1	−0.258
2	−0.312
3	−0.317
4	−0.266
5	−0.177
6	−0.021
7	0.178
8	0.426

Table 2.18 ANOVA Table for the Quadratic Model

Source of Variation	Sum of Squares	Degrees of Freedom	Mean Square
Model	3.307	1	3.307
Residual	8.2×10^{-5}	6	1.367×10^{-5}
Total	3.321		

Estimation of β_1 by an Unweighted Fit of a Quadratic Model. Let b_1 be an estimate of β_1 and define $\mathrm{SE} = \sum (y - b_1 x^2)^2$. Setting $d\mathrm{SE}/db_1 = 0$, gives

$$b_1 = \frac{\sum (y_i x_i^2)}{\sum x_i^4}. \tag{2.18}$$

For the data in Table 2.15, $b_1 = 10.04$. The model is

$$y = 10.04x^2 \tag{2.19}$$

or $i_d = 10.04[\mathrm{M}]^2$; i_d in μA and [M] in mM.

The analysis of variance for the above model [Eq. (2.19)] is shown in Table 2.18.

An F-test shows that the variance due to the model is highly significant (much more significant than that obtained with the linear model). The model accounts for 99.6% of the total variation. The sum of the model variation and the residual variation is very close (but not equal) to the total variation. The differences may be due to deviations from our assumption that ε_i values are distributed as $N(0, \sigma^2)$. Usually, when the model accounts for such a high portion (99.6%) of the total variance, the model need not be refined or changed (e.g., assuming higher degrees polynomials) because no *real* gain in accuracy is attained. In practice, the total variance and the variance due to the model are first calculated and the variance due to the residual is then calculated by subtraction.

Unweighted Linear Model in Two Variables. Assume that

$$y_i = \beta_1 x_{1i} + \beta_2 x_{2i}. \tag{2.20}$$

The above model can be used to describe the combined absorption (y) of two components (X_1 and X_2) at a certain wavelength, assuming that Beer's Law is additive for two interfering components. Using the method of least squares one can show that

$$b_1 \sum x_{1i}^2 + b_2 \sum x_{1i}x_{2i} = \sum y_i x_{1i} \tag{2.21a}$$

and

$$b_1 \sum x_{1i}x_{2i} + b_2 \sum x_{2i}^2 = \sum y_i x_{2i}. \tag{2.21b}$$

Equations (2.21a) and (2.21b), often referred to as the *normal equations*, can be solved simultaneously to calculate b_1 and b_2 (the estimates of β_1 and β_2, respectively).

Unweighted Nonlinear Models in Two Variables. Assume that

$$y_1 = \beta_1 x_{1i}^2 + \beta_2 x_{2i}^2 + \beta_3 x_{1i}x_{2i}. \tag{2.22}$$

Again, using the least squares approach, one can show that

$$b_1 \sum x_{1i}^4 + b_2 \sum x_{1i}^2 x_{2i}^2 + b_3 \sum x_{1i}^3 x_{2i} = \sum y_i x_{1i}^2, \tag{2.23a}$$

$$b_1 \sum x_{1i}^2 x_{2i}^2 + b_2 \sum x_{2i}^4 + b_3 \sum x_{1i} x_{2i}^3 = \sum y_i x_{2i}^2, \tag{2.23b}$$

and

$$b_1 \sum x_{1i}^3 x_{2i} + b_2 \sum x_{1i} x_{2i}^2 + b_3 \sum x_{1i}^2 x_{2i}^2 + b_3 \sum x_{1i}^2 x_{2i}^2 = \sum y_i x_{1i} x_{2i}. \tag{2.23c}$$

Equations (2.23a)–(2.23c) can be solved for b_1, b_2, and b_3 (the estimates β_1, β_2, and β_3, respectively).

Matrix Form Solution for Regression Parameters. The parameters in Equations (2.11), (2.17), (2.20), and (2.22) are known as regression parameters. The above method of solving for these parameters is also known as regression analysis. The above multivariate models can be expressed and solved for in a convenient manner using matrix notation and matrix algebra (5).

Let

$$y_i = \beta_1 x_{1i} + \beta_2 x_{2i} + \cdots + \beta_k x_{ki}.$$

The results, y_i's can be expressed in matrix form in terms of a coefficient matrix, $\boldsymbol{\beta}$, and a variable matrix, $\mathbf{X}$, as follows:

$$\mathbf{Y} = \mathbf{X}\boldsymbol{\beta}.$$

The coefficient matrix $\boldsymbol{\beta}$, can be estimated as $\mathbf{b}$ according to

$$\mathbf{b} = (\mathbf{X}^T\mathbf{X})^{-1}\mathbf{X}^T\mathbf{Y} \quad (2.24)$$

where $\mathbf{X}^T$ is the transpose of the matrix $\mathbf{X}$ and $(\mathbf{X}^T\mathbf{X})^{-1}\mathbf{X}^T$ is the so called *generalized inverse of* $\mathbf{X}$.

The parameter β_1 in the model given by Equation (2.17) can be estimated as b_1 from the data in Table 2.15 by Equation (2.24). After squaring each value of x in Table 2.15, b_1 is found by Equation (2.24) to be 10.04. This is the same value obtained by least squares [Eq. (2.19)]. This is also a special case of using the generalized inverse solution for a univariate (one *independent variable*, x, and one dependent *variable*, y) model. In this case, b becomes a one-dimensional vector (or a scalar) and y and x are vectors instead of matrices.

RESPONSE SURFACES

As mentioned earlier each experimental datum is referred to as a response. Consequently, the behavior of the responses as the factors vary (e.g., Fig. 2.1) is known as the *response surface*. Figure 2.1 is a two-dimensional response surface or a univariate, or one variable (factor), response surface. Generally, for N variables there is an $(N+1)$-dimensional space response surface. The relationship between the response and the factors is called the *response function*. Models, such as those discussed earlier, are usually used to approximate the response functions. If y denotes the response that is dependent on the factors $x_1, x_2, \ldots, x_k$, then y can be expressed as

$$y = f(x_1, x_2, \ldots, x_k) \quad (2.25)$$

and the above function, f, is referred to as the response function.

Response functions are the key to study and predict the different interactions leading to a certain phenomenon. For example, it is extremely important to know how the yield of a particular reaction varies as temperature and/or concentrations change. A knowledge of how the yield is affected (e.g., the combination of temperature and concentration that maximizes the yield) can prove to be crucial. In gas chromatography (GC) the resolution is dependent on the flow rate and column temperature. The analyst must choose the proper combination of these variables to maximize the resolution. In general, the maximum amount of information must be obtained from the experiment at the lowest cost. This is partly achieved by maximizing the *signal* (the systematic changes in the system which are caused by the controlled experimental factors and are related to the described information). Response surfaces are investigated in order to find the factors' levels at which the response (signal) is a maximum. It should also be said that the methods presented below can also find minima on the response surfaces. These points are elaborated on in Chapter 7.

Response Surfaces with Known or Estimable Shapes

Suppose that you are trying to determine the flow rate, f, and the molar ratio, r, of phase A to phase B at which the liquid chromatographic (LC) resolution, R, of two compounds is best. Assume that the resolution of these two compounds is known to be related to r and f as shown in Equation (2.26)

$$R = e^{-(r-1)^2} + e^{-(f-2)^2}, \tag{2.26}$$

where r = moles of A/moles of B and f is the flow rate in mL.

The above equation could have been either analytically derived or a fitted response surface (a model) obtained by a least squares fitting procedure of previous data.

To find the values r_{opt} and f_{opt} at which R is a maximum (R_{max}) we simply differentiate Equation (2.26) with respect to r and to f and set the derivative equal to zero:

$$\frac{\partial R}{\partial r} = \frac{\partial R}{\partial f} = 0.$$

Therefore,

$$r_{opt} = 1.0, \quad f_{opt} = 2.0, \quad \text{and} \quad R_{max} = 2.0. \tag{2.27}$$

Of all the optimization techniques, the above procedure is the simplest and most straightforward. Generally, however, the response function is *not* known beforehand and estimating a model of it is difficult if not practically impossible. It is extremely advantageous to be able to maximize a response function without any knowledge of the detailed shape of the surface other than knowing that the surface does (or is suspected to) have a maximum (a peak)—a single large peak with steep slopes.

We shall briefly demonstrate two methods of optimization that do not rely on the shape of the response function. We shall continue with the same LC example given above and use Equation (2.26) to simulate values of R at different r and f values. Remember that even though the R values are still given by Equation (2.26) we shall pretend that Equations (2.26) and (2.27) are unknown.

Response Surfaces with Unknown Shapes

The Method of Steepest Ascent

This method assumes that the response surface is similar to that shown in Figure 2.2. The method consists of employing a *linear model* away from the maximum and then "climbing" up the surface in the direction of the steepest slope (thus the name *steepest ascent*) as indicated by the linear model. The steps involved are as follows:

STEP 1. Conduct preliminary experiments to characterize the following linear model:

$$y = \beta_1 x_1 + \beta_2 x_2 + \cdots + \beta_k x_k + \beta_0, \tag{2.28}$$

where the x_i are factors.

The purpose of the preliminary experiment is to estimate the slopes $(b_1, b_2, \ldots, b_x)$ and the experimental random error. Two-level factorial

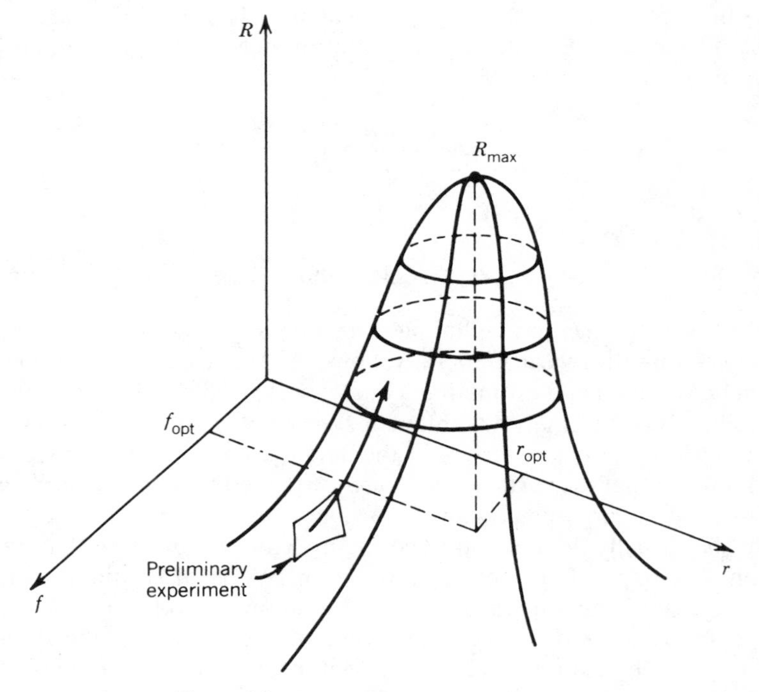

Figure 2.2 The method of steepest ascent.

Table 2.19 Value of R at Different r and f Values: $r = 0.5$, 0.6, and 0.7 (or −1, 0, and +1) and $f = 1.2$, 1.3, and 1.4 (or −1, 0, and +1)

Variable Space		Design Space		
r	f	r_c	f_c	R
0.5	1.2	−1	−1	1.31
0.7	1.4	+1	+1	1.61
0.5	1.4	−1	+1	1.48
0.7	1.2	+1	−1	1.44
0.6	1.3	0	0	1.47

designs are suitable at this stage. Table 2.19 shows the resolution obtained at five different combinations of r and f. The variables are *coded* (scaled) such that the low, average, and high values are given the values of −1, 0, and +1, respectively. These are the values of the variables in the *design space*. The coding makes the necessary calculations much easier than ordinary least squares procedure.

The point (0.6, 1.3), or (0, 0), is referred to as the *origin of the design*.

The linear model that needs to be fitted is

$$R = \beta_1 r + \beta_2 f + \beta_0. \tag{2.29}$$

The parameters β_1, β_2, and β_0 are estimated as b_1, b_2, and b_0, respectively (the origin of the design is excluded when b_1 and b_2 are calculated):

$$b_1 = \frac{1}{n} \sum R_i r_{ci}, \tag{2.30}$$

$$b_2 = \frac{1}{n} \sum R_i f_{ci}, \tag{2.31}$$

and

$$b_0 = \frac{1}{n} \sum R_i. \tag{2.32}$$

For data in Table 2.19 and the model given by Equation (2.29),

$$b_1 = \tfrac{1}{4}[-1.31 + 1.61 - 1.48 + 1.44] = 0.065,$$

$$b_2 = \tfrac{1}{4}[-1.31 + 1.61 + 1.48 - 1.44] = 0.085,$$

and

$$b_0 = \tfrac{1}{5}[1.31 + 1.61 + 1.48 + 1.44 + 1.47] = 1.46.$$

Therefore,

$$R = 0.065 r_c + 0.085 f_c + 1.46. \tag{2.33}$$

STEP 2. Examine the adequacy of the linear model (lack of curvature and interaction).

A quick test for the curvature of the plane is to measure the difference between the resolution at the origin of the design and the average of resolution at the other four points, ΔR.

$$\Delta R = 1.47 - \frac{1.31 + 1.61 + 1.48 + 1.44}{4} = 0.01.$$

ΔR should be expressed together with the experimental variance (as estimated by replication) in order to determine if it is significantly different from zero. Another test is to estimate the interaction parameter, β_{rf}. (The origin is excluded.)

$$b_{rf} = \frac{1}{n} \sum R_i r_{ci} f_{ci}$$

$$= \tfrac{1}{4}(1.31 + 1.61 - 1.48 - 1.44) = 0.0.$$

Again, this should be expressed together with the experimental error.

For the purpose of this discussion we shall assume that both ΔR and b_{rf} are negligible (not significantly different from zero) and that the linear model is adequate. If the surface shows large deviations from linearity in the region of the design different treatments are necessary and the method of steepest ascent is inapplicable.

STEP 3. The path of steepest ascent is determined by b_1 and b_2. For every $+1$ unit change of r in the design space, f must be changed by $+b_2/b_1$ (or 1.31) units in the design space. From Table 2.19, *every* $+1$ *unit of* r *in the design space is equivalent to* $+0.1$ *unit in the variable space*. Similarly, each $+1$ unit of f in the design space is equivalent to $+0.1$ unit in the variable space. A change of $+1.31$ units of f in the design space is equivalent to a change of $+0.131$ unit in the variable space. Therefore, every $+0.1$ unit change in r in the variable space must be accompanied by a change of $+0.131$ unit in f also in the variable space. It is important to realize that this method yields a direction or gradient in which to proceed. It does not tell us how far to climb in a single step.

Subsequently, a series of trials are suggested as follows:

r	f	R	Comments
0.6	1.30	1.47	Origin of design
0.7	1.43	1.64	A change of +0.1 unit in r is accompanied by a change of +0.131 unit in f
1.0	1.82	1.97	Moving up the surface
1.1	1.96	1.99	Near the peak
1.2	2.09	1.95	Moving down the surface
1.30	2.61	1.39	

The values we get for r_{opt} (1.1) and f_{opt} (1.96) are different from those of Equation (2.27). However, the maximum ($R_{max} = 1.99$) is in good agreement (within 0.5%) with the value of R_{max} in Equation (2.27). Note that the method of steepest ascent *does not* "know" the shape of the response surface.

There is a penalty, nevertheless, for our lack of knowledge of the shape of the response function. There is no guarantee that the above method will converge to the *global maximum*: the *highest* point on the response surface. Converging to a local maximum is not unusual. Unless strong reasons suggest that the surface indeed possesses several maxima, the maximum obtained by the method of steepest ascent is assumed to be the global maximum. The linear model given in Equation (2.33) can be used to predict the results during the "climbing" stage. If the predicted values are appreciably different from the experimental ones, a new model is postulated and a new path of steepest ascent is constructed from the new parameters. At the regions of maxima (global or local) the linear model will fail and a higher-order model *must* be implemented to investigate the nature of the maximal region.

Simple Simplex Method

A *simplex* is a geometrical figure in N-dimensional space with $N+1$ vertices. Each vertex represents a certain combination of variables and is considered a vector in N-dimensional space. The vertex associated with the least desirable response is eliminated and replaced with its reflection in the hyperspace defined by the other N vertices. The process is repeated until a maximum is reached. The process is schematized in Figure 2.3 when two variables are being considered.

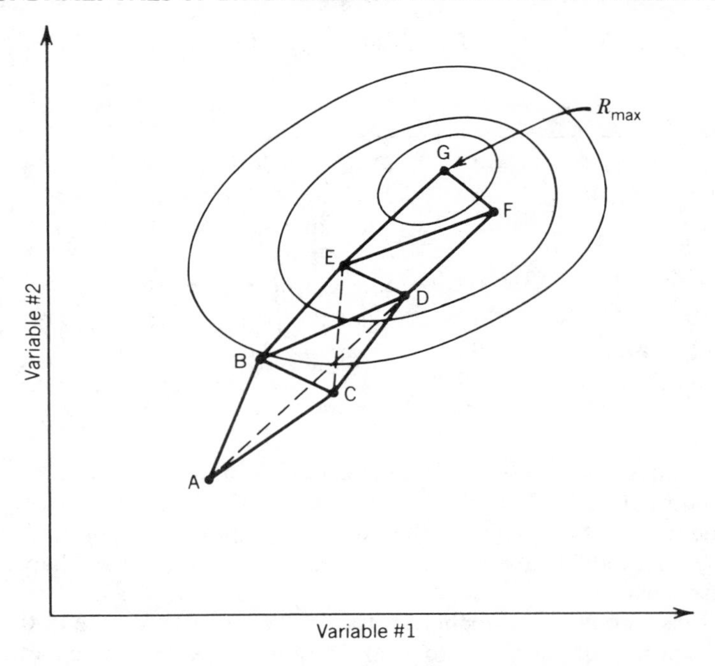

Figure 2.3 Schematic diagram of the simple simplex method.

Again using the chromatographic resolution optimization problem, initially three points usually at random A, B, and C are chosen. The point associated with the lowest resolution, A, is replaced by D. In the triangle BCD, vertex C is associated with the lowest resolution. Vertex C is thus replaced by vertex E. Of the three vertices B, D, and E vertex B represents the lowest resolution. It is replaced by vertex F. Finally, vertex D in the triangle EDF is replaced by G and no additional gain in resolution is obtained. The scheme by which a vertex is eliminated is summarized as follows:

Let $V_1, V_2, \ldots, V_k$ be the vertices of the simplex and let V_i be the vertex which should be eliminated.

(i) Calculate $\mathbf{P} = 1/(k-1)\sum_{j \neq i} V_j$ (vector addition).

(ii) Replace $\mathbf{V}_i$ by $\mathbf{V}'_i$ where $\mathbf{V}'_i = 2\mathbf{P} - \mathbf{V}_i$. (vector subtraction).

Let us apply the simple simplex method to the LC example.

STEP 1. Choose any three combinations and find R for each:

Vertex	r	f	R
1	0.4	0.5	0.80
2	0.7	0.7	1.10
3	0.3	0.8	0.85

Eliminate vertex 1 and replace with vertex 4. The coordinates of vertex 4 are given as follows:

$$\mathbf{P} = \tfrac{1}{2}[(0.7, 0.7) + (0.3, 0.8)] = (0.5, 0.75).$$

Therefore,

$$\mathbf{V}_4 = 2(0.5, 0.75) - (0.4, 0.5) = (0.6, 1.0).$$

STEP 2. Construct a new simplex with vertices 2, 3, and 4.

Vertex	r	f	R
2	0.7	0.7	1.10
3	0.3	0.8	0.85
4	0.6	1.0	1.22

Eliminate vertex 3 and replace it with vertex 5. The values of r and f at vertex 5 are calculated as follows:

$$\mathbf{P} = [(0.70, 0.70) + (0.60, 1.0)] = (0.65, 0.85).$$

$$\mathbf{V}_5 = 2(0.65, 0.85) - (0.3, 0.8) = (1.0, 0.9).$$

STEP 3. Construct a new simplex with vertices 2, 4, and 5.

Vertex	r	f	R
2	0.7	0.7	1.10
4	0.6	1.0	1.22
5	1.0	0.9	1.30

.
.
.

STEP 9.

r	f	R
0.8	1.5	1.74
1.1	1.7	1.90
0.7	1.8	1.87

STEP 10.

r	f	R
1.1	1.7	1.90
0.7	1.8	1.87
1.0	2.0	2.0

STEP 11.

r	f	R
1.1	1.7	1.90
1.0	2.0	2.0
1.4	1.9	1.84

STEP 12.

r	f	R
1.1	1.7	1.90
1.0	2.0	2.0
0.7	1.8	1.87

Step 12 is identical to Step 10. The simplex is oscillating in the maximal region. The vertex with the highest response is chosen. In this case

$$r_{opt} = 1.0, \quad f_{opt} = 2.0, \quad \text{and} \quad R_{max} = 2.0. \tag{2.34}$$

Purely by chance, these results are identical with the analytical solution and in good agreement with those of the steepest ascent method. The region of optimum response can subsequently be investigated by fitting a second- or a higher-order polynomial to data collected at the point where the simplex method stops. This is an important step because the simple simplex cannot often find the true maximum since its size is fixed.

The simple simplex method has been modified such that the displacement on the response surface can be accelerated or decelerated. In the *modified simplex* the reflection operation is accompanied by either simplex contraction or expansion and the convergence is expedited. The modified simplex method (MSM) as well as other methods used to control and/or optimize chemical systems will be introduced in Chapter 7.

REFERENCES

1. Herbert A. Laitinen and Walter E. Harris, *Chemical Analysis*, 2nd ed., McGraw-Hill, New York, 1975, pp. 545–546. Also, any book on Statistics.

   A brief description of the *F*-test and an example of its application.

2. R. Neill Carey, Svante Wold, and James O. Westgard, "Principal Component Analysis: An Alternative to 'Referee' Methods in Method Comparison Studies," *Anal. Chem.* **47**, 1824 (1975).

   Six methods of glucose determination are compared. Principal component analysis is used to examine their affinities to possible interferences.

3. Bo E. H. Saxberg and B. R. Kowalski, "Generalized Standard Addition Method," *Anal. Chem.* **51**, 1031 (1979).

   The generalized standard addition method (GSAM) is used to detect the presence of interfering analytes and to quantify their amounts.

4. Clemens Jochum, Peter Jochum, and B. R. Kowalski, "Error Propagation and Optimal Performance in Multicomponent Analysis," *Anal. Chem.* **53**, 85 (1981).

   A study of the effects of interferences on the experimental errors in multicomponent analysis.

5. Frank Ayres, Jr., *Matrices*, Schaum Outline Series in Mathematics, McGraw-Hill, New York, 1962.

   A discussion of matrices and the basic algebraic operations defined on them.

SUGGESTED READINGS

Edward L. Bauer, *A Statistical Manual for Chemists*, 2nd ed., Academic Press, New York, 1971, Chapter 3.

This is a good introduction to the subject. The fundamentals and nomenclature are explained and the examples should be helpful to those not yet familiar with the analysis of variance.

George E. P. Box, William G. Hunter, and J. Stuart Hunter, *Statistics for Experimenters*, Wiley, New York, 1978.

A valuable account of experimental designs, the analysis of variance, modeling, and optimization. The concepts are thoroughly explained and examples are given to further illustrate them.

William G. Cochran and Gertrude M. Cox, *Experimental Designs*, 2nd ed., Wiley, New York, 1957.

An extensive discussion of experimental design including simple and complex experimental plans.

Cuthbert Daniel and Fred S. Wood, *Fitting Equations to Data*, 2nd ed., Wiley, New York, 1980.

A detailed coverage of the theory and practice of mathematical modeling. Examples of fitting models to data with up to 11 variables are demonstrated.

Owen L. Davies, Ed., *The Design and Analysis of Industrial Experiments*, 2nd ed., Longman, London, 1956.

A detailed coverage of the theory and practice of experimental designs. Chapter 11 discusses the applications of the method of steepest ascent in optimization.

Stanley N. Deming and Stephen L. Morgan, "Simplex Optimization of Variables in Analytical Chemistry," *Anal. Chem.* **45**, 279A (1973).

A demonstration of the simple simplex and the modified simplex techniques.

Ronald A. Fisher, *The Design of Experiments*, 9th ed., Hafner Press, New York, 1971.

A treatment of the statistical theory behind the design of experiments including nonlinear designs.

Roland F. Hirsch, "Analysis of Variance in Analytical Chemistry," *Anal. Chem.* **49**, 6691A (1977).

A review of the analysis of variance and its possible applications in analytical chemistry. Examples of applications in atomic absorption, X-ray emission spectrometry, and ion selective electrode analyses are given.

D. L. Massart, A. Dijkstra, and L. Kaufman, *Evaluation and Optimization of Laboratory Methods and Analytical Procedures*, Elsevier, Amsterdam, 1978, pp. 213–302.

This section of the book (Part II) deals with the optimization of experiments. A discussion of several optimization methods and curve fitting techniques is presented. A good example on the use of factorial designs in atomic absorption is illustrated.

J. L. Myers, *Fundamentals of Experimental Design*, 3rd ed., Allyn and Bacon, Boston, 1979.

A thorough and detailed discussion of the analysis of variance and the most popular experimental plans. The text covers the necessary statistical background, matrix algebra, and linear regression.

CHAPTER

3

SIGNAL DETECTION AND MANIPULATION

In making qualitative and quantitative decisions, the analyst relies on one or more properties of the sample and/or the analyte(s). Instruments are often used to observe the properties of interest. These observations are usually in the form of an estimation (a measurement) of the amount of the desired property. An example is the use of a balance (an instrument) to determine the mass (a property) of a sample in grams (the measurement or an estimation of the amount of mass). Instruments are chosen such that the properties to be observed are "useful" stimuli to the instruments themselves. Consequently, the output (the response) of the instrument is expected to be related to (a function of) the property being observed. A balance, for example, is an inappropriate choice for measuring the refractive index of an unknown sample. The property of interest (the refractive index) *is not* a useful stimulus to the balance; the output of the balance is not related in any useful manner to the refractive index of the sample.

Any measurement obtained by the analyst may be referred to as a *signal*. The definition of the word *signal* varies from one discipline to another. From the analyst's point of view a signal can be defined as *the response of a device* (usually an instrument or a module in an instrument) *to certain stimuli*. In the following discussion we shall consider only those signals obtained as an instrumental response (in the form of a voltage or an electric current) to the presence (and the absence) of analytes. Signals can be measured over (as a function of) one or several parameters. In gas chromatography, for instance, the signal (e.g., the response of a flame ionization detector) is measured as a function of one parameter—time. The signal in this case is an electric current. In UV-Vis spectrometry the signal (the absorption) is a function of several parameters (all wavelengths between 190 and 800 nm).

Measurements of instrumental signals are the *sine qua non* of qualitative and quantitative inferences. Suppose that an instrument is known to respond only to an analyte, A, and is also known to yield no response in the absence of A. The presence of A in any sample is thus inferred whenever the response of the instrument is greater than zero. This is schematically shown in Figure 3.1 where replicate measurements are obtained for a blank (signal = 0) and a sample (signal = 2). Any amount of A that is present (no matter how small) is detectable (its presence can be verified) and the signal associated with it is precisely defined.

This situation is, unfortunately, an analyst's dream and is hardly ever realized. Any slight imperfections in the analytical procedure (reading errors, impurities, temperature changes, sampling inconsistencies, etc.) will add uncertainties to the signal. Introducing electronic modules and complex circuits to instruments has increased their analytical capabilities. However, additional uncertainties to the signals are also introduced due to detector noise and random fluctuations of frequencies and gains. A more realistic presentation of the situation described in Figure 3.1 is shown in Figure 3.2.

Note that the signal obtained in the absence of A is now referred to as the *background* (or *blank*) signal.* This is necessary because the response

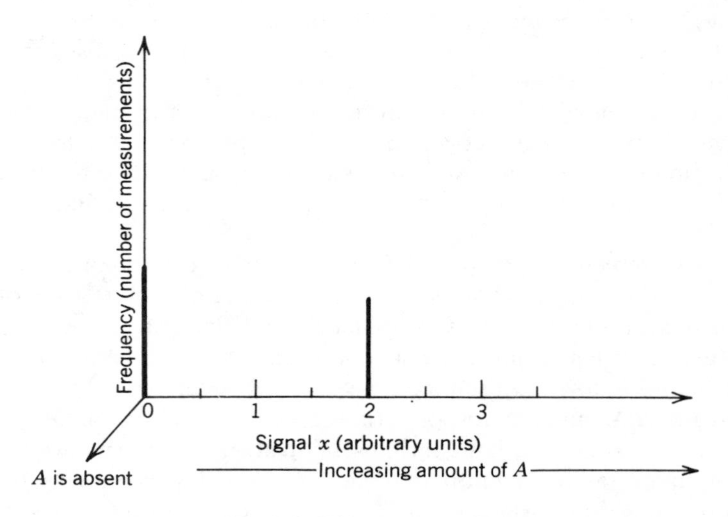

Fig. 3.1 The analyst's dream.

*A distinction is often made between background signals and blank signals. A background signal is the response of an instrument in the absence of any sample. A blank signal represents the response of an instrument obtained with a sample in which the analyte of interest is intentionally absent. Unless otherwise stated, the term background signal will conveniently refer to the signal which is obtained in the absence of the analyte (and is being compared to the signal obtained for the sample containing the analyte of interest).

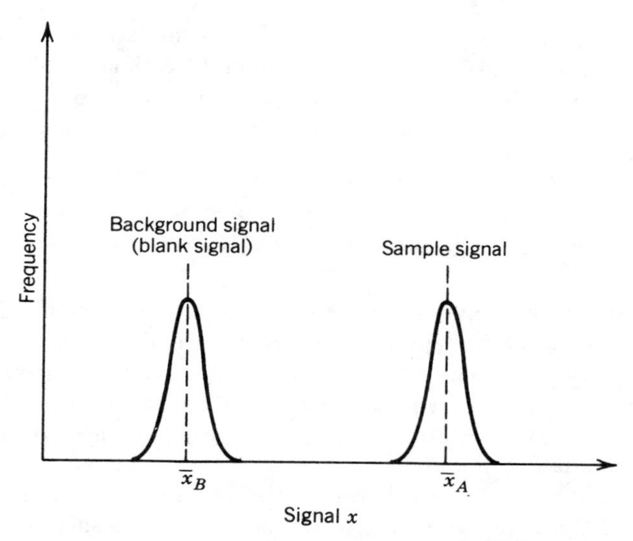

Figure 3.2 A more realistic picture of instrumental analysis.

of the instrument in the absence of A is no longer a fixed value and is not caused solely by the medium of the blank. The background signal may not even be distributed about zero. Thus in the absence of A the instrument shows a response that is distributed about a finite mean value, $\bar{x}_B$. The distribution about this average background signal is caused by random variations in the analytical procedure and by random processes in the various modules of the instrument itself. This combination of uncontrolled experimental and instrumental variations is referred to as *random noise*. The sources of random variations operate whether or not A is present. Consequently, both line signals are caused to widen (Fig. 3.1 versus Fig. 3.2).

The effect of random noise on the background and the sample signals is of special importance. The background signal distribution is generally obtained by experiment. Large numbers of measurements (≥ 20) are usually required to characterize the statistical parameters of the signal distribution. Generally, both the background and the sample signals are assumed to be normally distributed about μ_B and μ_A, respectively.* Both signals are also assumed to have a common variance, σ^2 ($\sigma_A^2 = \sigma_B^2 = \sigma^2$). When n measurements are obtained, the mean, μ_A, and the variance, σ_A^2,

*When counting statistics are pertinent (e.g., in X-ray spectroscopy) the background signals are assumed to have a Poisson distribution.

of the sample signal distribution can be estimated as $\bar{x}_A$ and s_A^2, respectively. Similarly, $\bar{x}_B$ and s_B^2 are unbiased estimates of the background signal distribution parameters μ_B and σ_B^2, respectively. The assumption of the equality of their population variances (σ_A^2 and σ_B^2) can be tested as shown in Chapter 2.

SIGNAL DETECTION

Comprehensive treatments of the theory and methods of signal detection are available (1, 2). This chapter will focus on a small area in which signal detection theory and analytical chemistry share common grounds. For convenience, a signal, x, will be represented by its probability density function, $P(x)$. All signals are considered Gaussian and are normalized to unit area. Population parameters, μ and σ^2, will be used. Separation between the means of the signals is expressed in units of σ. Experimental estimates of the population parameters, $\bar{x}$ and s_x^2, can be substituted in the text without loss of generality. The concepts discussed next can be applied to situations where the background (or the sample) signals are not Gaussian without loss of generality.

Figure 3.3 shows two signals whose means are μ_B (the background or the blank) and μ_A (the sample containing the analyte A). As shown, the two signals are completely separated (resolved). In such a situation the detection of the sample signal is achieved with absolute certainty. Though replicate measurements are often made to estimate μ_A and σ_A^2, *only one* measurement is required for qualitative inferences (to decide whether or not A is present).

Since the magnitude of the signal is dependent on (and usually directly

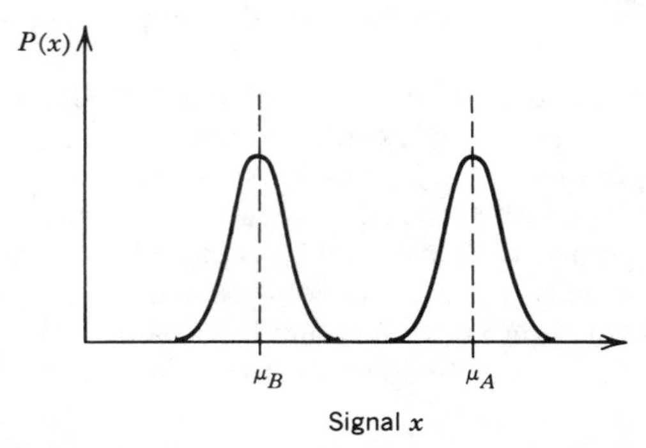

Figure 3.3 Probability density functions for signals A and B, where population means are μ_A and μ_B. Detection of the sample signal in this case is achieved with certainty.

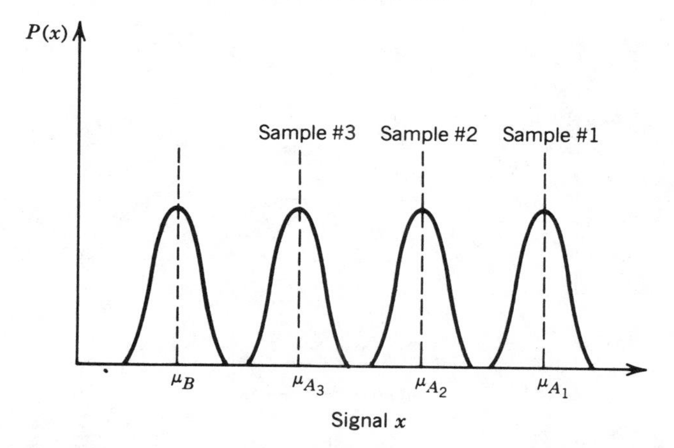

Figure 3.4 As the mount of A in the sample decreases, the separation between μ_A and μ_B decreases.

proportional to) the amount of A present in the sample, the separation between μ_A and μ_B decreases as the amount of A diminishes. This can be represented by translating the sample signal along the signal axes closer to μ_B (samples 2 and 3 in Fig. 3.4).

As the sample signal approaches the background signal, the two distributions begin to overlap (Fig. 3.5). The amount of this overlap (shaded area in Fig. 3.5) is a measure of the uncertainty associated with the

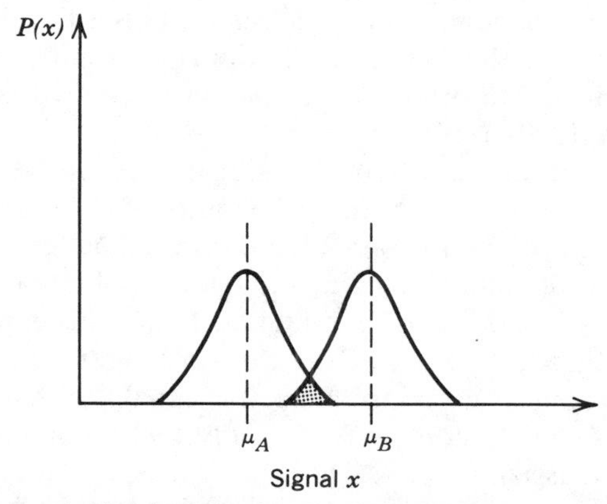

Figure 3.5 As the sample signal approaches the background, the two distributions begin to overlap.

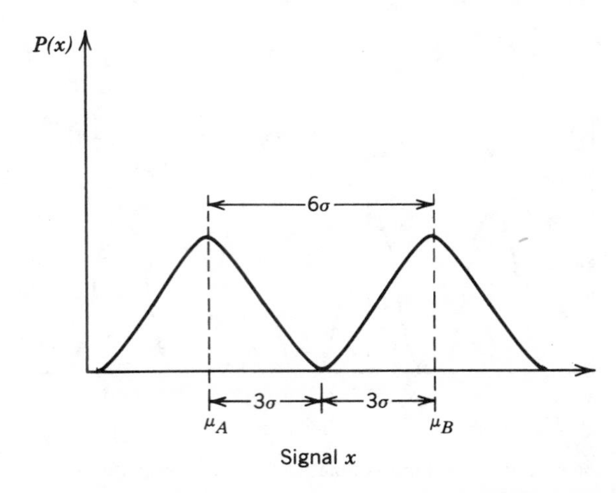

Figure 3.6 Means of normally distributed sample and background signals are separated by 6σ. Thus only a negligible 0.13% of the area of each signal is buried under the other.

detection. When the sample and the background are completely resolved (Fig. 3.3), the uncertainty in detection is near zero. In Figure 3.6 the means of the background and the sample signals are separated by 6σ. As mentioned earlier both signals are normally distributed and have a common variance σ^2 ($\sigma^2 = \sigma_A^2 = \sigma_B^2$). Thus, when the signal means are 6σ apart, only 0.13% of the area of each signal is overlapping with the other. This amount of overlap is considered negligible for most practical applications. At this resolution (degree of separation) the sample signal is considered "detectable with absolute certainty." A signal centered at $\mu_B + 6\sigma_B$ is often referred to as the *limit of guaranteed detection*. This is the closest the sample signal can approach the background signal and still be detected with "absolute certainty." In practice, the signals presented in Figure 3.6 are considered completely resolved.

A complication arises when the signals are *not* Gaussian. If the background signal is Gaussian, the probability of obtaining a background signal exceeding $\mu_B + 3\sigma_B$ is 0.0013. When the background signal is not Gaussian, this probability (as given by the Tschebyscheff inequality), $P(|x - \bar{x}| > k\sigma(x)) \leq 1/k^2$, can be as high as 11%. This implies that a signal of magnitude higher than $\mu_B + 3\sigma_B$ can no longer be attributed to the sample with "absolute" certainty. If such a signal is obtained the presence of A is inferred with a confidence of only 89% (compared to the near 100% for the Gaussian model). This is a serious deviation from the ideal. Unless experimental and/or theoretical grounds suggest otherwise, the Gaussian model is usually adopted.

If one defines the *limit of detection* as the smallest amount of analyte A that can be detected with "absolute certainty" then the limit of detection is the amount of the analyte that produces a signal whose mean has a magnitude of $\mu_B + 6\sigma_B$.

The detection limit is often expressed in units of concentration (e.g., 3 ppb, 5 ppm, etc.). This is plausible because the units are familiar and commonly used in practice. Strictly speaking, however, the detection limit is determined in the *signal domain* (e.g., 10^{-6} A, 10^{-8} V, etc.) The conversion from the signal domain to the concentration domain is accomplished by the use of a proper function (usually linear) that relates the magnitude of the signal to the amount of analyte. This process is known as *calibration* and will be treated in the next chapter together with *interval estimation* of the detection limit. This chapter is restricted to the signal domain and to *point estimation* of detection limit.

Assessing the detection limit of an instrument or an analytical procedure is *not* a simple matter. Several parameters have to be delineated concurrently. The International Union for Pure and Applied Chemistry (IUPAC) has recommended that the detection limit be taken as an amount of analyte giving a signal centered about $\mu_B + 3\sigma_B$ (3).

The sample signal distribution can not be allowed to approach that of the background signal without having the two distributions overlap at some point. This overlap is a measure of the probability of making errors in the detection (mistaking the sample for the background and vice versa). Definitions of the detection limit which allow μ_B and μ_A to be separated by less than $6\sigma_B$ have risks (i.e., errors in detection) associated with them.

Figure 3.7 shows two overlapping Gaussian signals. The background

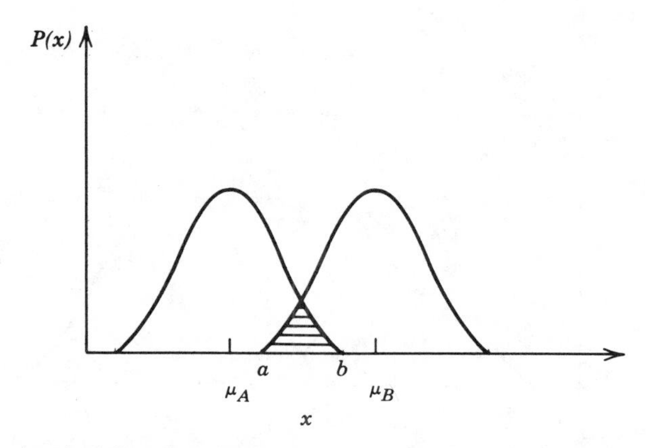

Figure 3.7 Overlapping Gaussian signals with thresholds defined as $a = \mu_A - 3\sigma_A$ and $b = \mu_B + 3\sigma_B$.

signal is centered about μ_B and the sample signal is centered about μ_A. Let the following two threshold points be defined on the signal axis:

$$b = \mu_B + 3\sigma_B \tag{3.1}$$

and

$$a = \mu_A - 3\sigma_A. \tag{3.2}$$

Assume that a signal x_i is obtained. If $x_i > b$ the analyst can "confidently" infer that analyte A is present (the probability of the background producing a signal greater than $\mu_B + 3\sigma_B$ is 0.0013). Also, if $x_i < a$ the analyst can "confidently" infer that analyte A is not present (the probability of the sample producing a signal less than $\mu_A - 3\sigma_A$ when A is present is 0.0013). If x_i is in the range $[a, b]$, decisions concerning the presence or the absence of A are not as confident as the above two situations. Both the sample containing A and the background give signals in the range $[a, b]$. The range $[a, b]$ is referred to as the *no-decision range*. The signal domain is thus divided into three distinct regions and three decisions can be made after a measurement has been obtained. If $x_i < a$ we accept the null hypothesis, H_0: A is not present. If $x_i > b$ we accept the alternate hypothesis, H_1: A is present. If x_i is in the range $[a, b]$ the analyst can not reject (or accept) H_0 or H_1 and more measurements must be made. The three regions of the signal domain are shown in Figure 3.8.

If x_i is in the range $[a, b]$, the measurement can be treated in two different manners. As has been implied already the experiment is simply

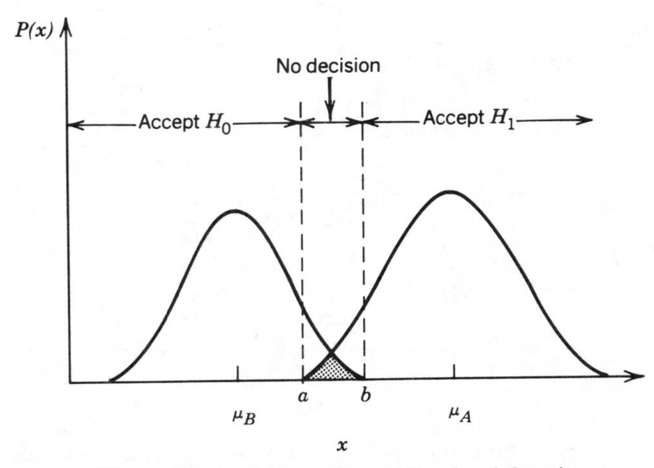

Figure 3.8 Decision regions of the signal domain.

repeated and more measurements are obtained until the signals are clearly out of the no-decision region. This route, however, can be costly and large numbers of measurements may be required. The method is known as *sequential detection* (4). This technique has found applications in industry and communication. It is not generally recommended in chemical analysis because of the potentially excessive number of measurements required. The second procedure is to "take a risk" and make a decision based on the probability of having each distribution produce measurement x_i. Figure 3.9 shows that the probability of having the sample signal distribution exhibits a measurement $\leq x_i$, $\int_{-\infty}^{x_i} P_A(x)\, dx$, is larger than the probability of having the background signal distribution exhibit a measurement $\geq x_i$, $\int_{x_i}^{\infty} P_B(x)\, dx$. Since it is more probable that the sample rather than the background produces a measurement $= x_i$, hypothesis H_1 (the presence of A) can be accepted. However, there is a finite probability (i.e., $\int_{x_i}^{\infty} P_B(x)\, dx \neq 0$) that H_0 is indeed correct. Accepting H_1 when H_0 is correct is known as Type I error. Deciding that A is present when it is indeed absent is sometimes referred to as a false alarm or a false detection. The probability of false alarm is equal to the area under the background signal density function to the right of x_i (shaded area in Fig. 3.9) and is denoted by P_{10}. The probability of Type I error is usually predetermined. Values between 2% and 3% are common. The preselected value of P_{10} determines the value of the threshold signal b. When $P_{10} = 0$, b is defined by Equation (3.1). As P_{10}

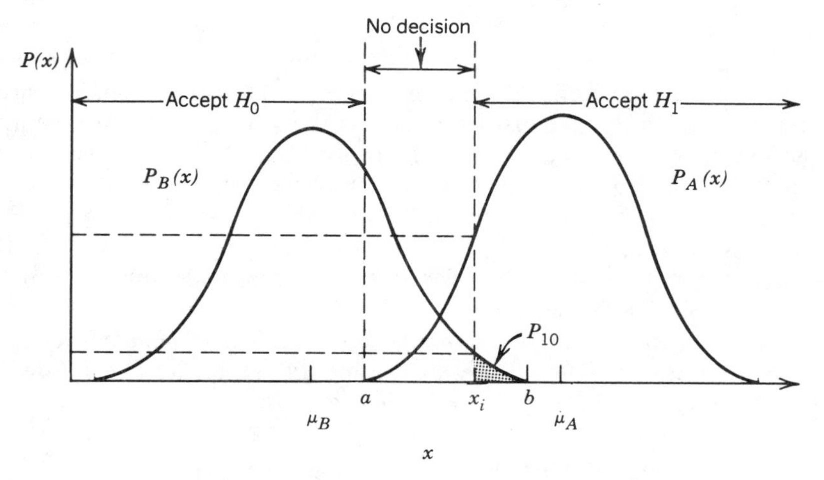

Figure 3.9 The probability of the sample exhibiting the measurement $\leq x_1$, given by $\int_{-\infty}^{x} P_A(x)\, dx$ is greater than the probability of the background exhibiting a measurement $\geq x_i : \int_{x_i}^{\infty} P_B(x)\, dx = P_{10}$ = shaded region = "false alarm."

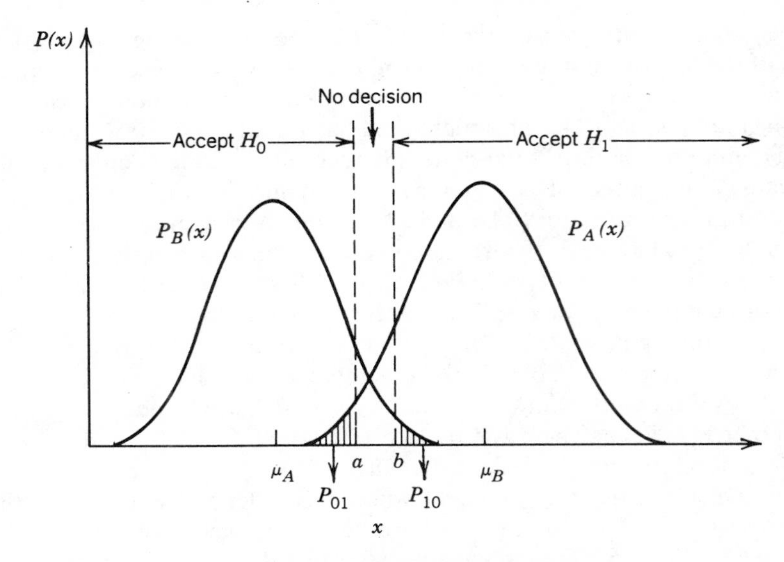

Figure 3.10 General detection model.

is allowed to take small positive values, point b in Figure 3.9 is, in effect, moving closer to μ_B and the no-decision region becomes smaller. This, in turn, increases the efficiency of detection. The disadvantage that accompanies this increased efficiency is, of course, the risk associated with the detection strategy.

The threshold signal a can also be moved closer to μ_A. By allowing this, a Type II error (accepting H_0 when H_1 is correct) is introduced. The area under the sample signal density function to the left of a is the probability of accepting H_0 when H_1 is true. The probability of making this error is denoted by P_{01}. It is the probability of inferring that the sample signal is the background. A value between 2% and 3% is also common for P_{01}. When P_{01} is zero, the value for a is as defined in Equation (3.2). By moving a closer to μ_A the no-decision region is made smaller and an additional risk must be accepted.

A general detection model is presented in Figure 3.10. The no-decision region becomes smaller only by increasing P_{10} and P_{01}. The following terms can be defined in reference to Figure 3.10:

$$P_{00} = \text{Probability of choosing } H_0 \text{ when } H_0 \text{ is true}$$

$$= \int_{-\infty}^{a} P_B(x)\,dx.$$

P_{11} = Probability of choosing H_1 when H_1 is true

$$= \int_b^{\infty} P_A(x)\, dx.$$

P_{10} = Probability of choosing H_1 when H_0 is true (probability of a false alarm)

$$= \int_b^{\infty} P_B(x)\, dx.$$

P_{01} = Probability of choosing H_0 when H_1 is true (probability of mistaking the sample signal for the background)

$$= \int_{-\infty}^{a} P_A(x)\, dx.$$

When the means of the sample and the background signals are separated by 6σ, and the threshold values are defined by Equations (3.1) and (3.2), a and b coincide at one point thus reducing the no-decision region into a single point. Therefore $P_{10} = P_{01} = 0$. More importantly, the sample signal lies completely in a decision area, as shown in Figure 3.11.

As the distributions of the sample and the background signals begin to

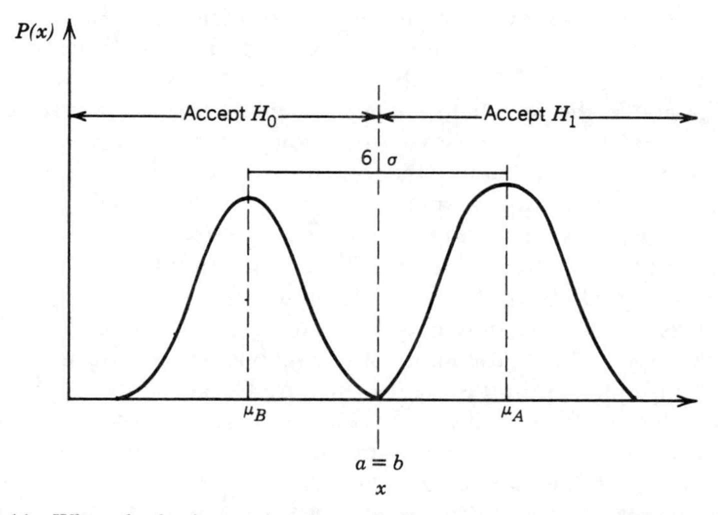

Figure 3.11 When the background and sample signal means are separated by 6σ, the no-decision region collapses to a single point, $P_{10} = P_{01} = 0$, and the sample signal lies completely in a decision region.

overlap, three important factors interact. These factors are (1) separation of the means, (2) allowable risks (P_{10} and P_{01}), and (3) the fractional area of the sample signal above the threshold value b. The interactions of these factors determine the smallest signal that can be detected—the detection limit. The following section demonstrates one approach to point estimation of detection limits based on preselected values for P_{10} and/or P_{01}. The analyst can preselect any proper value for P_{10} and/or P_{01} and examine any appropriate detection model by following the same steps.

POINT ESTIMATION OF THE DETECTION LIMIT

Assume the following experimental conditions:

1. $P_{01} = P_{10} = 0$ (this makes our inference always correct).
2. Define the threshold values for a and b as given in Equations (3.1) and (3.2).
3. The maximum allowable number of replicate analyses of a sample containing analyte A is 4.

Figure 3.12 shows the sample and the background signals at various degrees of separation. In Case A, the no-decision region has an area equal to zero and one measurement is sufficient to infer either H_0 or H_1. Since $P_{01} = P_{10} = 0$, the inferences are always correct. In Case B the area of the no-decision region is 2.3% of the total area. Only 97.7% of the sample signal is above the threshold value. The inferences are always correct; however, the efficiency of making decisions has been slightly decreased. There is about 9% probability that one measurement is in the no-decision region. In practice, such a small probability is not considered a serious inconvenience. As the sample signal moves closer to the background signal, the area of no-decision region increases and the fractional area of the sample signal above the threshold value b decreases. At a separation of 4σ (Case C) the area of the no-decision region is approximately 16% of the total area, and the probability of having one or two measurements out of the four replicates in it is 48.7%. In Case D, half of the measurements are expected to be *indecisive*. If the sample signal is moved any closer to μ_B, more than half of the measurements are expected to be indecisive. If the efficiency of detection is to be at least 50% (not more than half of the measurements are allowed to be indecisive), the sample signal *has to be* centered about $\mu_B + 3\sigma_B$. If only *one* measurement is allowed, then the smallest signal must be centered at $\mu_B + 6\sigma_B$ in order to guarantee its

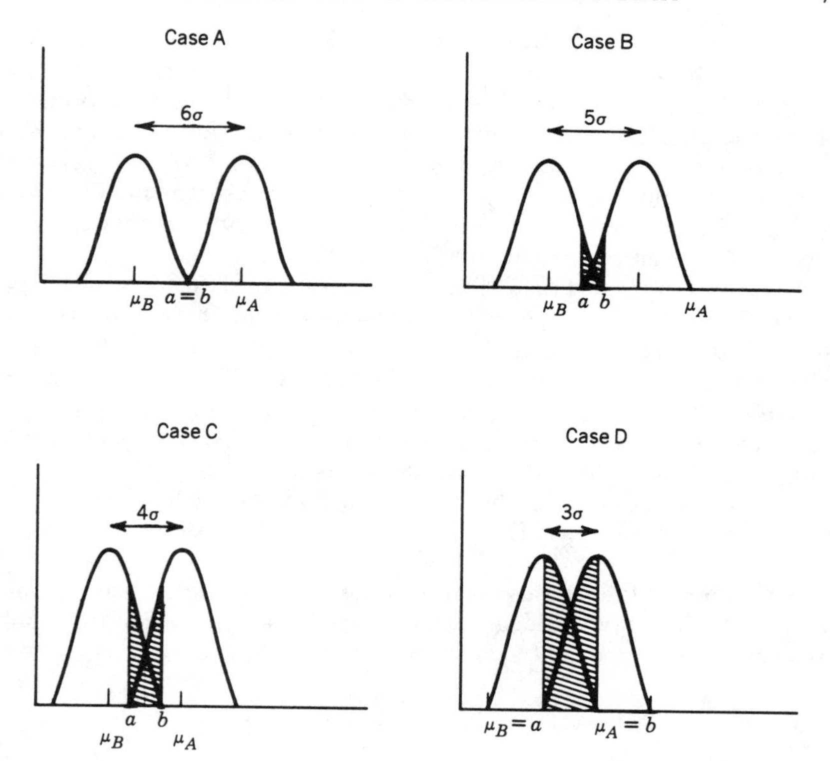

Figure 3.12 Effect of various degrees of separation on estimation of the detection limit assuming four replicates. Case A: inferences are always correct. Case B: 9% probability of one measurement falling in no-decision region. Case C: 48.7% probability of one or two measurements in no-decision region. Case D: half of measurements are indecisive.

detection. If four measurements are allowed, and the analyst can afford to allow *only two* measurements to be indecisive, then a sample signal centered about $\mu_B + 3\sigma_B$ is *detectable*.

It is clear that the detection limit is not independent of the number of replicates and the risks of detection. Indeed, the detection limit is a *working* concept and cannot be adequately postulated except in view of the experimental conditions.

Usually, only P_{10} is defined and no attention is given to P_{01}. In this case if the signal is greater than b (the only threshold value defined), the presence of A is *verified*. If the signal is less than b, the presence of A is *not verified* (note that the analyst cannot claim that A is absent).

Figure 3.13 shows the means of the background and the sample signal distributions separated by 2σ. The no-decision region is completely eliminated by defining only one threshold value (in this case $b = \mu_B + 3\sigma_B$). In Figure 3.13 only 16% of the sample signal distribution area is above the threshold value. The analyst must expect more than half of the measurements to be unable to verify the presence of A when A is actually present. The closer μ_A and μ_B are the more measurements the analyst needs in order to verify the presence of A. The detection limit depends on the *number* of measurements that can be obtained.

Kaiser (5) suggested that one threshold value be defined as $\mu_B + 3\sigma_B$ [b in Eq. (3.1)]. He also suggested that the detection limit is a signal centered at $\mu_B + 3\sigma_B$ (Fig. 3.12, Case D). If the signals are Gaussian then $P_{10} \simeq 0$. The probability of false detection can be as high as 11% if the background signal is not Gaussian. The value $\mu_B + 3\sigma_B$ is referred to as the *Kaiser detection limit*. Note, however, that half of our measurements are *not* expected to verify the presence of A at this detection limit.

A less conservative threshold value for b may be defined as

$$b = \mu_B + 2\sigma_B. \tag{3.3}$$

The probability of a false alarm in the above model is approximately 2.3%. The detection limit can be established again in view of the number of measurements needed to verify the presence of the analyte. As shown in

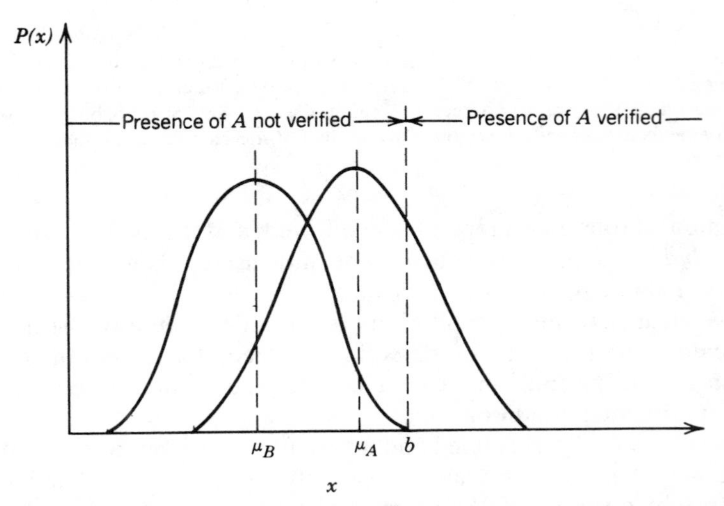

Figure 3.13 No-decision region is eliminated by defining only one threshhold value. Given 2σ separation, only 16% of sample distribution is above $b = \mu_B + 3\sigma_B$ (Kaiser's detection limit).

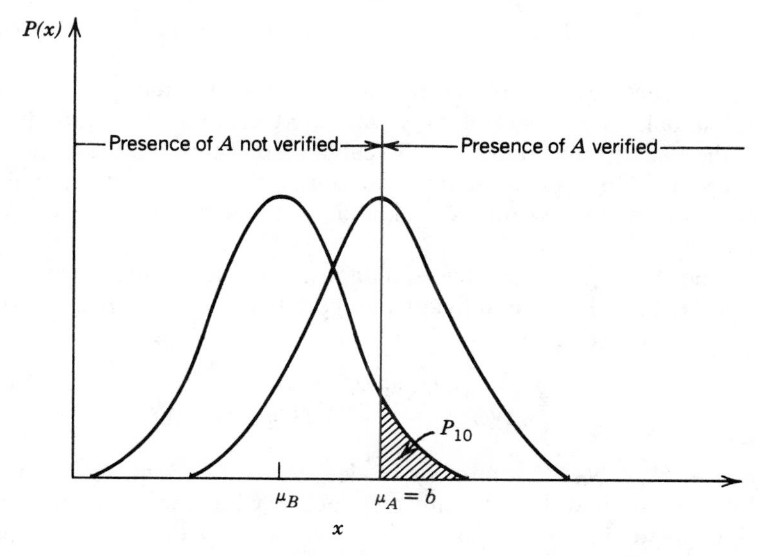

Figure 3.14 With a less conservative threshhold, $b = \mu_B + 2\sigma_B$, half of the measurements will be unable to verify A when A is present. Lower detection limit than Kaiser method.

Figure 3.14, at a separation of 2σ half of our measurements are expected to be unable to verify the presence of A when A is indeed present. When such a number of measurements is allowed the detection limit obtained according to Equation (3.3) is lower than that obtained according to Kaiser. However, when the signal distributions are Gaussian, the probability of false alarms in the model expressed by Equation (3.3) is 2.3% while that in the Kaiser model is virtually zero.

When the background signal distribution is *not* Gaussian, P_{10} can be as high as 25% in the model shown by Equation (3.3) compared to a maximum of about 11% in the Kaiser model.

When the threshold value is chosen according to a preselected value for P_{10}, the detection criterion is called the Neyman–Pearson criterion. Several working definitions of the limit of detection has been discussed and compared (6).

POINT ESTIMATION OF THE DETECTION LIMIT BY *t*-TESTS

t-Tests based on the difference between the estimated measurements $(x_A - x_B)$ are demonstrated in Chapter 2. The same concept can be used to point estimate the detection limit by straightforward application of the test.

Assume that several measurements, N_B, are obtained for the background signal. The experimentally estimated average, $\bar{x}_B$, and variance, s_B^2, of the background signal can be calculated. Similarly, if N_A measurements are available for the sample containing the analyte, it is possible to calculate $\bar{x}_A$, average sample signal, and s_A^2, the experimentally estimated variance.

Assume that σ_A^2 and σ_B^2 are equal and choose a confidence level (P_{10}) equal to α, then μ_A is significantly larger than μ_B (the presence of the analyte can be verified) if

$$\frac{(\bar{x}_A - \bar{x}_B)(N_A + N_B - 2)^{1/2}}{(1/N_A + 1/N_B)^{1/2}[(N_A - 1)s_A^2 + (N_B - 1)s_B^2]^{1/2}} > t_{\alpha,\nu}, \tag{3.4}$$

where $\nu = N_A + N_B - 2 =$ number of degrees of freedom. Consequently, the detection limit at the α significance level is when the left-hand side and the right-hand side of inequality (3.4) are equal. The condition for the smallest significant difference is thus

$$\frac{(\bar{x}_A - \bar{x}_B)(N_A + N_B - 2)^{1/2}}{(1/N_A + 1/N_B)^{1/2}[(N_A - 1)s_A^2 + (N_B - 1)s_B^2]^{1/2}} = t_{\alpha,\nu}. \tag{3.5}$$

The detection limit is clearly a function of N_A, N_B, s_A^2, s_B^2, x_A, x_B, and α (the level of confidence or P_{10}).

If it is shown that $\sigma_A^2 \neq \sigma_B^2$, the t-test for the populations with unequal variances can be used (see Chapter 2) and the same conclusion can be found.

Signal-to-Noise Ratio

The difference between $\bar{x}_A$ and $\bar{x}_B$, $\bar{D}$, is an estimate of $\mu_A - \mu_B$. The variance of $\bar{D}$ is given as

$$\sigma^2(\bar{D}) = \sigma^2(\bar{x}_A) + \sigma^2(\bar{x}_B) = \frac{\sigma_A^2}{N_A} + \frac{\sigma_B^2}{N_B}. \tag{3.6}$$

Assuming that $\sigma_A^2 = \sigma_B^2 = \sigma^2$, then

$$\sigma^2(\bar{D}) = \sigma^2\left(\frac{1}{N_A} + \frac{1}{N_B}\right).$$

An estimate of σ^2 can be calculated from s_A^2 and s_B^2 and is given as

$$s^2 = \frac{(N_A - 1)s_A^2 + (N_B - 1)s_B^2}{N_A + N_B - 2}. \tag{3.7}$$

Therefore $\sigma^2(\bar{D})$ can be estimated as $s^2(\bar{D})$ according to

$$s^2(\bar{D}) = \frac{[1/N_A + 1/N_B][(N_A - 1)s_A^2 + (N_B - 1)s_B^2]}{N_A + N_B - 2}. \tag{3.8}$$

Inequality (3.4) can now be rewritten

$$\frac{\bar{D}}{s(\bar{D})} > t_{\alpha,\nu}, \tag{3.9}$$

where $\nu = N_A + N_B - 2$.

The left-hand side of inequality (3.9) is referred to as the *signal-to-noise* (S/N) ratio. The detection limit can thus be expressed as *a minimum signal-to-noise ratio*. Given a particular α and a certain ν, the minimum significant difference is obtained when both sides of inequality (3.9) are equal. If, for example, α is set at 0.05 and $\nu = 10$, the detection limit is the amount of analyte whose signal has a signal-to-noise ratio of 1.812.

Table 3.1 lists the measurements obtained for the background and a sample containing analyte A. Ten replicates are obtained for each.,

An F-test at the 95% level shows that the assumption of equality of σ_A^2 and σ_B^2 is valid.

If α is set at 0.05, the threshold value of detection as given in the t-table is $t_{0.05,18} = 1.734$.

Table 3.1 Current Output (Arbitrary Units) of an Instrument Obtained for the Background and a Sample

Background	Sample
0.031	0.051
0.010	0.032
0.025	0.043
0.042	0.054
0.051	0.070
0.050	0.072
0.032	0.050
0.021	0.039
0.054	0.072
0.041	0.060
$x_B = 0.036$	$x_A = 0.054$
$s_B = 0.014$	$s_A = 0.014$
$N_B = 10$	$N_A = 10$

When the proper values in Table 3.1 are substituted in the left-hand side of inequality (3.4), we obtain a value of 2.875 which is greater than $t_{0.05,18}$. Therefore the difference between $\bar{x}_A$ and $\bar{x}_B$ is significant and the signal is "detectable" at the 0.05 significant level. The detection limit, given that $\alpha = 0.05$ and $\nu = 18$, is a value of 1.734 for the left-hand side of Equation (3.4). From inequality (3.9) and S/N obtained for the sample is 2.875. By the same reasoning, the signal is detectable and the detection limit at $\alpha = 0.05$ and $\nu = 18$ is a signal with an S/N = 1.734. If a sample signal with an S/N = 2.875 is obtained under different conditions, for example, $\alpha = 0.005$ and $\nu = 10$, it would *not* be considered significantly different from the background signal $t_{0.005,10} = 3.169$). If such a sample indeed contains A, the presence of the analyte cannot be verified under these conditions. Note that if $\alpha = 0.005$ the sample signal in Table 3.1 is near the detection limit ($t_{0.005,18} = 2.878$). For $\alpha = 0.005$ and $\nu = 10$, the detection limit corresponds to a S/N = 3.169. If more measurements are allowed, say $\nu = 18$, the minimum S/N becomes 2.878. By making more measurements the analyst can reduce the detection limit to a very small value.

t-Test Based on the Difference Between Individual Measurements ($x_A - x_B$)

Assume that the data in Table 3.1 were obtained by taking a measurement of the background signal followed by a measurement of the sample signal. One can now subtract the background measurement from the sample measurement and evaluate the N differences ($N = N_A = N_B$). Each difference, d_i, is given as

$$d_i = x_{A,i} - x_{B,i}. \tag{3.10}$$

The variance of the differences can be calculated from the propagation of error principle (see Chapter 1) and is given by

$$\sigma^2(d) = \sigma^2(x_A) + \sigma^2(x_B) \tag{3.11}$$

or

$$\sigma_d^2 = \sigma_A^2 + \sigma_B^2. \tag{3.12}$$

Since $\sigma_A^2 = \sigma_B^2$,

$$\sigma_d^2 = 2\sigma_B^2. \tag{3.13}$$

An estimate of σ_d^2, s_d^2, can be calculated using an estimate of σ_B^2, s_B^2. Equation (3.13) can be rewritten in terms of these estimates:

$$s_d^2 = 2s_B^2. \tag{3.14}$$

An average difference, $\bar{d}$, is calculated according to Equation (3.15) and its significance,

$$\bar{d} = \frac{1}{N} \sum_{i=1}^{N} d_i, \tag{3.15}$$

can be tested; $\bar{d}$ is considered significantly different from zero if

$$\frac{\bar{d}\sqrt{N}}{s_d} > t_{\alpha,N-1}. \tag{3.16}$$

Inequality (3.16) can be expressed in terms of s_B as

$$\frac{\bar{d}\sqrt{N}}{s_B\sqrt{2}} > t_{\alpha,N-1}. \tag{3.17}$$

Treating the data in Table 3.1 as described above and substituting the proper values in the left-hand side of inequality (3.17) a value of 2.87 is obtained. Therefore, $\bar{d}$ is significantly different from zero at the 0.05 significant level ($t_{0.05,9} = 1.833$).

Note that the differences in this example are *not* treated as those obtained from paired observations (see Chapter 2). The measurements obtained are not considered "paired," and the test used for paired comparisons is not generally recommended.

Inequality (3.17) can be written

$$\frac{\bar{d}}{s_B} > t_{\alpha,N-1} \sqrt{\frac{2}{N}}. \tag{3.18}$$

The left-hand side of inequality (3.18) may also be referred to as the signal-to-noise (S/N) ratio. In our example the S/N obtained is 1.29. The detection limit is a signal for which the left-hand side of inequality (3.18) is equal to the right-hand side. If $\alpha = 0.005$ and $N = 10$, the detection limit is a signal with S/N = 1.45. The conditions for the significance of $\bar{d}$ can be further simplified if we express d in units of s_B.

Let

$$\bar{d} = ks_B, \tag{3.19}$$

where k is a constant. Then $\bar{d}$ is significantly different from zero if

$$k > t_{\alpha,N-1} \sqrt{\frac{2}{N}}. \tag{3.20}$$

The conditions at the detection limit are

$$k = t_{\alpha,N-1} \sqrt{\frac{2}{N}}. \tag{3.21}$$

Given a particular α, it is possible to estimate the number of measurements needed to detect a signal centered at $\bar{x}_B + ks_B$. For example, at $\alpha = 0.005$, a signal centered at $\bar{x}_B + s_B$ ($k = 1$) requires a minimum of 18 measurements yielding a value for $\bar{d}$ equal to s_B in order to be detected (differentiated from background).

THE WILCOXON TEST

Problems in detection arise when the signal can no longer be approximated by a "well-defined" distribution (e.g., Gaussian, Poisson). Some of the statistical tests discussed earlier can tolerate deviation from normality. However, pertinent signal parameters cannot be estimated with a high degree of confidence when the signal distribution is severely skewed. Box and Cox have shown that a skewed distribution can be transformed into a Gaussian-like distribution (7). The choice of a suitable transformation, however, can be a difficult task. Nonparametric tests offer an easier solution (8). The Wilcoxon test can be applied to the data directly. The Wilcoxon test is used when the background and sample signals can no longer be considered normally distributed. The test relies on the differences, d_i's, between the sample and the background measurements ($d_i = x_{A,i} - x_{B,i}$). The test assumes that the d_i's are independent and identically distributed. Two hypotheses are tested:

H_0: The distribution of d_i's is symmetric about zero.
H_1: H_0 is untrue.

The steps involved in the test are as follows:

1. Calculate the differences, $d_i = x_{A,i} - x_{B,i}$.
2. Calculate the absolute value of each d_i, $|d_i|$.
3. Rank the n nonzero absolute values in ascending order from 1 to n.
4. Attach the sign of each difference as calculated in 1 to the rank.
5. Calculate the sum of positive rank, T^+.
6. Compare T^+ to the critical values in the Wilcoxon T^+ tables (Table A.5).

If two or more differences are equal, they are given the average of the ranks that would have been given to them if the differences were slightly different.

Table 3.2 Signed Rank Values for the Wilcoxon Test

		(1)	(2)	(3)	(4)
Background	Signal	d_i	$\lvert d_i \rvert$	Rank	Signed Rank
0.031	0.051	+0.020	0.020	8	+8
0.010	0.032	+0.022	0.022	9.5	+9.5
0.025	0.043	+0.018	0.018	3.5	+3.5
0.042	0.054	+0.012	0.012	1	+1
0.051	0.070	+0.019	0.019	6.5	+6.5
0.050	0.072	+0.022	0.022	9.5	+9.5
0.032	0.050	+0.018	0.018	3.5	+3.5
0.021	0.039	+0.018	0.018	3.5	+3.5
0.054	0.072	+0.018	0.018	3.5	+3.5
0.041	0.060	+0.019	0.019	6.5	+6.5
					$T^+ = 55$
					$n = 10$

Table 3.2 shows the results of applying steps 1–4 described above to the data in Table 3.1. The value for T^+ is greater than the critical value in Table A.5 for $\alpha = 0.05$ and $n = 10$. Therefore, H_0 is rejected. The mean of the sample signal is significantly larger than the mean of the background signal. As can be seen from Table A.5, for $\alpha = 0.05$ and $n = 10$ the detection limit corresponds to a signal for which $T^+ = 45$.

PRECISION AT THE DETECTION LIMIT

The precision of the average of the sample signal can be expressed in terms of the relative standard deviation, s_r. An estimate of s_r is calculated as the reciprocal of the signal-to-noise ratio and is sometimes expressed as a percentage:

$$s_r = \frac{100}{\text{S/N}}. \tag{3.22}$$

At the detection limit S/N is equal to the selected value $t_{\alpha,\nu}$. Therefore, at the limit of detection

$$s_r = \frac{100}{t_{\alpha,\nu}}. \tag{3.23}$$

When $\alpha = 0.005$ and ten measurements are obtained ($\nu = 9$) the s_r of the average of the ten measurements is

$$s_r(\bar{x}_A) = \frac{1}{3.25} \times 100 = 30.8\%. \tag{3.24}$$

Therefore, $\bar{x}_A$ is not a highly precise value. Experimentally determined detection limits are *imprecise*. Since precision is the criterion for quantification, quantitative analysis at or near the detection limit can be meaningless.

Equation (3.22) can be used to estimate the minimum S/N needed for a certain quantitative precision. If $s_r(\bar{x}_A)$ is not to exceed say, 0.1%, the signal must have a minimum S/N of 10^3.

INCREASING THE S/N RATIO

The significance of the S/N ratio associated with any measurement has been demonstrated in the preceding section. The analyst must attempt to obtain the analytical signal with the *highest* S/N ratio possible within the practical limits of the analysis. Some measures must be taken to keep the effect of noise to a minimum. Several techniques of increasing the S/N ratio will be briefly outlined.

Optimization

The response of any instrument depends on many factors other than the amount of analyte present in the sample. Generally, by the time the sample is ready for analysis, the amount of the analyte cannot be changed; thus one parameter of the instrumental response is invariant. The analyst must therefore *optimize* the remaining experimental factors in order to obtain the highest S/N ratio possible. Optimization of experimental variables depends on the nature of the analysis. It varies from manipulating one parameter (e.g., choosing the proper pH in complexometric titrations) to testing several factors (e.g., flame temperature, type of auxiliary fuel, and flow rate in flame spectrometry). Since the experimental factors may be correlated, a proper optimization scheme must involve simultaneous changes of their levels. Any of the optimization techniques demonstrated in Chapter 2 may be implemented for the particular analysis at hand.

Signal Averaging

If the analytical signal is *repeatable*, the analyst can obtain several estimates of the signal and simply calculate an average value (9, 10). When the measurement is repeated the random noise is expected to have an average value of zero. When n measurements are averaged, the S/N ratio is increased by a factor of $n^{1/2}$. In practice, a large number of measurements may be required to reduce the average noise level near zero. The number of replicate measurements is determined by the degree of allowable uncertainty, the nature of the analytical instrument, and the cost of replicate analysis and sample preparation.

Nondestructive analyses are ideal situations for performing signal averaging. Repetitive scanning of some emission and absorption spectra, for example, is inexpensive and efficient for routine analysis. The sample remains virtually intact during replicate measurements of IR, NMR, and UV-Vis spectra.

Boxcar Integration

A boxcar integrator is a single-channel signal averager. Several measurements are obtained for *only one portion* of the signal, and an average amplitude is calculated. The detector is turned ON and the signal is measured (sampled). Following the measurement, the detector is turned OFF. The detector is synchronized such that when it is turned ON again, it measures the same portion of the signal. After n measurements are obtained for a particular portion, the detector is allowed to sample other portions of the signal in the same manner (9, 10). The process is continued until the whole signal is sampled. Again, if n measurements are

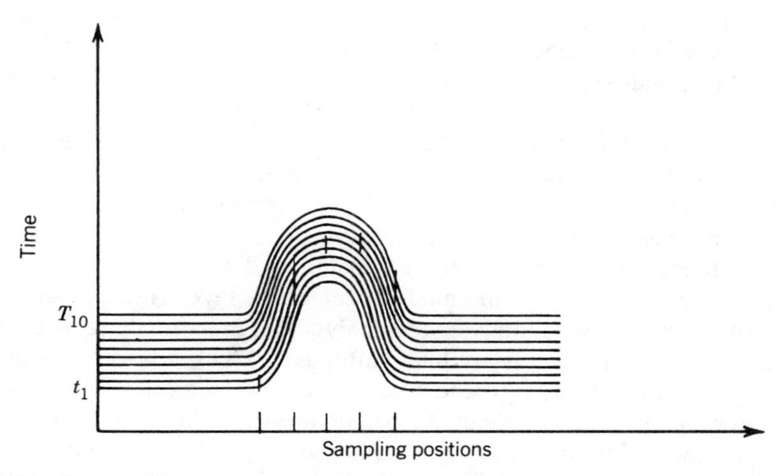

Figure 3.15 The operation of a boxcar integrator. Five portions of signal are sampled and two measurements obtained per portion.

obtained for each portion of the signal, the S/N ratio is increased by a factor of $n^{1/2}$. Figure 3.15 depicts the operation of a boxcar integrator where five portions of a signal are sampled, and two measurements are obtained for each portion before moving to the next sampling position.

Clearly, the signal must be *repeatable*. Synchronization is achieved by employing a proper reference signal. As mentioned earlier, average amplitudes are the only information acquired by such a device. It is not recommended for detecting complex waveforms. However, when the only information needed is an average amplitude of a pulse train, a boxcar integrator is a good choice. Compared to a signal averager, a boxcar integrator is a slow detector. The advantages of a boxcar integrator are attributed to its instrumental simplicity. Unlike a signal averager, a boxcar integrator is turned ON for a short duration only—the time needed to sample the signal. Consequently, the boxcar integrator is exposed to random noise for a much shorter amount of time than a signal averager. This further decreases the effects of random noise.

Signal Filtering and Modulation

A significant increase in the S/N ratio can be achieved by modifying the signal sources and/or the detection strategy such that the noise is ignored by the detector. The noise components, for example, can be filtered by employing a capacitor across the voltage output terminals of the instrument. *Signal modulation* is by far the most popular means of separating an analytical signal from random noise (11–13). Noise signals originate from different sources (14). In terms of their power spectra (10), there are three common types of noise—white, flicker, and interference. The power spectra of these three types of noise are shown in Figure 3.16. The noise level becomes significant in analytical techniques where the energy of the signal source is limited (e.g., IR and NMR spectroscopy).

The noise factor can also be important when the signals are generated near the detection limit. In these cases, the magnitude of the signal is not much higher than that of the noise. It is therefore important that the signal becomes easily distinguished from noise.

In flame spectrophotometry, for instance, flicker noise is important. As shown in Figure 3.16, minimal flicker noise levels are present at high frequencies. Unfortunately, many analytical signals are dc (zero frequency) signals. Small dc signals are badly affected by flicker noise. It is therefore necessary to *shift* the signal to a high-frequency region where flicker noise is low. Such a signal is said to be *modulated*. This can be achieved by simply switching the signal ON and OFF repeatedly; the signal is chopped. The resultant signal is a train of pulses with a frequency equal to the

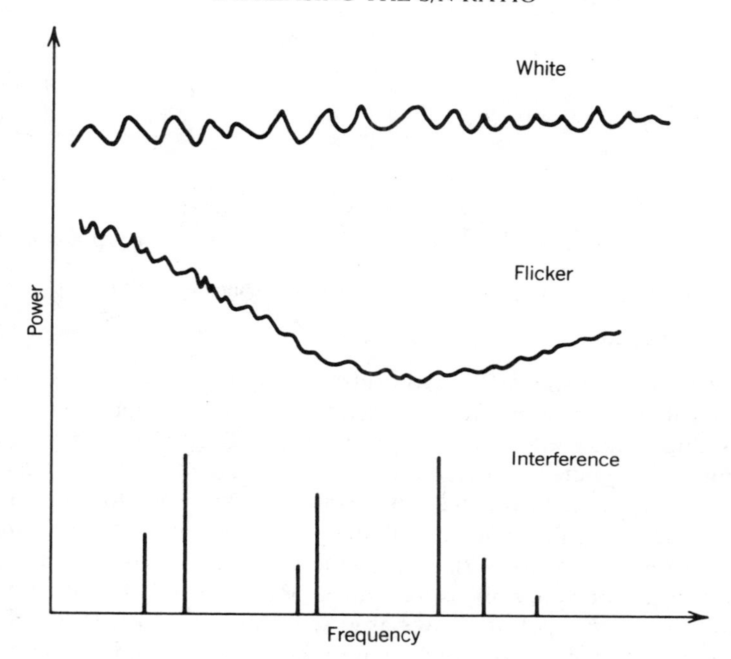

Figure 3.16 Three common types of noise.

frequency of chopping the signal. The chopping frequency is chosen such that the signal is shifted beyond the region of high flicker noise.

The choice of the modulation frequency depends on other modules in the instrument such as the detector's response time. A typical modulation frequency is about 1000 Hz and can be achieved by an electromechanical chopper. By modulating the signal we have not only reduced the effect of noise but also "coded" the signal such that any other accompanying noise component can be filtered out during the detection phase—the demodulation step. Lock-in amplifiers are used to convert the modulated waveform back to a dc signal (11, 14). A lock-in amplifier discriminates against noise on the basis of both frequency and phase. Only those signals that are phase synchronized with a reference signal will produce a net dc output. This further reduces the noise components and increases the S/N ratio significantly.

Multiplex Spectroscopy

Multiplex spectroscopy is a technique based on detecting several wavelengths simultaneously. The advantage of this practice over conventional

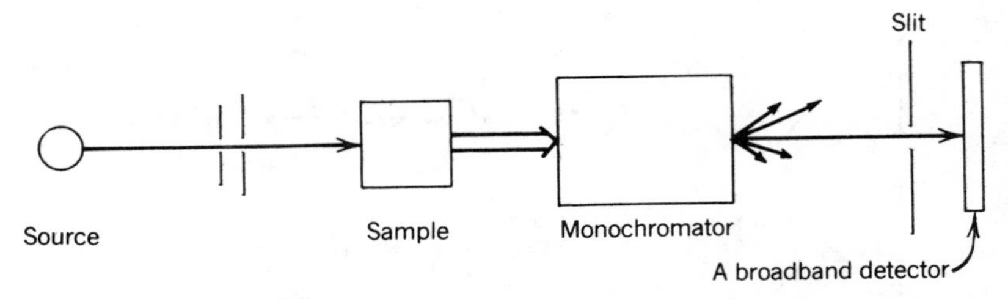

Figure 3.17 Schematic diagram of a single-slit scanning spectrometer.

"one-wavelength-at-a-time" detection is that during the same detection time several measurements are obtained for several wavelengths. The S/N ratio is thus improved (15). Figures 3.17 and 3.18 depict a single-slit scanning spectrometer and a multichannel detector based spectrometer.

If a finite amount of time, Δt, is required to measure the response at each of n wavelengths, a single-slit spectrometer (Fig. 3.17) requires an amount of time equal to $n\,\Delta t$ to scan the n channels. At the end of the scanning cycle, only *one* measurement is available for each channel. In contrast, a multichannel detector based spectrometer (Fig. 3.18) requires an amount of time equal to Δt to record the n channels with one measurement per channel. Thus the same amount of information is acquired in a shorter time. More importantly, during a scanning cycle of duration $n\,\Delta t$, n measurements are obtained for *each* channel. Consequently, the S/N ratio is, in principle, increased by a factor of $n^{1/2}$ for the same measurement time.

The main problem with multichannel spectrometers is the prohibitively

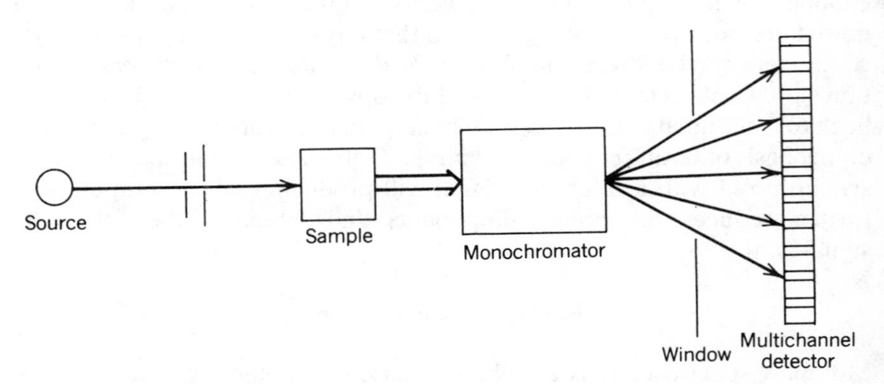

Figure 3.18 Schematic diagram of a multichannel detector based spectrometer.

large number of channels required. In IR spectroscopy, for example, the typical range of the spectrum is 1.5×10^{14} Hz. If a resolution of 3×10^9 Hz is desired, one needs 50,000 channels (15). It is necessary to resort to indirect approaches in order to perform multiplex spectroscopy. Hadamard and Fourier transform spectroscopy offer two such approaches.

Hadamard Transform Spectroscopy

The Hadamard multiplex scheme consists of using a broad band detector with a "mask" placed between the detector and the window of Figure 3.18. This is schematically shown in Figure 3.19. The mask contains several openings with each opening as wide as a single slit. Half the slits in the mask are left open while the other half remain closed during each measurement interval. By opening and closing the various slits different channels are monitored. The response obtained during each measurement interval can be expressed as

$$R_i = a_1 X_1 + a_2 X_2 + \cdots + a_n X_n, \tag{3.25}$$

where

$$a_i \begin{cases} = 0 & \text{when slit is closed} \\ = 1 & \text{when slit is opened} \end{cases}$$

X_j = individual response of the jth channel,

R_i = response of the detector during the ith measuring interval.

By changing the pattern of opening and closing the various slits, N *independent* masks can be obtained. The N observations can be expressed

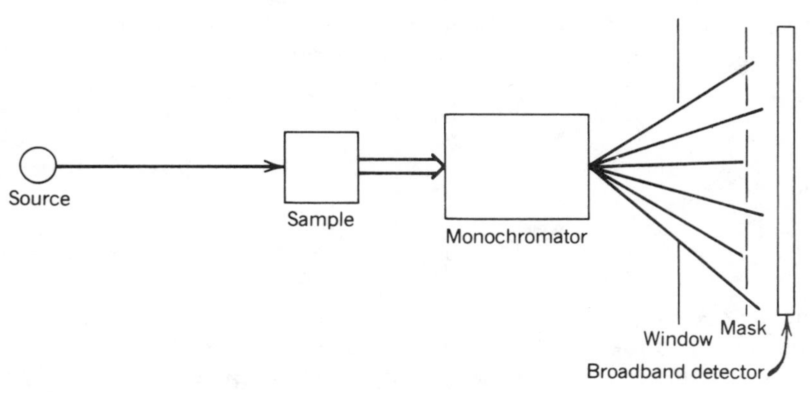

Figure 3.19 Hadamard multiplex scheme.

as N independent equations:

$$\begin{aligned} R_1 &= a_{11}X_1 + a_{12}X_2 + \cdots + a_{1n}X_n, \\ R_2 &= a_{21}X_1 + a_{22}X_2 + \cdots + a_{2n}X_n, \\ &\vdots \\ R_N &= a_{N1}X_1 + a_{N2} + \cdots + a_{Nn}X_n, \end{aligned} \tag{3.26}$$

where $a_{ij} = 1$ or 0.

Using linear algebraic notation, Equation (3.26) can be expressed as

$$\mathbf{R} = \mathbf{AX} \tag{3.27}$$

where **R** is a column matrix containing the responses of the detector
A is an $N \times n$ matrix whose rows represent the various patterns of the mask used during the different measurements
X is a column matrix containing the responses of the individual channels

The elements of **R** and **A** are known. Assuming that the inverse of **A**, $\mathbf{A}^{-1}$, exists, **X** can be solved for according to

$$\mathbf{X} = \mathbf{A}^{-1}\mathbf{R}. \tag{3.28}$$

The response of each channel, x_i, is measured $N/2$ times over the same period of time needed to scan the spectrum conventionally. If the noise is limited only by the detector, the above scheme offers an increase in the S/N ratio by a factor of $(N/2)^{1/2}$, the multiplex advantage.

Fourier Transform Spectroscopy

The Fourier transform is a mathematical relationship that relates two functions, $f(t)$ and $F(\nu)$. The relationships can be written

$$f(t) = \int_{-\infty}^{\infty} F(\nu)e^{2\pi jt\nu}\, d\nu \tag{3.29}$$

and

$$F(\nu) = \int_{-\infty}^{\infty} f(t)e^{-2\pi jt\nu}\, dt, \tag{3.30}$$

where $j = \sqrt{-1}$.
$f(t)$ is usually a function of time (or distance) and is said to be in the *time domain*. Similarly, $F(\nu)$ is usually a function of frequency and is said to be in the *frequency domain*. Thus $f(t)$ is a *waveform* while $F(\nu)$ is a *spectrum*.

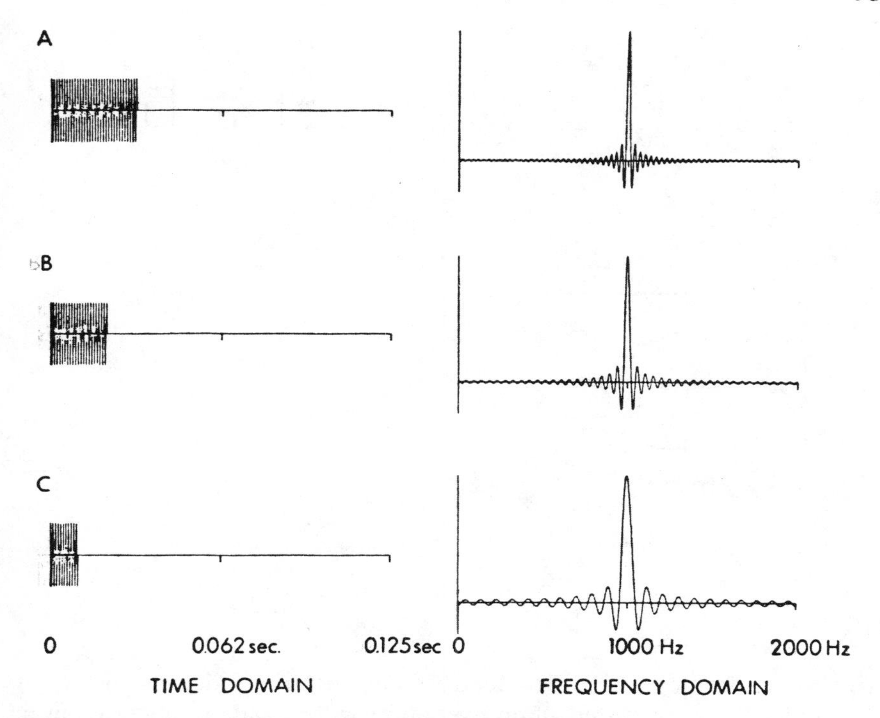

Figure 3.20 Pictorial representations of the Fourier transformation of (A) 32 cycles, (B) 21 cycles, and (C) 10 cycles of a 1000-Hz cosine wave. Note the inverse dependence of the width of the frequency domain function on the length of the time domain function. [From Horlick, "Fourier Transform Approaches to Spectroscopy," *Anal. Chem.* **43**(8), p. 62A (July 1971), Fig. 1.]

Figure 3.20 shows three time domain functions, $f(t)$, and their frequency domain transforms, $F(\nu)$.

In conventional spectroscopy the spectrum, $F(\nu)$, is monitored directly by observing the radiations of different wavelengths emerging from the monochromator. By contrast, in Fourier transform spectroscopy a waveform, $f(t)$, is measured and then decomposed to obtain the spectrum, $F(\nu)$. The process is depicted in Figure 3.21. Note that *all* wavelengths are monitored simultaneously, the advantage being that the resultant spectrum has a better S/N ratio than the spectrum obtained by conventional scanning over the same amount of time. If N measurements are obtained, the S/N ratio is increased by a factor of $N^{1/2}$, provided the noise is limited only by the detector.

The Fourier transform of a spectrum, $f(t)$, is detected via different

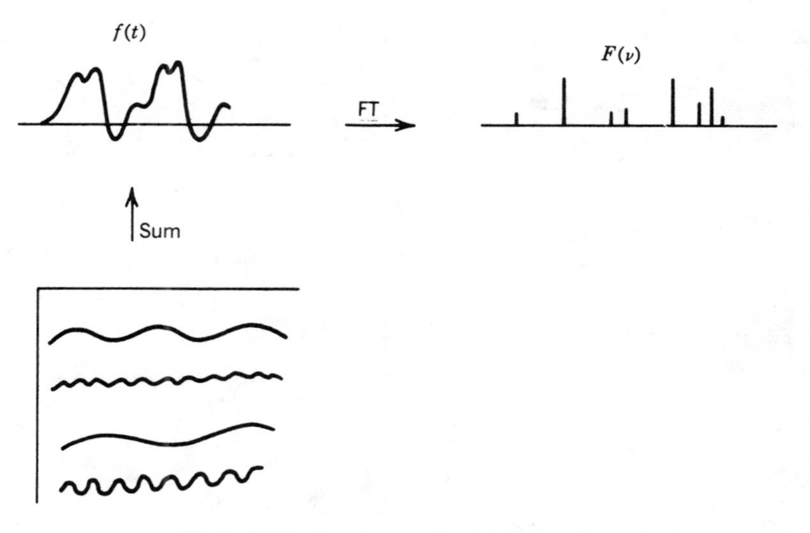

Figure 3.21 Fourier transform spectroscopy.

detection schemes. In NMR spectroscopy, the free induction decay of the molecule is measured upon excitation with a radiofrequency pulse. The obtained waveform is the Fourier transform of the common NMR spectrum. As shown in Figure 3.21, the waveform being detected is composed of the individual frequencies in the spectrum; these frequencies are detected simultaneously. In IR spectroscopy, the Fourier transform of the spectrum is obtained by using a Michelson interferometer. The common IR spectrum, $F(\nu)$, is then calculated from the Fourier transform of $f(t)$.

Decoding f(t) [Computing F(ν)]

The time domain signal, $f(t)$, is sampled at N equally spaced intervals. Each signal, $f(t_i)$, can be expressed as a series of frequency domain amplitudes:

$$
\begin{aligned}
f(t_1) &= a_{11}X(\omega_1) + a_{12}X(\omega_2) + \cdots + a_{1N}X(\omega_N) \\
f(t_2) &= a_{21}X(\omega_1) + a_{22}X(\omega_2) + \cdots + a_{2N}X(\omega_N) \\
&\vdots \\
f(t_n) &= a_{N1}X(\omega_1) + a_{N2}X(\omega_2) + \cdots + a_{NN}X(\omega_N),
\end{aligned}
\qquad (3.31)
$$

where $t_n = \text{nT/N},\ n = 1, 2, \ldots, N$
$T = \text{observation time}$
$\omega_m = 2\pi m/T,\ m = 1, \ldots, N$
$a_{nm} = e^{2\pi jnm/N}$
$j = \sqrt{-1}$

Equation (3.31) can be solved to determine the amplitudes of the various frequencies, $X(\omega_m)$. Equation (3.31) can be rewritten

$$f(t_n) = \sum_{m=1}^{N} e^{2\pi jnm/N} X(\omega_m). \tag{3.32}$$

Note the similarity between Equations (3.32) and (3.29). $X(\omega_m)$ can be solved for by computing the Fourier transform of $f(t)$ in a fashion similar to Equation (3.30). This is often referred to as the *discrete Fourier transform*. $X(\omega)$ is also referred to as the reverse Fourier transform of $f(t)$.

$$X(\omega_m) = \sum_{n=1}^{N} f(t_n) e^{-2\pi jnm/N}, \qquad m = 1, 2, \ldots, N. \tag{3.33}$$

Equation (3.33) can be simplified by expressing $X(\omega_m)$ and $f(t_n)$ as $X(m)$ and $f(n)$, respectively, and by replacing $e^{-2\pi j/N}$ by the operator W. Equation (3.35) can therefore be written

$$X(m) = \sum_{n=1}^{N} f(n) W^{nm}, \quad m = 1, 2, \ldots, N. \tag{3.34}$$

Direct solution of the above equation requires N^2 multiplication operations. The computation of the inverse Fourier transform was therefore once considered a disadvantage. Cooley and Tukey (16) have developed a fast method for computing inverse Fourier transforms. Appropriately, the process is referred to as the *fast Fourier transform* (FFT). This will be briefly demonstrated next.

Fast Fourier Transform

The following is a brief description of how the FFT technique is used to calculate inverse Fourier transforms. A more detailed discussion can be found elsewhere (17). Consider the simple case of $N = 4$. Equation (3.34) can be written

$$X(n) = \sum_{k=0}^{N-1} f(k) W^{nk}, \qquad n = 0, 1, 2, \ldots, N-1. \tag{3.35}$$

(A slight change in notation is made here to conform with the notation used in ref. 17 from which this example is taken with some modifications.) Equation (3.33) can therefore be written

$$\begin{bmatrix} X(0) \\ X(1) \\ X(2) \\ X(3) \end{bmatrix} = \begin{bmatrix} W^0 & W^0 & W^0 & W^0 \\ W^0 & W^1 & W^2 & W^3 \\ W^0 & W^2 & W^4 & W^6 \\ W^0 & W^3 & W^6 & W^9 \end{bmatrix} \begin{bmatrix} f(0) \\ f(1) \\ f(2) \\ f(3) \end{bmatrix}. \tag{3.36}$$

It can be shown that

$$W^{nk} = w^{\ell}, \tag{3.37}$$

where ℓ is the remainder of the division of nk by $N[e^{-j\theta} = \cos\theta - j\sin\theta]$.

Equation (3.36) can now be expressed as

$$\begin{bmatrix} X(0) \\ X(1) \\ X(2) \\ X(3) \end{bmatrix} = \begin{bmatrix} 1 & 1 & 1 & 1 \\ 1 & W^1 & W^2 & W^3 \\ 1 & W^2 & W^0 & W^2 \\ 1 & W^3 & W^2 & W^1 \end{bmatrix} \begin{bmatrix} f(0) \\ f(1) \\ f(2) \\ f(3) \end{bmatrix}. \tag{3.38}$$

The matrix containing the 1's and W^{ℓ}'s can be factored into two matrices, and Equation (3.38) can be rewritten

$$\begin{bmatrix} X(0) \\ X(2) \\ X(1) \\ X(3) \end{bmatrix} = \begin{bmatrix} 1 & W^0 & 0 & 0 \\ 1 & W^2 & 0 & 0 \\ 0 & 0 & 1 & W^2 \\ 0 & 0 & 1 & W^3 \end{bmatrix} \begin{bmatrix} 1 & 0 & W^0 & 0 \\ 0 & 1 & 0 & W^0 \\ 1 & 0 & W^2 & 0 \\ 0 & 1 & 0 & W^2 \end{bmatrix} \begin{bmatrix} f(0) \\ f(1) \\ f(2) \\ f(3) \end{bmatrix}. \tag{3.39}$$

Note the interchange of $X(1)$ and $X(2)$ in the matrix on the left-hand side of Equation (3.39). Computing $X(m)$ from Equation (3.39) requires *four* complex multiplications and *eight* complex additions. In contrast, computing the same elements from Equation (3.36) requires *16* complex multiplications and *12* complex additions. The number of necessary multiplication and addition operations is substantially reduced. This fact becomes more appreciated when N is much larger than 4. The only problem, however, is that the $X(n)$'s are scrambled. The resultant vector can be unscrambled by replacing the subscript n by its binary equivalent (17).

SIGNAL MANIPULATION

Analytical signals are often subjected to several manipulations after they are acquired. In spectroscopy, for example, the maximum of the signal must be determined. In chromatography, the maximum amplitude and the area of the signal are desired. Frequently, the signal is "processed" such that its information content is more easily recognizable. The various objectives of signal processing can be grouped into two major categories: (1) to estimate pertinent signal parameters (e.g., maximum amplitude, area, and general shape) and (2) to enhance its information content. The first category utilizes the techniques of *curve fitting* while the second relies on *least-squares polynomial smoothing*, *deconvolution*, and *differentiation*.

The applications of these and other techniques to the signal is often referred to as *signal processing*. Some popular methods of signal processing will be briefly outlined.

Curve Fitting

As has already been demonstrated (Chapter 2) least-squares curve fitting is employed to describe the relationship between analyte concentration and instrumental response. Calibration represents only one application of curve fitting. It is an example of curve fitting of linear functions. The technique can be used to describe experimental data in terms of an appropriate model and/or to examine the agreement between theory and experiment.

The fields of chromatography and spectroscopy have found numerous applications for curve fitting methodologies. Modeling chromatographic peaks and spectral signals accounts for a large portion of the use of curve fitting in analytical chemistry. The analyst can apply the method(s) to *raw* or smoothed data. (Smoothing of raw data will be discussed in the next section).

Curve Fitting of Nonlinear Functions

Figure 3.22 shows the digital output of a detector obtained during a measurement interval over a certain parameter x. The ordinate, R, is the

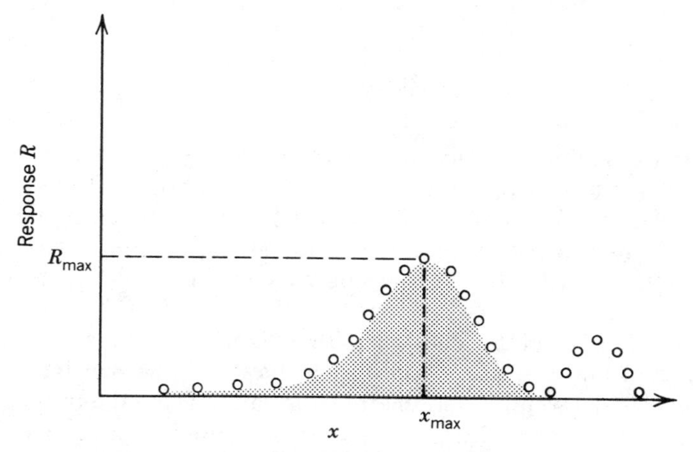

Figure 3.22 Digital output of a detector.

response of the instrument (e.g., magnitude of the current or voltage output). The abscissa is the measurement parameter [e.g., time (in chromatography), wavelength (in spectroscopy), or mass-to-charge ratio (in mass spectrometry)]. The parameters R_{max} and x_{max} are the values of R and x at the maximum of the signal. Three of the most important characteristics of such a signal are R_{max}, x_{max} and the area under the signal (the shaded region in Fig. 3.22). The spread of the signal about x_{max} (the variance or the width of the signal) is also important.

Estimating Peak Parameters

The values of R_{max}, x_{max}, and the variance, s^2, of the signal can be estimated from an appropriate model. If the signal shown in Figure 3.22 can be adequately described by a Gaussian model, such a model can be used to estimate the signal's parameters. Let the following model describe the signals shown in Figure 3.22:

$$R = R_{max} e^{-(x_{max}-x)^2/s}. \tag{3.40}$$

The direct application of least-squares curve fitting yields nonlinear equations in the parameters (R_{max}, x_{max}, and s) whose solutions can be an involved task. This problem can be easily overcome by *linearizing* the model (18). Equation (3.40) can be expressed as

$$\ln R = \ln R_{max} - \frac{(x_{max} - x)^2}{s} \tag{3.41}$$

or

$$\ln R = Cx - Ax^2 + D, \tag{3.42}$$

where $A = 1/s$, $C = 2x_{max}/s$, and

$$D = \ln R_{max} - \frac{x_{max}^2}{s}.$$

Equation (3.42) conforms to the general model ($Y = \beta_0 + \beta_1 X_1 + \beta_2 X_2 + \cdots + \beta_n X_n$) and its parameters ($C$, A, and D) can be estimated by linear least-squares curve fitting as shown in Chapter 2. If the model is adequate and can account for the systematic variance in the data (this is also demonstrated in Chapter 2), the analyst can then proceed to estimate R_{max}, x_{max}, and s.

The reader is hereby cautioned that linearization also changes the weighting in curve fitting. Appropriate new weights can be calculated from the original weights and the transformation using the propagation of error principle. Curve fitting can be implemented in several variations:

1. The data can be smoothed *before* the fitting.
2. The *whole curve* can be fitted at once. This can be employed only when one peak is present. If small peaks (see Fig. 3.22) are present, they can be completely overlooked and the fitting will produce severe errors in the estimated parameters.
3. Only a small portion of the data (the part suspected of containing a peak) is fitted at a time. If the background is varying slowly under the peak (and can be interpolated from fitting on both sides) it is recommended that it is fitted separately (as a straight line or a quadratic function) on both sides of the peak. The region of the peak is then modeled separately after subtracting the contribution of the background.
4. The signal and the background are fitted simultaneously. This is a more complex variation as it often requires *nonlinear least-squares curve fitting*. If a signal like that shown in Figure 3.22 is superimposed on a quadratic function representing the background, Equation (3.40) becomes

$$R = R_{\max} e^{-(x_{\max}-x)^2/s} + \underbrace{a_1 + a_2 x + a_3 x^2}_{R_b}, \qquad (3.43)$$

where $R_{\max}$ and $x_{\max}$ are as defined above and R_b is the contribution of the background expressed as a second-order polynomial in x.

Note that Equation (3.43) *cannot* be linearized. The parameters in the model [e.g., Eq. (3.43)] are estimated by minimizing χ^2 with respect to the various parameters. The measure of goodness of fit, χ^2, is given by

$$\chi^2 = \sum \left\{ \frac{1}{\sigma_i^2} [R_i - R_{\max} e^{-(x_{\max}-x_i)^2/s} - a_1 - a_2 x_i - a_3 x_i^2]^2 \right\}, \qquad (3.44)$$

where σ_i^2 is the variance of the corresponding datum. This is, in principle, an optimization problem. Several methods are available to find the minimum value of χ^2 as a function of the various parameters (19). Figure 3.23 shows the results of fitting a Gaussian signal superimposed on a quadratic background. The single most serious source of errors in curve fitting is an inadequate model (i.e., when the proposed model does not match the experimental data). It is of utmost importance that any model be statistically validated. Fitting Gaussian profiles to tailing chromatographic peaks is one example of such a mismatch between the model and the data. The tailing section of the peak contributed most to the lack-of-fit error.

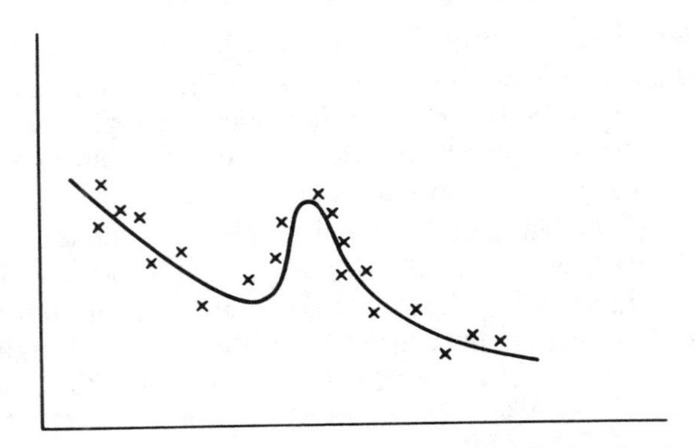

Figure 3.23 Results of fitting a Gaussian signal superimposed on a quadratic background. [From Bevington, *Data Reduction and Error Analysis for the Physical Sciences*, McGraw-Hill, New York, 1969, Fig. 11-5, p. 215.]

Tailing peaks are better modeled as gamma functions or as a combination of a Gaussian profile and an exponentially decaying function (20).

Estimating the Area Under the Peak

In several applications (e.g., chromatography, spectroscopy, and thermogravimetry) the area under the signal is proportional to the amount of analyte. This makes estimating the area under the peak a crucial step in the analysis. The simplest method of calculating the area is to add all ordinate values [e.g., $R(X_i)$ in Fig. 3.22] due only to the peak (i.e., after the background contributions have been eliminated). Thus, for a peak defined by N equally spaced data points, the area of the peak, A_p, is given by

$$A_p = \sum_{i=1}^{N} R(X_i), \tag{3.45}$$

where $R(X_i)$ is the response obtained at X_i. Equation (3.45) implies that data points in the "middle" of the signal are given equal importance (weight) as the data points near the "edges" of the signal. There are, however, several factors that suggest that the middle points be weighted more than the edge points. Random errors, for instance, are apt to average out at the middle, where each data point has more "neighbors," than at the end where neighboring points are excluded from the signal distribution. Consequently, the middle points are often given more weight than the edge points when the area of the signal is estimated. This practice is familiar to those who have experience with numerical integration using the trapezoidal or Simpson rules (21). For a signal with N equally spaced data points, the areas as calculated by the trapezoidal (A_p^T) and the Simpson (A_p^S) rules are given by

$$A_p^T = \tfrac{1}{2}[R(X_1) + R(X_N)] + \sum_{i=2}^{N-1} R(X_i) \tag{3.46}$$

and

$$A_p^S = \tfrac{1}{3}[R(X_1) + 4R(X_2) + 2R(X_3) + \cdots + 4R(X_{n-1}) + R(X_N)]. \tag{3.47}$$

Integration formulas are derived by modeling portions (or intervals) of the function being integrated. Their accuracy, therefore, is dependent on the accuracy of the model being employed. Generally, n data points are chosen as an interval and a polynomial of degree $n-1$ is fitted to the data. The trapezoidal integration formula is derived by assuming that the shape of the signal between every two adjacent ordinate values [the interval consists of $R(X_i)$ and $R(x_{i+1})$] can be approximated as a straight line. The formula known as the Simpson rule is derived by assuming that the shape of the signal between every three adjacent ordinate values [the interval consists of $R(X_i)$, $R(X_{i+1})$, and $R(X_{i+2})$] can be modeled as a parabola. The final formulas [Eqs. (3.46) and (3.47)] are obtained by adding the areas of all individual intervals. Clearly, the trapezoidal formula yields incorrect results when the shape of the signal in one or more intervals is nonlinear. This shortcoming has limited the application of the trapezoidal rule in analytical chemistry where the shape of most signals is nonlinear. By contrast, the Simpson rule is applicable in situations where the shape of the signal is parabolic. It is also accurate for linear segments. The Simpson rule is therefore a common and widely used integration procedure. As mentioned earlier, the accuracy of any integration formula depends on the underlying assumptions of its derivation. The analyst is encouraged to investigate the underlying assumptions involved in any "built-in" integration procedure before subjecting the data to processing by that procedure.

A successful application of the Simpson rule requires that the number of data points, n, is odd (i.e., the number of intervals is even). If the number of intervals is not even, the Simpson rule can be employed for an even number of the available intervals and some other method can be employed to integrate the remaining interval(s). One such method is the *Simpson* (or Newton) *three-eights rule* (21). If the number of points is even, the Simpson rule is used for the interval(s) between $R(X_1)$ and $R(X_{n-3})$. The last interval is made up of $R(X_{n-3})$, $R(X_{n-2})$, $R(x_{n-1})$, and $R(X_n)$

and its area is computed by fitting a third-degree polynomial. The area of the last interval, A_w, is given by

$$A_w = \tfrac{3}{8}[R(X_{n-3}) + 3R(X_{n-2}) + 3R(X_{n-1}) + R(X_n)]. \qquad (3.48)$$

Using the three-eights rule in conjunction with the Simpson rule covers any number of data points. Generally, the smaller the interval the more accurate the integration.

Alternatively, the area can be calculated using an appropriate model that has been previously fitted to the data. In this case the area calculated using the model will differ from the area calculated using the raw data by an amount equivalent to the minimum χ^2 value used to estimate the parameters [Eq. (3.44)].

There is no unique recipe for modeling or integrating an analytical signal. The choice of the model, the treatment of background noise, and the strategy of curve fitting and computing the area of the signal are dependent on the kind of experiment being performed, the cost of analysis, the allowable error, and the analyst. There are situations where peak height (i.e., R_{max} in Fig. 3.22) is used for calibration instead of peak area. In these situations computing R_{max} is more pertinent than the integration procedure. If the background signal is negligibly small compared to the analytical signal, the area may be calculated from the raw data points rather than from a model.

Smoothing of Data

Smoothing is a means of data manipulation by which small variations are discarded and large ones are retained. While large variations are usually expected to contain useful information, small variations are often attributed to random errors. Smoothed data *do not* contain any additional information and they are by no means a substitute for "good" data. In general, the original measurements are divided into several subsets and each subset is used to calculate a smoothed average value for the data. Some common methods of smoothing (a subset of digital filtering) will be outlined.

Boxcar Averaging

As shown in Figure 3.24, the data are divided into separate groups (boxcars). Each group of data is replaced by its centroid. Clearly, some resolution is lost (see Fig. 3.24). An important requirement for a successful boxcar averaging application is that the rate of variations in the signal is smaller than the sampling rate. Random noise is expected to have an

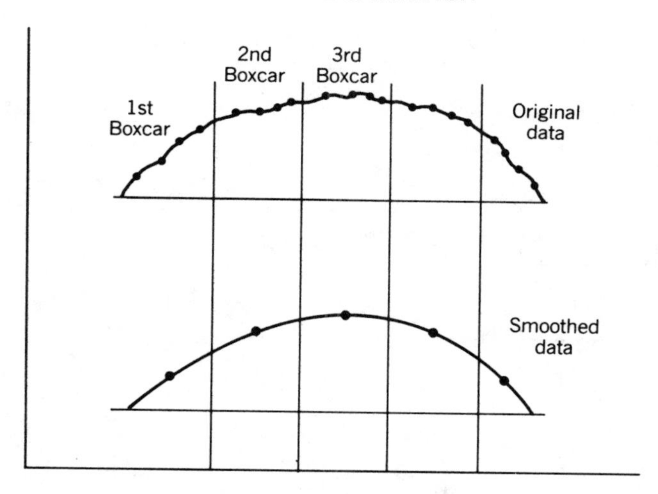

Figure 3.24 Boxcar averaging.

average value of zero in each boxcar. The larger the number of points in every boxcar the higher the probability of having an average value of zero for random noise. This also results in a higher degree of resolution loss and more distortion of the original data. Determining the span of the boxcar (i.e., the number of points in each boxcar) is thus dependent on the particular application. One of the attractive features of this technique is the ease of its implementation.

Moving Window Averaging

This is a variation of boxcar averaging and a special case of the general technique known as least-squares polynomial smoothing (to be discussed in the next section). As shown in Figure 3.25, in the moving window averaging method, the spans of the different boxcars overlap.

Assume that a boxcar is made of five data points, P_1, P_2, P_3, P_4, and P_5. In the moving window averaging technique, the centroid of the boxcar, P_3, is replaced by the arithmetic average of P_1, P_2, P_3, P_4, and P_5. The following boxcar is formed by dropping P_1 and adding the following datum, P_6. The centroid of the subsequent boxcar, P_4, is replaced by the arithmetic average of P_2, P_3, P_4, P_5, and P_6. A third boxcar is then formed by dropping P_2 and adding P_7. The process is repeated until all the data points are included. At the end, the original N data points are replaced by $N-2r$ smoothed data points, where $2r+1$ is the span of the moving

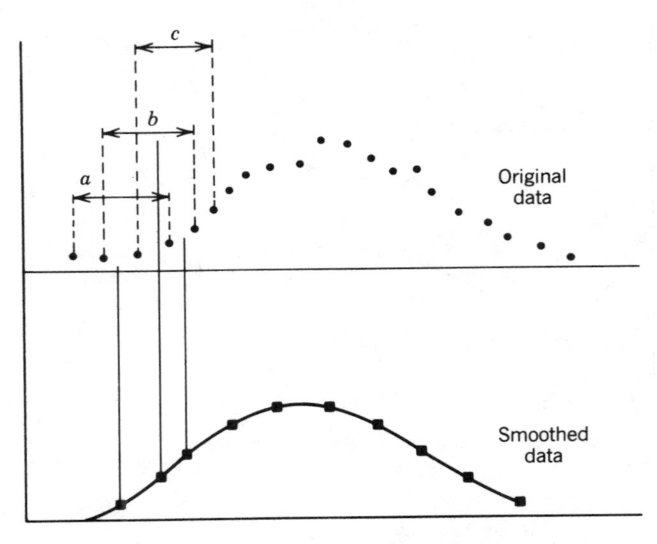

Figure 3.25 Moving window averaging.

window. Loss of resolution is much less in this technique than in boxcar averaging, and the effective sampling rate is higher.

Least-Squares Polynomial Smoothing

In this method a curve (usually linear or quadratic) is used to model a subset of the data. The criterion of least squares (thus the name) is used to estimate the parameters of the model. Consequently, a smoothed average value is computed for the subset. The process is similar to the moving window averaging procedure. As mentioned earlier, the moving window averaging process is a special case of least-squares polynomial smoothing—namely, the case where the polynomial used in fitting the data is linear. The following discussion will illustrate this point and will demonstrate how the smoothing formulas are derived.

Consider a subset (or a window) of three data points. The three points (the first, the central, and the last) will be denoted R_{-1}, R_0, and R_{+1}, respectively. The centroid of these three points, R_0, is to be replaced by a smoothed value, R_0^*. Assume that the following model fits the three data points:

$$R = \beta_0 + \beta_1 x, \tag{3.49}$$

where x is the measurement parameter.

Using the method of least squares, the normal equations are

$$\left.\begin{aligned} b_0 \sum_{i=-m}^{m} (1) + b_1 \sum_{i=-m}^{m} x_i &= \sum_{i=-m}^{m} R_i \\ \text{and} \quad b_0 \sum_{i=-m}^{m} x_i + b_1 \sum_{i=-m}^{m} x_i^2 &= \sum_{i=-m}^{m} R_i x_i \end{aligned}\right\}, \tag{3.50}$$

where b_0 and b_1 are the estimates of β_0 and β_1, respectively, and $2m+1$ is the span of the window. The solution of Equation (3.50) is simplified by assuming that the x_i's are equally spaced and can be expressed as

$$x_i = ih,\ i = -m, -m+1, \ldots, 0, 1, 2, \ldots, m, \tag{3.51}$$

where h is the interval between the values of x. Equation (3.50) reduces to

$$b_0 = \frac{R_{-1} + R_0 + R_{+1}}{3} \quad \text{and} \quad b_1 = \frac{R_{+1} - R_{-1}}{2h}. \tag{3.52}$$

Smoothed values (modeled values) can now be calculated for all R_i's by substituting Equation (3.52) in the model [Eq. (3.49)]. The smoothed value, R_i^*, is given by

$$R_i^* = \tfrac{1}{3}(R_{-1} + R_0 + R_{+1}) + (R_{+1} - R_{-1})\frac{x_i}{2h}. \tag{3.53}$$

The centroid of the three-point window, R_0, is replaced by R_0^*.

$$R_0^* = \tfrac{1}{3}(R_{-1} + R_0 + R_{+1}). \tag{3.54}$$

This equation is the arithmetic average needed for the moving window averaging procedure. A new window is then formed in exactly the same fashion of the moving window averaging process, and the centroid of the new window is replaced by a smoothed value. Equation (3.54) is used to compute the smoothed value for all windows. Note, however, that unlike the moving window averaging strategy, it is now possible to compute smoothed values for *all* points in the window [Eq. (3.53)]. Smoothed values for R_{-1} and R_{+1} are given by

$$R_{-1}^* = \tfrac{1}{6}(5R_{-1} + 2R_0 - R_{+1}) \quad \text{and} \quad R_{+1}^* = \tfrac{1}{6}(-R_{-1} + 2R_0 + 5R_{+1}). \tag{3.55}$$

Two important features must be realized in Equations (3.54) and (3.55). First, for equally spaced variables and an odd number of points, the independent variable, x, does not appear in the smoothing formulas. Second, the smoothed value, R^*_i, is a *weighted* average. The centroid, R_0, is replaced by the arithmetic average. The end points are replaced by smoothed values that are heavily weighted in each of the corresponding datum. In practice, a moving window with a span of $2m-1$ data points, $R_{-m}, R_{-m+1}, \ldots, R_0, R_1, R_2, \ldots, R_m$, scans the measurements in the fashion shown in Figure 3.25. In the first window, smoothed values for $R_{-m}, R_{-m+1}, \ldots, R_0$ are calculated. In the subsequent windows only, R_0 is replaced by R^*_0. In the last window, smoothed values for $R_0, R_1, \ldots, R_m$ are computed. Often, the first and last m points of the original N data points are discarded and only $N-2m$ smoothed measurements are retained. The smoothed data can further be subjected to smoothing. This is known as multiple smoothing or multiple-pass smoothing.

Several smoothing formulas can be derived by changing the window span and/or the model employed to fit the data. Quadratic models are quite common. The quadratic model is given by Equation (3.56):

$$R = \beta_0 + \beta_1 x + \beta_2 x^2. \tag{3.56}$$

Application of the least-squares criterion yields the following normal equations:

$$\left.\begin{aligned}
&b_0 \sum_{i=-m}^{m} (1) + b_1 \sum_{i=-m}^{m} x_i + b_2 \sum_{i=-m}^{m} x_i^2 = \sum_{i=-m}^{m} R_i,\\
&b_0 \sum_{i=-m}^{m} x_i + b_1 \sum_{i=-m}^{m} x_i^2 + b_2 \sum_{i=-m}^{m} x_i^3 = \sum_{i=-m}^{m} R_i x_i,\\
\text{and}\quad &b_0 \sum_{i=-m}^{m} x_i^2 + b_1 \sum_{i=-m}^{m} x_i^3 + b_2 \sum_{i=-m}^{m} x_i^4 = \sum_{i=-m}^{m} R_i x_i^2.
\end{aligned}\right\} \tag{3.57}$$

where b_0, b_1, and b_2 are the estimates of β_0, β_1, and β_2, respectively, and $2m+1$ is the span of the moving window. For a window with a five-point span, and assuming that $x = ih$, $i = -2, -1, 0, 1, 2$, the normal equations become

$$\left.\begin{aligned}
5b_0 + 10h^2 b_2 &= \sum_{i=-m}^{m} R_i,\\
10hb_1 &= -2R_{-2} - R_{-1} + R_1 + 2R_2,\\
\text{and}\quad 10b_0 + 34h^2 b_2 &= 4R_{-2} + R_{-1} + R_1 + 4R_2.
\end{aligned}\right\} \tag{3.58}$$

The above equations yield

$$b_0 = \frac{1}{35}(-3R_{-2} + 12R_{-1} + 17R_0 + 12R_1 - 3R_2),$$

$$b_1 = \frac{1}{10h}(-2R_{-2} - R_{-1} + R_1 + 2R_2),$$

and

$$b_2 = \frac{1}{14h^2}(2R_{-2} - R_{-1} - 2R_0 - R_1 + 2R_2). \tag{3.59}$$

The first equation of (3.59) is used to calculate a smoothed value for R_0, R_0^* ($R = \beta_0$ at $x = 0$). It is also possible to compute smoothed values for R_{-2}, R_{-1}, R_1, and R_2.

A quadratic model and a window with a five-point span yields the following smoothing formulas (22):

$$R_{-2}^* = \tfrac{1}{35}(31R_{-2} + 9R_{-1} - 3R_0 - 5R_1 + 3R_2),$$

$$R_{-1}^* = \tfrac{1}{35}(9R_{-2} + 13R_{-1} + 12R_0 + 6R_1 - 5R_2),$$

$$R_0^* = \tfrac{1}{35}(-3R_{-2} + 12R_{-1} + 17R_0 + 12R_1 - 3R_2),$$

$$R_1^* = \tfrac{1}{35}(-5R_{-2} + 6R_{-1} + 12R_0 + 13R_1 + 9R_2),$$

and

$$R_2^* = \tfrac{1}{35}(3R_{-2} - 5R_{-1} - 3R_0 + 9R_1 + 31R_2). \tag{3.60}$$

Similarly, a seven-point window and a cubic model give the following smoothing formulas (22):

$$R_{-3}^* = \tfrac{1}{42}(39R_{-3} + 8R_{-2} - 4R_{-1} - 4R_0 + R_1 + 4R_2 - 2R_3),$$

$$R_{-2}^* = \tfrac{1}{42}(8R_{-3} + 19R_{-2} + 16R_{-1} + 6R_0 - 4R_1 - 7R_2 + 4R_3),$$

$$R_{-1}^* = \tfrac{1}{42}[-4R_{-3} + 16R_{-2} + 19R_{-1} + 12R_0 + 2R_1 - 4R_2 + R_3],$$

$$R_0^* = \tfrac{1}{42}[-4R_{-3} + 6R_{-2} + 12R_{-1} + 14R_0 + 12R_1 + 6R_2 - 4R_3],$$

$$R_1^* = \tfrac{1}{42}[R_{-3} - 4R_{-2} + 2R_{-1} + 12R_0 + 19R_1 + 16R_2 - 4R_3],$$

$$R_2^* = \tfrac{1}{42}[4R_{-3} - 7R_{-2} - 4R_{-1} + 6R_0 + 16R_1 + 19R_2 + 8R_3],$$

and

$$R_3^* = \tfrac{1}{42}[-2R_{-3} + 4R_{-2} + R_{-1} - 4R_0 - 4R_1 + 8R_2 + 39R_3]. \tag{3.61}$$

The studies of Savitzky and Golay (23) are among the early developments of least-squares polynomial smoothing in analytical chemistry. They have demonstrated that the noise in every window is reduced by a factor approximately equal to the square root of the span of the window, provided the noise is normally distributed. Note, however, that the larger the window the higher the loss of resolution. They have also tabulated the coefficients needed for the smoothing formulas using various models (e.g., quadratic and higher-order polynomials and their derivatives) and various window spans. Some of these tables have been corrected by Steinier et al. (24).

Least-squares polynomial smoothing has found several applications in analytical chemistry. A detailed study of the effects of smoothing on signal parameters and the extent of noise reduction has been advanced (25).

Fourier Transform Smoothing

Spectral signals are distorted by several processes. Finite monochromator resolution, collisional broadening, and Doppler effects are common sources of noise in spectroscopy. The experimentally obtained spectrum, $F(\nu)$, can be expressed as

$$F(\nu) = \int_{-\infty}^{\infty} F^*(\nu) N(\nu)\, d\nu, \tag{3.62}$$

where $F^*(\nu)$ is the true spectrum and $N(\nu)$ is a noise function. The process of extracting $F^*(\nu)$ from $F(\nu)$ and $N(\nu)$ is referred to as *deconvolution*. The experimental spectrum, $F(\nu)$, and the noise function, $N(\nu)$, are Fourier transformed into $F(t)$ and $N(t)$, respectively. The ratio of $F(t)$ to $N(t)$ is the inverse Fourier transform of $F^*(\nu)$, $F^*(t)$.

$$F^*(t) = \frac{F(t)}{N(t)}. \tag{3.63}$$

The true spectrum, $F^*(\nu)$, is obtained by computing the inverse transform of $F^*(t)$. Obviously, $N(\nu)$ [or $N(t)$] must be estimated a priori. Inappropriate choice of the noise function can produce undesirable artifacts in the resultant signal. Applications of deconvolution in analytical chemistry have been demonstrated (26–29).

Alternately, the Fourier transform of the experimentally obtained spectrum $F(t)$ is *convoluted* with a *smoothing function*. The inverse Fourier transform of the resultant signal represents a *smoothed* spectrum. The process is depicted in Figure 3.26 and is analogous to applying smoothing formulas to the original spectrum (30).

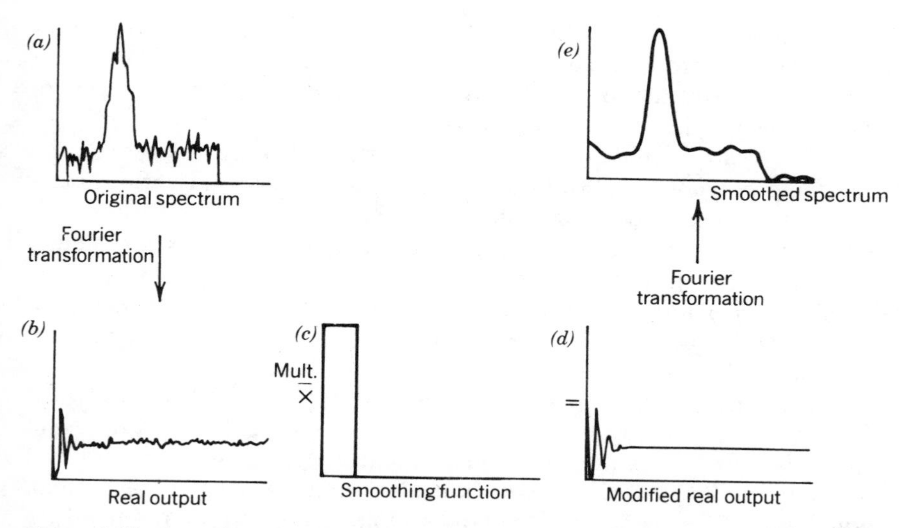

Figure 3.26 Smoothing function applied to a spectrum using Fourier transformations. [From Horlick, "Digital Data Handling of Spectra Utilizing Fourier Transformations," *Anal. Chem.* **44**(6), p. 945 (May 1972), Fig. 2.]

As in deconvolution, some artifacts can be produced in the smoothed signal if the choice of the smoothing function is unsuitable. Indeed, the choice of the smoothing function (also referred to as the *Fourier domain weighting function*) is the most difficult step in implementing Fourier domain smoothing methods. Discussions of the techniques and the effects of several smoothing and noise functions are available (27, 28, 30, 31).

Differentiation of the Signal

Useful information can often be "hidden" in the usual R versus x response plot. The combined chromatographic response, for example, of two poorly resolved components may not indicate the presence of two coeluting components; the R versus x representation of the signal may be unimodal and may show no "shoulders." Such hidden features can be enhanced in the derivative of the signal (i.e., dR/dx versus x). The derivative of the signal is sensitive to subtle features of its distribution and is therefore effective in detecting important, yet subtle, details such as weak shoulders. Unfortunately, noise can also be enhanced and the analyst must be careful when the results are interpreted. The effects of noise enhancement are more pronounced when higher derivatives are being generated.

The ability (and the limitation) of signal differentiation to detect overlapping spectral signals has been demonstrated as early as 1955 (32). Indeed, the realization of the potential of signal differentiation is responsible for techniques such as derivative spectroscopy (33, 34).

The derivative of the signal can be computed using a least-squares smoothing polynomial (23). In derivative spectroscopy and various electrochemistry techniques, the instrumental response is in the form of the first (or higher) derivative of the "usual" signal. Clearly, the instrumental design and the strategy of detection must be modified. In the Fourier domain, a *differentiating function* is convoluted with the Fourier transform of the signal to produce the inverse Fourier transform of the derivative of the signal (30).

Signal differentiation has been successfully employed in analytical chemistry to achieve better data manipulation and analysis; more useful information can be extracted from analytical signals. In spectroscopy and chromatography, the signal can be differentiated to detect any weak shoulders that may be indicative of overlapping bands. In electrochemistry signal differentiating enhances the end points of titration curves and polarographic waves.

The derivative of the signal can be numerically computed in a fashion similar to that of the moving window averaging process. A window with a $(2m+1)$-point span is allowed to scan the signal curve. The derivative of the signal curve, $R'(x_0)$, at the centroid of the moving window, x_0, is calculated. The formula used to *approximate* the first derivative at the centroid of each window will be derived.

Assume that the window span is five points, (R_{-2}, x_{-2}), (R_{-1}, x_{-1}), (R_0, x_0), (R_1, x_1), and (R_2, x_2), and express x_i as

$$x_i = x_0 + ih, \qquad i = -2, -1, 0, 1, 2,$$

where h is the spacing between adjacent points. Let the following relationship approximate $R'(x_0)$:

$$R'(x_0) \simeq C_{-2}R_{-2} + C_{-1}R_{-1} + C_0R_0 + C_1R_1 + C_2R_2, \qquad (3.64)$$

where R_i is the response at x_i and the C_i's are coefficients needed to calculate the derivative at x_0, $R'(x_0)$.

If the C_i's can be estimated, Equation (3.64) can be used to compute the first derivative of the signal at x_0 in every window. Since the window span is five points, it is desirable that Equation (3.64) be valid up to when $R(x)$ is a fourth-degree polynomial in x. This is the key for estimating the C_i's.

Let $R(x) = \text{constant}$ (a polynomial of degree zero in x). Then

$$\because \qquad R(x_i) = \text{constant};$$

$$\therefore \qquad R'(x) = 0.$$

Substituting in Equation (3.64) gives

$$0 = C_{-2} + C_{-1} + C_0 + C_1 + C_2. \tag{3.65}$$

Now, let $R(x) = x - x_0$ (a linear function). Then $R(x_i) = ih$ and $R'(x) = 1$. Substituting in Equation (3.64) gives

$$1 = -2hC_{-2} - hC_{-1} + hC_1 + 2hC_2. \tag{3.66}$$

Next, let $R(x) = (x - x_0)^2$ (a quadratic function). Then $R(x_i) = (ih)^2$ and $R'(x) = 2(x - x_0)$. Substituting in Equation (3.64) yields

$$0 = 4h^2C_{-2} + h^2C_{-1} + h^2C_1 + 4h^2C_2. \tag{3.67}$$

Continuing the above process with $R(x) = (x - x_0)^3$ and $R(x) = (x - x_0)^4$, results in two additional equations:

$$0 = -8h^3C_{-2} - h^3C_{-1} + h^3C_1 + 8h^3C_2 \tag{3.68}$$

and

$$0 = 16h^4C_{-2} + h^4C_{-1} + h^4C_1 + 16h^4C_2. \tag{3.69}$$

Simultaneous solution of Equations (3.65)–(3.69) yields the estimates of C_i. Substituting the values of C_i in Equation (3.64) yields

$$R'(x_0) \simeq \frac{1}{12h}(R_{-2} - 8R_{-1} + 8R_1 - R_2). \tag{3.70}$$

The above procedure can be used to derive approximate equations for $R'(x_0)$ in larger windows (more than five-point span) and/or to derive approximate equations for higher derivatives (23, 35). For a five-point span window, the second, third, and fourth derivatives at x_0 are given by

$$R''(x_0) = \frac{1}{12h^2}(-R_{-2} + 16R_{-1} - 30R_0 + 16R_1 - R_2), \tag{3.71}$$

$$R'''(x_0) = \frac{1}{2h^3}(-R_{-2} + 2R_{-1} - 2R_1 + R_2), \tag{3.72}$$

and

$$R''''(x_0) = \frac{1}{h^4}(R_{-2} - 4R_{-1} + 6R_0 - 4R_1 + R_2). \tag{3.73}$$

REFERENCES

1. J. C. Hancock and P. A. Wintz, *Signal Detection Theory*, McGraw-Hill, New York, 1966.

   A discussion of the theory of signal detection and its application in digital communication. Chapter 3 covers hypothesis testing and its application to signal detection in the presence of noise. Chapter 4 treats sequential detection and the Wald's sequential probability ratio test.

2. C. W. Helstrom, *Statistical Theory of Signal Detection*, 2nd ed., Pergamon Press, Oxford, U.K., 1975.

   The elements of parametric, nonparametric, and sequential detection of signals are demonstrated. Chapter 2 is devoted to the characterization of noise. Filters, digital communication, and signal resolution are also covered.

3. International Union for Pure and Applied Chemistry (IUPAC), "Nomenclature, Symbols, Units and Their Usage in Spectrochemical Analysis—II. Data Interpretation," *Pure and Appl. Chem.* **45**, 101 (1976).

   Recommendations on data handling, reporting of detection limits, and standardization of terms and symbols.

4. O. L. Davies, Ed. *The Design and Analysis of Industrial Experiments*, 2nd ed., Longman, New York, 1978, Chap. 3.

   A thorough discussion with examples of sequential detection and tests of significance.

5. H. Kaiser, "Quantitation in Elemental Analysis," Part I: *Anal. Chem.* **42**(2), 24A (1970); Part II: *Anal. Chem.* **42**(4), 26A (1970).

   The principles of information theory and mathematical statistics are used to demonstrate some important concepts such as the information content and the detection limit of analytical procedures.

6. L. A. Currie, "Limits for Qualitative Detection and Quantitative Determination," *Anal. Chem.* **40**, 586 (1968).

   Several definitions for the limit of detection are compared at three levels—the critical level, the detection limit, and the determination limit.

7. G. E. P. Box and D. R. Cox, "An Analysis of Transformation," *J. R. Statist. Soc. Ser. B.* **26**, 211 (1964).

   This is a discussion of mathematical transformations. The λ transformation is proposed and its applications are demonstrated.

8. J. Neter, W. Wasserman, and G. A. Whitmore, *Applied Statistics*, Allyn and Bacon, Boston, 1978, Chap. 15.

   Nonparametric statistical procedures such as the Wilcoxon and Mann–Whitney tests are demonstrated.

9. R. L. Rowell, "Signal Averagers," *J. Chem. Ed.* **51**, A71 (1974).

   A discussion of signal averaging and its use in analytical instruments.

10. G. M. Hieftje, "Signal-to-Noise Enhancement Through Instrumental Tech-

niques," Part I: *Anal. Chem.* **44**(6), 81A (1972); Part II: *Anal. Chem.* **44**(7), 69A (1972).

Part I covers noise power spectra, signal-to-noise ratio, signal filtering, and modulation techniques. Noise reduction methods such as signal averaging and boxcar integration are covered in Part II.

11. T. C. O'Haver, "Lock-in Amplifier," Part I: *J. Chem. Ed.* **49**, A131 (1972); Part II: *J. Chem. Ed.* **49**, A211 (1972).

Signal modulation is discussed in Part I. Lock-in amplifiers and signal demodulation are covered in Part II.

12. G. M. Heiftje, B. E. Holder, A. S. Maddox, and R. Lim, "Selection of Source-Modulation Waveform for Improved Signal-to-Noise Ratio in Atomic Absorption Spectrometry," *Anal. Chem.* **45**, 238 (1973).

Several modulation waveforms (e.g., sine and square waves) are studied and compared in terms of detection efficiency and signal-to-noise ratio.

13. M. Marinkovic and T. J. Vickers, "Novel Sample Modulation Device for Flame Spectrometry and Evaluation of Its Utility for Noise Reduction," *Anal. Chem.* **42**, 1613 (1970).

The analytical signal is modulated by periodically varying the amount of sample being introduced to the flame.

14. T. Coors, "Signal-to-Noise Optimization in Chemistry," Part I: *J. Chem. Ed.* **45**, A533 (1968); Part II: *J. Chem. Ed.* **45**, A583 (1968).

Noise sources are characterized in Part I. Part II demonstrates how a high S/N ratio can be obtained by proper system design.

15. A. Marshall and M. B. Comisarow, "Fourier and Hadamard Transform Methods in Spectrometry," *Anal. Chem.* **47**, 491A (1975).

A lucid explanation of the principles of Hadamard and Fourier transform spectroscopy.

16. J. W. Cooley and J. W. Tukey, "An Algorithm for the Machine Calculation of Complex Fourier Series," *Math. Comput.* **19**, 297 (1965).

The fast Fourier transform algorithm is described.

17. E. O. Brigham, *The Fast Fourier Transform*, Prentice-Hall, Englewood Cliffs, N.J., 1974.

A comprehensive treatment of Fourier transformation and its application. Chapter 20 covers the fast Fourier transform computation.

18. R. L. LaFara, *Computer Methods for Science and Engineering*, Hayden, Rochelle Park, N.J., 1973.

Chapter 8 discusses curve fitting of linear and nonlinear functions.

19. P. R. Bevington, *Data Reduction and Error Analysis for the Physical Sciences*, McGraw-Hill, New York, 1969, Chap. 11.

Brief descriptions of several nonlinear least-squares curve fitting strategies are given. The results of different algorithms are compared.

20. J. L. Glajch, D. C. Warren, M. S. Kaiser, and L. B. Rogers, "Effects of

Operating Variables on Peak Shape in Gel Permeation Chromatography," *Anal. Chem.* **50**, 1962 (1978).

Chromatographic peaks are modeled using a Gaussian and an exponentially decaying function. Experimental factors that influence the shape of the peak are identified and related to the parameters of the model—the peak parameters.

21. Reference 18, Chap. 11.

An overview of numerical integration. The coverage includes error analysis, Gaussian quadrature, and integration rules for up to a fifth-order interpolating polynomial.

22. Reference 18, Chap. 9.

A discussion of polynomial smoothing. The effects of smoothing and multiple-pass smoothing are demonstrated.

23. A. Savitzky and M. J. E. Golay, "Smoothing and Differentiation of Data by Simplified Least Squares Procedures," *Anal. Chem.* **36**, 1627 (1964).

A treatment of the theory and application of digital filters utilizing least-squares polynomial smoothing. The authors have tabulated several smoothing and differentiation weighting functions.

24. J. Steinier, Y. Termonia, and J. Deltour, "Comments on Smoothing and Differentiation of Data by Simplified Least Squares Procedure," *Anal. Chem.* **44**, 1906 (1972).

Corrections of the equations given in Ref. 23 are reported.

25. C. G. Enke and T. A. Nieman, "Signal-to-Noise Ratio Enhancement by Least-Squares Polynomial Smoothing," *Anal. Chem.* **48**, 705A (1976).

A detailed study of the theory and practice of polynomial smoothing. Effects of experimental noise, multiple-pass smoothing, and window span are demonstrated. The extents of signal distortion and S/N ratio enhancement are discussed.

26. R. W. Dwyer, Jr., "Isolation of Column Phenomena in Gas Chromatography," *Anal. Chem.* **45**, 1380 (1973).

Fourier transform techniques are used to deconvolute gas chromatographic elution curves. The parameters of column processes (e.g., diffusion coefficients) can be estimated from the elution profile.

27. D. W. Kirmse and A. W. Westerberg, "Resolution Enhancement of Chromatographic Peaks," *Anal. Chem.* **43**, 1035 (1971).

Overlapping chromatographic peaks are resolved and sharpened by deconvolution. A synthetic noise function is used. The difficulties associated with the choice and the implementation of the noise function are reported.

28. T. A. Maldacker, J. E. Davis, and L. B. Rogers, "Applications of Fourier Transform Techniques to Steric-Exclusion Chromatography," *Anal. Chem.* **46**, 637 (1974).

The factors giving rise to peak broadening are "stored" in the signal of an unretained peak. The elution profile of the unretained peak is used as a noise function, and a sample peak is consequently corrected by deconvolution.

29. D. E. Smith, "Data Processing in Electrochemistry," *Anal. Chem.* **48**, 517A (1976).

    The applications of convolution and deconvolution techniques in electrochemistry are overviewed.

30. G. Horlick, "Digital Data Handling of Spectra Utilizing Fourier Transformations," *Anal. Chem.* **44**, 943 (1972).

    Several signal processing techniques in the Fourier domain (e.g., smoothing and differentiation) are demonstrated. The effects of several smoothing functions are compared.

31. C. A. Bush, "Fourier Methods for Digital Data Smoothing in Circular Dichroism Spectrometry," *Anal. Chem.* **46**, 890 (1974).

    Application of digital smoothing in the Fourier domain to CD data is demonstrated.

32. A. T. Giese and C. S. French, "The Analysis of Overlapping Spectral Absorption Bands by Derivative Spectrophotometry," *Appl. Spectrosc.* **9**, 78 (1955).

    One of the earliest studies of the subject. The ability of signal differentiation to extract pertinent information from the parent signal is demonstrated by resolving overlapping spectral bands. The authors illustrate graphically how the separation of the bands, their relative areas, and their shape influence the total signal and its derivative.

33. R. N. Hager, Jr., "Derivative Spectroscopy with Emphasis on Trace Gas Analysis," *Anal. Chem.* **45**, 1131A (1973).

    The principles and instrumentation of derivative spectroscopy are introduced.

34. T.C. O'Haver, "Derivative and Wavelength Modulation Spectrometry," *Anal. Chem.* **51**, 91A (1979).

    A review of the theory and applications of derivative spectroscopy.

35. Reference 18, Chap. 10.

    A discussion of numerical differentiation including the methods of smoothing and interpolating polynomials.

SUGGESTED READINGS

E. Bruninx and A. van Eenbergen, "Computerized Detection and Evaluation of Peaks in Survey Spectra from Photoelectron Spectroscopy," *Anal. Chim. Acta Comp. Tech. Opt.* **133**, 339 (1981).

Data are smoothed using least-squares polynomial smoothing techniques with a five-point window span. The second derivative of the signal curve is used to estimate peak parameters. Detection of overlapping peaks is also described.

R. G. Dromey, M. J. Stefik, T. C. Rindfleisch, and A. M. Duffield, "Extraction of Mass Spectra Free of Background and Neighboring Component Contributions from Gas Chromatography/Mass Spectrometry Data," *Anal. Chem.* **48**, 1368 (1976).

The modeling of gas chromatographic peaks allows background subtraction from the signal of interest.

R. R. Ernst, Sensitivity Enhancement in Magnetic Resonance, in *Advances in Magnetic Resonance*, Vol. 2, J. S. Waugh (Ed.), Academic Press, New York, 1966, p. 2.

A comprehensive report on signal filtering and resolution enhancement and their applications. The coverage includes signal differentiation, least-squares polynomial smoothing, and Fourier transform spectroscopy.

P. R. Griffiths, Ed., *Transform Techniques in Chemistry*, Plenum Press, New York, 1978.

The principles and applications of transform spectroscopy are described. The coverage includes the computational aspects of Fourier methods and data processing in the Fourier domain.

J. D. Ingle, Jr., "Sensitivity and Limit of Detection in Quantitative Spectrometric Methods," *J. Chem. Ed.* **51**, 100 (1974).

A report on the definition of detection limit and sensitivity. The statistical concepts pertaining to the evaluation of the limit of detection are covered.

R. B. Lam and T. L. Isenhour, "Equivalent Width Criterion for Determining Frequency Domain Cutoffs in Fourier Transform Smoothing," *Anal. Chem.* **53**, 1179 (1981).

Several methods of selecting the Fourier domain smoothing function are briefly described. The authors propose a different criterion for defining the filter function (the smoothing function) and demonstrate its applications to synthetic Gaussian signals and Lorentzian infrared bands.

J. R. Morrey, "On Determining Spectral Peak Positions from Composte Spectra with a Digital Computer," *Anal. Chem.* **40**, 905 (1968).

A detailed study of signal differentiation. Overlapping bands are resolved by generating the first and higher derivatives of the parent spectrum. Profiles being studied are Gaussian, Lorentzian, and Student T_3.

J. Pitha and R. N. Jones, "A Comparison of Optimization Methods for Fitting Curves to Infrared Band Envelopes," *Can. J. Chem.* **44**, 3031 (1966).

A detailed study of the performances of several iterative procedures of nonlinear least-squares curve fitting. Among the tested techniques are conjugated gradient, Marquardt, Fletcher and Powell, and multiplet methods.

A. Proctor and P. M. A. Sherwood, "Data Analysis Techniques in X-ray Photoelectron Spectroscopy," *Anal. Chem.* **54**, 13 (1982).

The application of many numerical techniques in data analysis are described. The discussion includes background correction, nonlinear least-squares curve fitting, and differentiation.

A. W. Westerberg, "Detection and Resolution of Overlapped Peaks for an On-Line Computer System for Gas Chromatographs," *Anal. Chem.* **41**, 1770 (1969).

Chromatographic data are smoothed and resolved by curve fitting of Gaussian and non-Gaussian models. Curve fitting methods are compared to geometric techniques of resolution. The choice of the Fourier domain smoothing function is described. Some criteria for signal sampling rate and signal detection are discussed.

H. P. Yule, "Study of Gamma Ray Spectrum Distortion by Mathematical Smoothing," *Anal. Chem.* **44**, 1245 (1972).

The effects of single- and multiple-pass smoothing on signal distortion are reported for single and overlapping peaks.

CHAPTER

4

CALIBRATION AND CHEMICAL ANALYSIS

Calibration is the process by which the response of a measurement system is transformed to (or expressed in terms of) a quality or a quantity of interest. A familiar calibration process is that involving the arbitrary assignments of 0°C and 100°C to the heights of a mercury column in a thermometer to indicate the freezing and boiling points of water, respectively. The purposes of calibration can be classified in two major categories. The first category contains calibration processes in which *a measurement domain is either defined or determined*. Graduating a thermometer is an example of a calibration process where a measurement domain is *defined*. An example of a calibration process where a measurement domain is *determined* is the use of a standard to establish the "mass axis" in mass spectrometry. The assignments along the mass axis are *not* arbitrary since the ratio of the molecular weight of an ion to its charge is related to the scanning parameter (e.g., the magnetic field).

The second category of calibration processes includes *the estimation of one or more measurement parameters*. These are crucial processes that rely on the use of *standard materials* and are performed under *highly controlled conditions*. An example is the estimation of the molarity of an acidic solution by titration with a standard base. The estimation of the amount of a particular analyte in a given sample matrix (i.e., quantitative chemical analysis) is of a particular interest to the analyst. It is often difficult, if not impossible, to define the boundaries between the second category of calibration processes and quantitative chemical analysis. The two operations are strongly related. It can be argued, indeed, that quantitative chemical analysis is a calibration process; the analyst is asked to estimate the *caliber* of the sample vis-a-vis a particular analyte.

This chapter will discuss the influence of calibration methods on quantitative chemical analysis and the effects of the sample's matrix and the estimation model on the accuracy and precision of the analysis.

COMPARISON WITH STANDARDS

The estimation of a sample's property by comparing it with a standard is a common operation. Determining the pH of a sample solution using pH paper and a color chart is a familiar example. Comparison with preselected standards is one of the simplest methods used to estimate the amount of a particular analyte in a sample matrix. The selected standard must fulfill two important requirements: (1) the *content* of the analyte of interest must be known accurately, and (2) its *medium* must be identical (or as similar as practically possible) to the sample's medium. This minimizes the effects of the matrix and interfering analytes on the accuracy of estimation (these effects will be discussed later in the chapter). By observing a particular signal (e.g., pH, and IR spectrum, or the weight of a precipitate), the analyst can determine the significance of the differences between the sample and the standard, and upper and lower bounds for the amount of analyte in the sample. This method of analysis is fast but it lacks precision. It is usually utilized in situations where it is sufficient to know that the analyte is present in the sample within a prespecified range (e.g., 5–10% or <10 ppm) such as in industrial quality control. One or more key constituents in an industrial product may be monitored by sampling the production stream. By comparing the sample product with a "standard product," the analyst can determine the acceptability of the product. Consequently, this information may be used for feedback purposes.

CONSTRUCTING A CALIBRATION CURVE

It is often desirable to estimate the amount of a particular analyte in a sample with a high degree of precision. While it is conceivable to attempt to "trap" the response of the sample between the responses of two standards, this technique can be time consuming, especially as the required degree of precision increases. Direct comparisons with standards are therefore not recommended in these cases. An alternative method, and a more general one than direct comparison with standards, is to determine the relationship between the concentration of the analyte and the property being observed (i.e., the response or the analytical signal of the instrument chosen for the analysis). When this response function (or transformation) is known, the correspondence between the concentration domain and the analytical signal domain becomes feasible.

Response Function

The instrumental signal (e.g., output, voltage or current) is determined by many experimental parameters. The concentration of the analyte of interest is only one of these parameters. Important variables that influence the instrumental signal include the sample's matrix, interfering analytes, random perturbations in the instrumental modules, the sample preparation process, signal acquisition conditions, and many other experimental parameters. Indeed, the instrumental signal is affected by the *whole* experimental procedure. The relationship between the response of the analytical system and the amount of analyte depends also on the nature of the analysis. In gravimetric analysis, for example, the weight of a precipitate (e.g., $BaSO_4$) is linearly related to the weight of the analyte (e.g., SO_4^{2-}) in the sample. In potentiometry there is a logarithmic relationship between the electrode potential and the concentration of the analyte of interest. In many other situations the response functions are difficult, if not impossible, to determine from theoretical considerations and must be determined empirically.

Instrumental and experimental conditions (e.g., appropriate sampling strategy, signal stability, high S/N ratio, minimal or no sample loss, and/or contamination) are usually selected for *optimal* performance. These variables are determined by proper experimental designs. The analyst's main concern then becomes the amount of analyte (which is unknown), random perturbations or random noise (which are uncontrollable), and the effects of the sample's matrix and interfering analytes. By keeping the instrumental and experimental conditions constant, the analyst can determine the response function experimentally by observing the signals obtained with several standards. The standards are prepared such that different analyte concentrations are present in the same matrix. Any possible interfering analyte should be either absent or present in known equal amounts in *all* standards (and in the sample being analyzed). Under these conditions the instrumental signal is determined by two variables: (1) the amount of analyte and (2) random noise.

Figure 4.1 shows the response function obtained when the absorbance at 212 nm is monitored for several standard biphenyl solutions. It can be seen that as the concentration of biphenyl increases, the response (i.e., the absorbance) rises fairly linearly up to a certain limit (approximately until $c = 2.5\ \text{m}M$). A curvature then occurs and the response becomes almost independent of (or varying very slowly with) the concentration. The *overall* response function is *nonlinear*. The behavior shown in Figure 4.1 is common to many applications. Clearly, the whole response function is

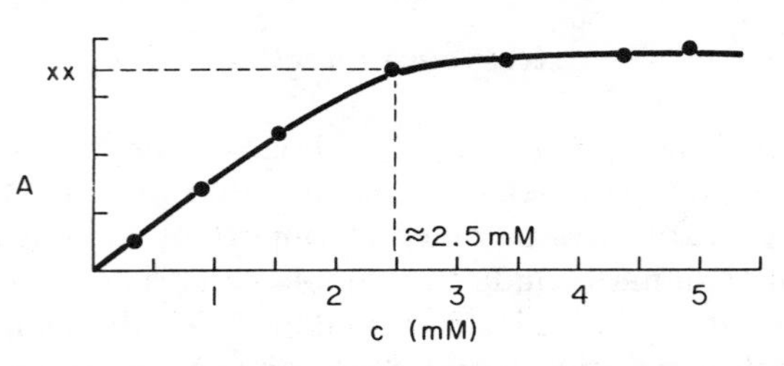

Figure 4.1 Example of response function for biphenyl: Absorbance, A, at 212 nm versus concentration. Taken in part from L. M. McDowell et al., *Anal. Chem.* **53**, 1373 (1981).

unsuitable for the analysis of an unknown sample. For example, any sample whose content of biphenyl is greater than about 2.5 mM will yield approximately the same absorbance. The analyst is therefore interested only in the portion of the response function where the variations in the response with respect to the concentration of the analyte contain *useful information*. In Figure 4.1 this portion is the part of the response curve between $c = 0$ mM and $c \cong 2.5$ mM. Variations in absorbance along other segments of the curve *do not* contain useful information pertaining to the analyte concentration. The segment of the response function that the analyst chooses for estimating the caliber of an unknown sample is referred to as the *calibration curve*. The contents of any sample showing an absorbance value between zero and xx (see Fig. 4.1) can be estimated by interpolation. This technique is also known as the *method of external standards*. Figure 4.1 also indicates that serious errors can result when the sample's caliber is estimated using the calibration curve by extrapolation. For example, if a sample shows an absorbance greater than xx and its concentration is determined by extrapolation using the linear segment of the response curve, the true concentration will be greatly underestimated. The range of concentrations in which the response varies linearly with concentration is referred to as the *linear dynamic range*. Linear response functions are preferred. When the response function is *not* linear, it is customary to invoke a suitable transformation to linearize the function.

Linear Calibration

The experimental observations within the linear dynamic range are used to estimate the parameters of the linear calibration functions. The technique of least-squares curve fitting offers an unbiased method to construct such a calibration curve based on the points $(c_1, R_1)(c_2, R_2), \ldots, (c_N R_N)$. The following requirements must be satisfied for successful application of unweighted least-squares regression:

1. The only errors of measurements are in the dependent variable (the response of the analytical system, R). If the independent variable (the concentration of analyte, c) is subject to errors, they must be much smaller than those of the dependent variable (i.e., $\sigma_c^2 = 0$ or $\sigma_c^2 \ll \sigma_R^2$).
2. The variances in the values of R are equal (i.e., $\sigma_{R_1}^2 = \sigma_{R_2}^2 = \cdots = \sigma_{R_N}^2 = \sigma_R^2$).
3. The errors in the values of R are independent.
4. $\sigma_R^2 = N(0, \sigma^2)$.

The following model can be used to compute the parameters of the calibration curve:

$$R = \beta_0 + \beta_1 c + \epsilon, \tag{4.1}$$

where R = instrumental response
c = analyte concentration
β_0, β_1 = model parameters
ϵ = random error

The model states that the analytical response is determined by the amount of analyte present in the sample in addition to some background β_0. The parameters β_0 and β_1 can be estimated as b_0 and b_1, respectively, according to

$$b_0 = \frac{(\sum c_i^2)(\sum R_i) - (\sum c_i)(\sum c_i R_i)}{N(\sum c_i^2) - (\sum c_i)^2} \tag{4.2}$$

and

$$b_1 = \frac{N(\sum c_i R_i) - (\sum c_i)(\sum R_i)}{N(\sum c_i^2) - (\sum c_i)^2}. \tag{4.3}$$

The effect of the variance in R, σ_R^2, on b_0 and b_1 can be found by direct application of the propagation of error principle. It can be shown that the variance of b_0, $\sigma_{b_0}^2$, and that of b_1, $\sigma_{b_1}^2$ can be estimated as $s_{b_0}^2$ and $s_{b_1}^2$, respectively, according to

$$s_{b_0}^2 = \frac{s_R^0 \sum c_i^2}{N \sum (c_i - \bar{c})^2} \tag{4.4}$$

and

$$s_{b_1}^2 = \frac{s_R^2}{\sum (c_i - \bar{c})^2}, \tag{4.5}$$

where s_R^2 is an unbiased estimator of σ_R^2 and is given by

$$s_R^2 = \frac{\sum (R_i - \bar{R})^2 - b_1^2[\sum (c_i - \bar{c})^2]}{N-2}, \tag{4.6}$$

and $\bar{c}$ and $\bar{R}$ are the averages of the N concentrations and N responses, respectively.

Note that the estimation of β_0, the background, is improved by having large N and low concentration standards and the estimation of β_1 is improved by a large range of standard concentrations.

The confidence intervals for β_0 and β_1 associated with a probability of P are given by

$$b_0 - |t_{\alpha/2,n-2}|S_{b_0} < \beta_0 < b_0 + |t_{\alpha/2,n-2}|S_{b_0} \tag{4.7}$$

and

$$b_1 - |t_{\alpha/2,n-2}|S_{b_1} < \beta_1 < b_1 + |t_{\alpha/2,n-2}|S_{b_1}, \tag{4.8}$$

where $\alpha = 1 - P$.

The uncertainties associated with estimating β_0 and β_1 imply that the regression line is *not* unique. Instead, there is a *regression band*. For any value of c, a confidence interval, c^*, can be computed for the average value of corresponding R^* at the $(1-\alpha)$ confidence level according to

$$b_0 + b_1 c^* - t_{\alpha/2} s_R \left[\frac{1}{N} + \frac{(c^* - \bar{c})^2}{\sum (c_i - \bar{c})^2}\right]^{1/2}$$

$$< R^* < b_0 + b_1 c^* + t_{\alpha/2} s_R \left[\frac{1}{N} + \frac{(c^* - \bar{c})^2}{\sum (c_i - \bar{c})^2}\right]^{1/2}, \tag{4.9}$$

where $t_{\alpha/2}$ is the value of the t-distribution with $N-2$ degrees of freedom

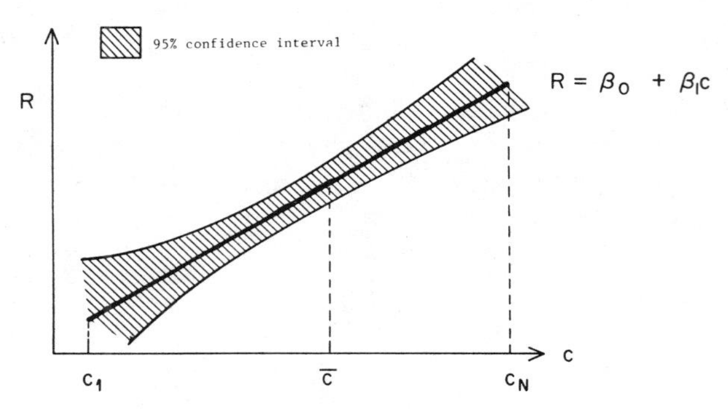

Figure 4.2 Regression band. Confidence intervals are smallest near $\bar{c}$.

at the $(1-\alpha)$ confidence level. The regression band is schematically shown in Figure 4.2. Note that the confidence interval for R^* is smallest at $c^* = \bar{c}$. Equation (4.9) can be derived by direct application of the propagation of error theory. (Set $R^* = \bar{R} + b_1[c^* - \bar{c}]$.) Ideally, the calibration curve must pass through the origin; where $c = 0$, there is no significant analytical response. This implies that the instrumental and experimental background noise as well as any possible interferences have no significant effect on the signal [i.e., $E(\beta_0) = 0$]. When this is the case, the model $R = \beta_1 c$ must be used instead of the model given in Equation (4.1). It is generally better to include β_0 in the model and to evaluate its significance (test if it is statistically different from zero) subsequently.

Examination of the Residuals

A residual is defined as the difference between the observed response, R_i, and the fitted response (i.e., the response calculated from the model, $\hat{R}_i$).

$$e_i = R_i - \hat{R}_i, \qquad i = 1, 2, \ldots, N. \tag{4.10}$$

The behavior of e_i can indicate if any of the requirements mentioned earlier are violated or if the model itself is inappropriate (1). An overall plot of the residuals (i.e., a histogram) should resemble observations from a Gaussian distribution centered about zero. Usually the residuals are plotted against the independent variable, c. If our assumptions about the data and the model are correct, a plot of e versus c results in a "horizontal band." This is schematically shown in Figure 4.3.

Abnormalities in either the model or the data are indicated when a different band is obtained. Figure 4.4 shows three cases where a plot of e versus c does not give a horizontal band.

Case I in Figure 4.4 suggests that the variances in R_i are *not* constant, thus necessitating a weighted least-squares curve fitting procedure (2). Case II indicates that the independent variable has a linear effect. Finally, Case III shows that the model needs extra terms (e.g., c^2). The residual can also be plotted against the modeled values of R, $\hat{R}_i$. The adequacy of the model and the assumptions about R_i are supported by a horizontal band

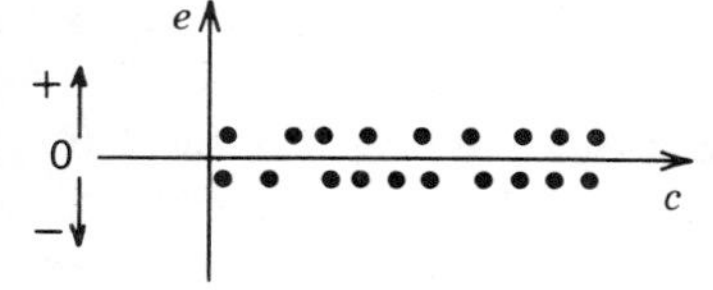

Figure 4.3 Residuals plotted against the independent variable, c, ideally should resemble observations from a Gaussian distribution centered about zero.

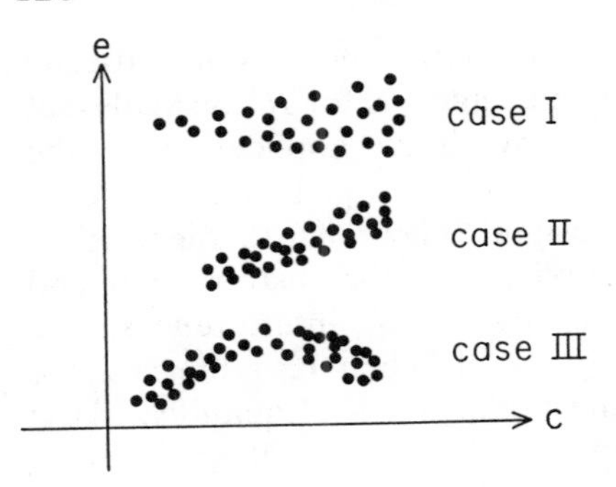

Figure 4.4 A poor model or abnormalities in the data give rise to nonrandom structure in the residuals plot.

(e.g., Fig. 4.3). If bands similar to these shown in Figure 4.4 are obtained when e is plotted against $\hat{R}$, the following discrepancies are indicated.

Case I — Same as for e versus c.

Case II — The plot shows systematic deviations from the model (negative at low R and positive at high R). This can be caused by experimental errors or missing terms in the model.

Case III — The plot suggests the need for more terms or a transformation of R.

The residuals can be examined in terms of any factor that may affect the experiment. If the data have been acquired over a long period of time or in different laboratories, the analyst may examine the residuals as a function of time or other experimental procedures.

This discussion is not restricted to calibration curves. The residual examination procedures that have been described are applicable to any modeling experiment. The construction of a calibration curve is a special case of general mathematical modeling and parameter estimation used in many areas of science and engineering.

Utilizing the Calibration Curve for Chemical Analysis

Having constructed the calibration curve properly, the analyst estimates the amount of an analyte in a sample, c_s, using the instrumental response obtained for the sample, R_s, by interpolation. This process is also known as *inverse regression*. Figure 4.5 shows the effect of the regression band on the estimation of c_s. As shown in Figure 4.5, the amount of analyte giving rise to R_s can be calculated from the model as c_s^m. But the true value for c_s is bounded by c_s^l and c_s^u. These values are given by the intersection of the two regression boundaries and the straight line $R = R_s$

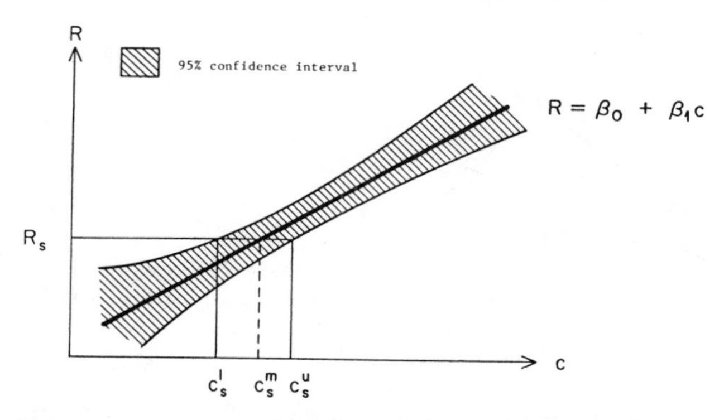

Figure 4.5 Effect of the regression band on the estimation of c_s.

(3). If n observations are used to construct the calibration curve, and m responses are obtained for an unknown sample showing an average response of R_s, the amount of analyte in the sample, c_s, is given by

$$c_s = \bar{c} + \frac{b_1(R_s - \bar{R})}{K} \pm \frac{t_{1-\alpha/2,\nu}}{K} s'_R \left[\frac{(R_s - \bar{R})^2}{\sum (c_i - \bar{c})^2} + \left(\frac{1}{n} + \frac{1}{m} \right) K \right]^{1/2}. \quad (4.11)$$

where $\bar{R}$ = the average response of the n observations
$\bar{c}$ = the average concentration for the n observations
b_1 = estimator of β_1
$\nu = n + m - 3$ = the number of degrees of freedom
$1 - \alpha$ = confidence level
$K = b_1^2 - (t_{1-\alpha/2,\nu})^2 s_{b_1}^2$

$$s'^2_R = \frac{(n-2)s_R^2 + \sum_{i=1}^{m} (R_{s,i} - R_s)^2}{n + m - 3}$$

Note that the smallest interval is obtained when $R_s = \bar{R}$. Equation (4.11) is more accurate than the "classical" estimation of c_s. By setting $S_{b_1}^2 = 0$, Equation (4.11) reduces to the classical expression (if $m \ll n$, s_R can be used instead of s'_R).

Constructing a Calibration Curve with Heteroscedastic Data

The behavior shown in Case I of Figure 4.4 indicates that the variances in R_i are *not* equal (i.e., data are *heteroscedastic*). Some values of R are more precise than others. Application of "regular" least-squares pro-

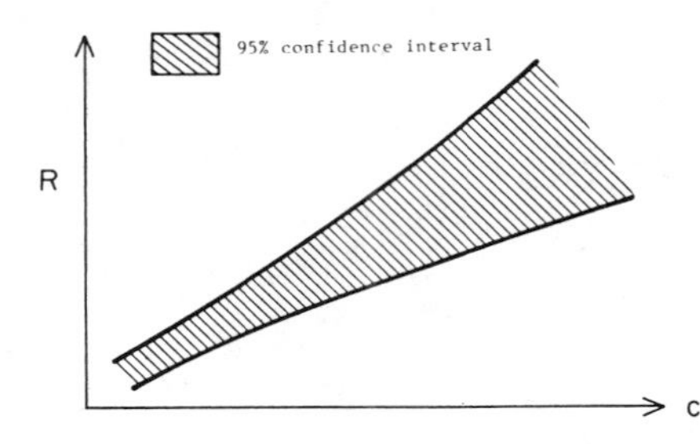

Figure 4.6 Regression band obtained when the variance in R increases as R increases.

cedures is invalid under these conditions. The sum of squares that is being minimized must be weighted (2). The function to be minimized becomes

$$s = \sum \frac{(R_i - \beta_0 - \beta_1 c)^2}{\sigma^2_{R_i}}. \tag{4.12}$$

The values of β_0 and β_1 that minimize the above expression are estimated as b_0 and b_1 and are given according to

$$b_1 = \frac{(\sum w_i)(\sum w_i c_i R_i) - (\sum w_i c_i)(\sum w_i R_i)}{(\sum w_i)(\sum w_i c_i^2) - (\sum w_i c_i)^2} \tag{4.13}$$

and

$$b_0 = \frac{(\sum w_i R_i) - b_1(\sum w_i c_i)}{\sum w_i}, \tag{4.14}$$

where $w_i = 1/s^2_{R_i} \simeq 1/\sigma^2_{R_i}$. The regression band obtained when the variance in R increases as R increases is schematically shown in Figure 4.6. Confidence intervals are obtained by weighting the appropriate terms in the equations used for unweighted least squares. The weights calculated as indicated above may be normalized such that the average weight is 1. This can be accomplished by transforming each "old" weight, w_i, into a "new" weight, w'_i, according to

$$w'_i = \frac{w_i}{N^{-1} \sum w_i}.$$

Estimating the Detection Limit from a Linear Calibration Curve

As shown in Figure 4.5 a confidence interval for c_s is found for any response value R_s. The intersection of the regression band and the vertical axis defines the range of responses obtained for an analyte-free

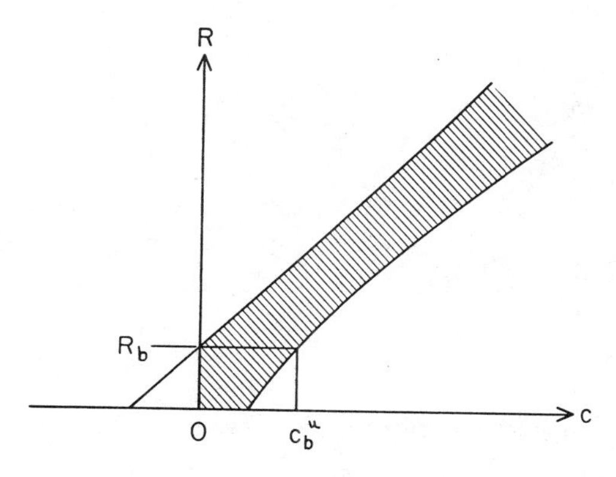

Figure 4.7 The response of a blank, R_b, is associated with the interval $[0, c_b^u]$. c_b^u is the detection limit.

sample ($c = 0$) at the specified confidence level (responses less than the point of intersection result in negative concentration bounds).

As shown in Figure 4.7, a response of R_b (e.g., a blank solution) is associated with the interval $[0, c_b^u]$, where c_b^u is the upper bound for the true value for c_b which corresponds to R_b. At a confidence level of $(1-\alpha)$, the probability that a blank will produce a signal $\geq R_b$ is $\alpha/2$. If this is an acceptable risk, then the lowest detectable amount of analyte is c_b^u (i.e., c_b^u is the detection limit). The presence of any amount less than c_b^u can not be verified. The confidence interval for c_b is dependent on the value R_b. The analyst can restrict the values of the detection limits to "low" values of c with a prespecified precision. In such a case the detection limit can be considered as the amount of c_b associated with a response, R_b, giving rise to a confidence interval [Eq. (4.11)] of a particular size.

At this point we suggest that a *very* useful learning aid for the material covered in the section on constructing a calibration curve is to write a computer program to accept calibration data, do the appropriate least-squares analysis, and report values for the equations given in this chapter.

Intersection of Two Regression Lines

In many analytical applications (e.g., photometric titrations), it is necessary to locate the intersection of two regression lines (4). The point of intersection of two regression bands, c_I, is schematically shown in Figure 4.8.

Let $R_{1i} = (\beta_0)^1 + (\beta_1)^1 c_{1i}$ and $R_{2i} = (\beta_0)^2 + (\beta_1)^2 c_{2i}$ be two regression

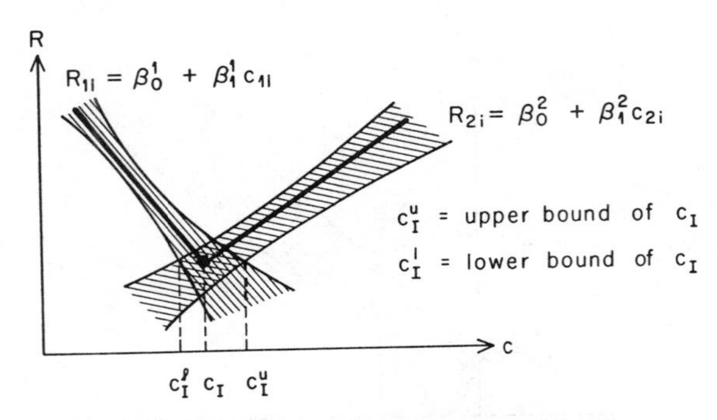

Figure 4.8 Intersection of two regression lines.

lines determined by n_1 and n_2 observations, respectively. The point of intersection of the two lines, c_I, can be shown to be

$$c_I = \frac{(b_0)^2 - (b_0)^1}{(b_1)^1 - (b_1)^2}. \tag{4.15}$$

Verification of this equation is recommended as an exercise. This is only an estimate of the true point of intersection, ϕ, given by

$$\phi = \frac{(\beta_0)^2 - (\beta_0)^1}{(\beta_1)^1 - (\beta_1)^2}. \tag{4.16}$$

The confidence interval for c_I is computed by solving the following quadratic equation:

$$c_I^2[(\Delta b)^2 - t_{\alpha,\nu}^2 S_{\Delta b}^2] - 2c_I(\Delta a \Delta b - t_{\alpha,\nu}^2 S_{\Delta a \Delta b}) + [(\Delta a)^2 - t_{\alpha,\nu}^2 S_{\Delta a}^2] = 0, \tag{4.17}$$

where

$$\Delta a = (b_0)^2 - (b_0)^1,$$

$$\Delta b = (b_1)^1 - (b_1)^2,$$

$$S_{\Delta a}^2 = S_p^2\left[\frac{1}{n_1} + \frac{1}{n_2} + \frac{\bar{c}_1^2}{\sum(c_{1i} - \bar{c}_1)^2} + \frac{\bar{c}_2^2}{\sum(c_{2i} - \bar{c}_2)^2}\right],$$

$$S_{\Delta b}^2 = S_p^2\left[\frac{1}{\sum(c_{1i} - \bar{c}_1)^2} + \frac{1}{\sum(c_{2i} - \bar{c}_2)^2}\right],$$

$$S_{\Delta a \Delta b} = S_p^2\left[\frac{\bar{c}_1}{\sum(c_{1i} - \bar{c})^2} + \frac{\bar{c}_2}{\sum(c_{2i} - \bar{c}_2)^2}\right],$$

and S_p^2, the pooled variance, is given by

$$S_p^2 = \frac{\sum (R_{1i} - \hat{R}_{1i})^2 + \sum (R_{2i} - \hat{R}_{2i})^2}{(n_1 + n_2 - 4)}.$$

The roots of Equation (4.17) define the appropriate interval for the point of intersection.

Linear Model when Both Variables Are Subject to Error

The general case for a linear model relating two variables (e.g., R and C) comes about when both variables are subject to error. Such a situation arises when, for example, two measured properties are correlated. In this case the calibration curve is used to predict one property after measuring the other (e.g., estimating the concentration of the chloride ion from the measurement of the concentration of the sodium ion in brine samples). Assume that the following model is used to find the relationship between property Y and property X:

$$y_i = \beta_0 + \beta_i x_i. \tag{4.18}$$

The parameters β_0 and β_1 are estimated as b_0 and b_1, respectively, by minimizing the sum of squares, S, given by

$$S = \sum \frac{(y_i - b_0 - b_i x_i)^2}{S_{y_i}^2 + b_1^2 S_{x_i}^2}. \tag{4.19}$$

Note that when $S_{x_i}^2 = 0$ and $S_{y_i}^2 = s^2$ (homoscedastic data) we obtain the simple case of the least-squares solution. Also, if $S_{x_i}^2 = 0$ or $b^2 S_{x_i}^2 \ll S_{y_i}^2$ and the variances of the y_i's are not equal we have the case of weighted least-squares solution.

Solving Equation (4.19) for b_0 and b_1 by minimizing S is complicated by the fact that the normal equations obtained are nonlinear in the parameters. Wald (5) has proposed a simple solution for b_0 and b_1 based on the grouping of data points. If N (assume that N is even for simplicity) points are available, then b_0 and b_1 can be calculated according to

$$b_1 = \frac{(y_1 + y_2 + \cdots + y_m) - (y_{m+1} + y_{m+2} + \cdots + y_N)}{(x_1 + x_2 + \cdots + x_m) - (x_{m+1} + x_{m+2} + \cdots + x_N)} \tag{4.20}$$

and

$$b_0 = \bar{y} - b_1 \bar{x}. \tag{4.21}$$

where $m = N/2$. Iterative solutions for b_0 and b_1 have also been proposed (6).

There are many cases where regression analysis is performed with errors associated with both the independent and the dependent variable. Indeed, it can be said that all calibrations fit into this class as the standards are prepared by weighing, dilution, and so on, and their concentrations are rarely known exactly. In these cases, the errors in the independent variables are often simply ignored. Perhaps in the future more use will be made of methods such as the above.

Nonlinear Calibration

Quite often the analyst needs to model data using a nonlinear model. Such cases can be found for calibration in emission spectroscopy and X-ray fluorescence. Schwartz (7) has discussed three methods to deal with nonlinear calibration. The simplest approach is to assume that the relationship between the variables is linear over short intervals. The response curve is thus divided into regions of more or less linear segments (the process is referred to as the *method of linear segments*) and parameter estimation proceeds in a *piecewise* manner. The success of this method depends on the curvature of the response curve in the region where an unknown sample is being analyzed. It is sometimes impossible to estimate a confidence interval for the amount of analyte in the sample, c_s. A more appropriate method is to model the data using a second-order polynomial. A suitable model may be

$$R = \beta_0 + \beta_1 c + \beta_2 c^2. \tag{4.22}$$

The parameters β_0, β_1, and β_2 may be estimated using the criterion of least squares. It is also possible to estimate a confidence interval for the model (i.e., a regression band) similar to that given for the linear case (7). The response function can also be generalized and expressed as a higher-order polynomial.

EFFECTS OF THE SAMPLE'S MATRIX

When standards are used to construct a calibration curve, they must be prepared such that the matrix of the standard is identical to the sample's matrix. The values of the parameters β_0 and β_1 associated with a linear calibration curve are matrix dependent. Examples of calibration problems involving matrix effects can be found in emission and absorption spectroscopy and in many other areas of chemical analysis. Analysis of trace metals in complex natural product matrices such as crude oils is subject to a large matrix effect.

It is often difficult, if not impossible, to duplicate the sample matrix when preparing external standards. Since it is desirable to eliminate any matrix effects, it is attractive to perform the analysis in the sample matrix itself. The method of standard addition offers such an opportunity.

Standard Addition Method (SAM)

For this method, the standards are added to the sample matrix and the response of the analyte plus the standard is monitored as a function of the added amount of the standard. Assume that the initial response (i.e., the response of the sample *before* any additions are made) is given as R_0. Let the relationship between the response and the concentration of the analyte be given as

$$R_0 = kC_0 = k\frac{n_0}{V_0}, \qquad (4.23)$$

where n_0 = initial number of moles in the analyte
V_0 = initial volume of the sample

Here, the model with no intercept $R = \beta_1 c$ is assumed for simplicity. The parameter β_1 is replaced here by k. Assuming the model holds, after the ith addition,

$$R_i = k\left[\frac{n_0 + n_i}{V_0 + V_i}\right], \qquad (4.24)$$

where R_i = the response after the ith addition
n_i = the total number of moles *added* at the ith addition
V_i = the total volume *added* at the ith addition

Equation (4.24) can be written $Q_i = kn_0 + kn_i$, where $Q_i = R_i(V_0 + V_i)$. (Q is often referred to as the volume-corrected response.) A plot of Q_i versus n_i yields a straight line with a slope of k and an intercept of kn_0. The initial number of moles, n_0, is found by *extrapolation*. At $Q_i = 0$ we have

$$kn_0 = -kn_i \quad \text{or} \quad n_0 = -n_i. \qquad (4.26)$$

This scheme is illustrated in Figure 4.9 [set $R_s = 0$ and eliminate the m term in Equation (4.11)].

The initial number of moles, n_0, is estimated away from $\bar{Q}$. This is indeed one disadvantage of SAM because the confidence interval for n_0 is larger at this value of Q [see Eq. (4.11)].

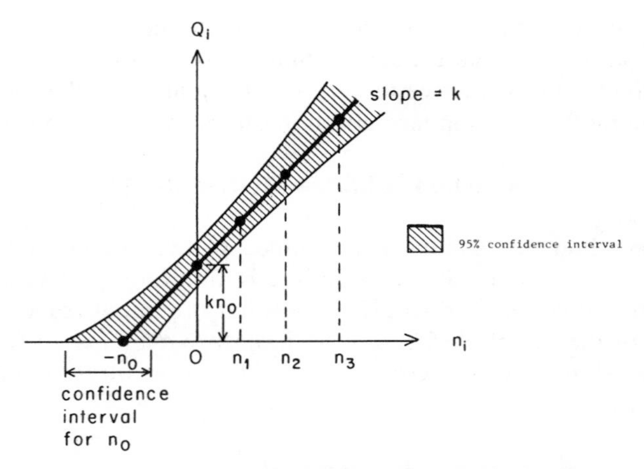

Figure 4.9 The Standard Addition Method.

The method of standard additions is a tool well known to electrochemists and spectroscopists. The analyst has no choice but to employ matrix matching or SAM when matrix effects are suspect.

ERROR PROPAGATION IN CALIBRATION AND ANALYSIS

The effect of the calibration and analysis procedures on the estimation of an unknown amount of an analyte in a sample can be evaluated from the error propagation principles. Consider the following calibration model:

$$R = \beta_0 + \beta_1 c. \tag{4.27}$$

The variance of c caused by variations in R is given from the relationship

$$c = \frac{R - \beta_0}{\beta_1} \quad \text{as} \quad \sigma_c^2 = \frac{\sigma_R^2}{\beta_1^2}. \tag{4.28}$$

The *sensitivity* of an analytical system is generally defined as the ratio of the change in response to the change in concentration (i.e., sensitivity = dR/dc). Thus b_1, an estimate of β_1, is referred to as the *sensitivity factor*. [A more practical, and justifiable, definition for sensitivity should be $(dR/dc)/\sigma_R^2$.]

From Equation (4.28) it is clear that as the analytical system becomes more sensitive, error propagation becomes less pronounced. The propagation of error in the SAM calculation yields some very important results. To illustrate, we shall, for simplicity, assume that the volumes added are negligible compared to the initial volume, V_0. Equations (4.23) and (4.24) can then be combined to give

$$n_0 = n_1 R_0/(R_1 - R_0), \tag{4.29}$$

where n_1 = amount of standard added
R_1 = response obtained after the addition (assume only *one* standard addition).

By the propagation of error theory

$$\sigma_{n_0}^2 = n_0^2 \left[\frac{\sigma_{R_0}^2}{R_0^2} + \frac{\sigma_{R_1}^2}{R_1^2}\right]\left[1 + \frac{n_0}{n_1}\right]^2. \tag{4.30}$$

Equation (4.30) can be simplified if we assume that a common coefficient of variation (8) exists for all measurements (i.e., σ/μ = constant for all R). Let θ denote the coefficient of variation. Then

$$\sigma_{n_0}^2 = n_0^2 \left[1 + \frac{n_0}{n_1}\right]^2 \left[\frac{1}{N_0} + \frac{1}{N_1}\right] \theta^2, \tag{4.31}$$

where N_0 = number of replicate analyses used to estimate R_0
N_1 = number of replicate analyses used to estimate R_1

According to Equation (4.31) the largest amount of standard possible should be added while remaining in the linear range (large n_1). For a fixed number of analyses (i.e., $N_1 + N_0$ = a constant) and a fixed n_1, the variance in n_0 is a minimum when $N_1 = N_0$.

This result is an example of the advantage the analyst can obtain by a careful examination of the propagation of experimental errors. Later in this chapter, another type of error analysis will be introduced that can also be used to provide improved experimental design.

MULTICOMPONENT ANALYSIS

Thus far we have discussed calibration and analysis schemes where one analyte is being determined and one sensor is used to determine one analyte. Quite often the analyst is asked to analyze for several analytes in the same sample. The fundamentals pertaining to response functions remain unchanged. If external standards are to be used (i.e., the analyte concentrations are to be determined using a calibration curve), the analyst must find the relationships between concentration and response for all analytes prior to analysis. The linear model [Eq. (4.23)] can be generalized for multicomponent systems, where p different responses (e.g., p different wavelengths) are used to determine r analytes, for n different samples, and expressed in matrix notation as

$$\mathbf{R}_{n,p} = \mathbf{C}_{n,r}\mathbf{K}_{r,p}, \tag{4.32}$$

where p = the number of sensors (i.e., different responses) used
r = the number of analytes of interest
n = the number of standard solutions used
$\mathbf{K}_{r,p}$ = matrix of linear response constants whose elements represent the sensitivity of each sensor to each analyte

To be able to find a solution for the systems of equations expressed in Equation (4.32), the number of sensors must be at least equal to the number of analytes. As we have discussed for single-component analysis using a calibration curve, the analyst must first determine the sensitivity factors using external standards. This is equivalent to solving Equations (4.32) for $\mathbf{K}$ using known $\mathbf{C}$ and $\mathbf{R}$. Since $\mathbf{C}$ is generally *not* a square matrix, Equation (4.32) is solved by the *generalized inverse method*. $\mathbf{K}$ is given by

$$\mathbf{K} = (\mathbf{C}^T\mathbf{C})^{-1}(\mathbf{C}^T\mathbf{R}), \tag{4.33}$$

where $\mathbf{C}^T$ = transpose of $\mathbf{C}$
$(\mathbf{C}^T\mathbf{C})^{-1}$ = inverse of $\mathbf{C}^T\mathbf{C}$ (assuming it exists)

Also, the standard solutions must *span the space* of analytes and $n > r$.

The matrix $(\mathbf{C}^T\mathbf{C})^{-1}\mathbf{C}^T$ is called the generalized inverse of $\mathbf{C}$. Having estimated the sensitivity factors matrix $\mathbf{K}$, the analyst can then estimate the amounts of analytes in an unknown sample. This is equivalent to the interpolation step in single-component analysis. If the number of sensors is equal to the number of analytes, $\mathbf{K}$ is a square matrix. If $\mathbf{K}^{-1}$ exists then

$$\mathbf{C}_s = \mathbf{R}_s\mathbf{K}^{-1}, \tag{4.34}$$

where $\mathbf{C}_s$ = matrix containing the concentration of analytes in unknown samples
$\mathbf{R}_s$ = matrix containing the responses obtained for the unknown samples

If the number of sensors is greater than the number of analytes (the significance of this condition will be discussed shortly) $\mathbf{K}$ is *not* a square matrix. $\mathbf{C}_s$ is solved for using the generalized inverse of $\mathbf{K}$ according to

$$\mathbf{C}_s = \mathbf{R}_s\mathbf{K}^T(\mathbf{K}\mathbf{K}^T)^{-1}. \tag{4.35}$$

The interested reader can show that the generalized inverse method simply amounts to another formulation of multilinear least-squares analysis. All the usual assumptions involved with least squares apply, including the specification that the independent variable be independent. To the extent that they are not, a colinearity problem arises—the subject of much study in statistics. In Chapter 6, we will introduce a superior algorithm to alleviate this problem.

The sensitivity of the analytical system in the case of multicomponent analysis with a square **K** matrix may be defined as the absolute value of the determinant of **K**.

$$\text{Sensitivity} = |\det \mathbf{K}|, \tag{4.36}$$

where **K** is a square matrix.

Ideally, when only specific sensors are used, the **K** matrix should have nonzero entries along the diagonal and zero entries everywhere else (i.e., $k_{ij} = 0$ when $i \neq j$, and $k_{ii} \neq 0$). In this case *all* sensors are *specific* vis-a-vis the analytes. In other words, each sensor responds only to *one* analyte. If any sensor is not specific (i.e., responds to at least two analytes) but rather *selective*, then the appropriate k_{ij} will have a finite nonzero value, the sign and magnitude of which will, of course, depend on the nature of the interference. The more selective the sensor, the smaller the absolute value for this k_{ij}. If interferences are present, it is essential to include them in the calibration process and evaluate their sensitivity factors. For example, when two interfering analytes (A and B) are determined with no regard to their effects on one another, serious errors occur. This is demonstrated in Figure 4.10.

By including interfering species in the calibration and analysis processes these errors can be eliminated. In the case of two interfering analytes,

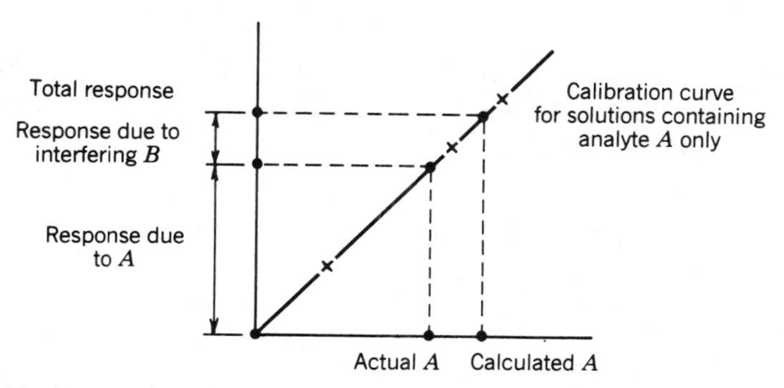

Figure 4.10 Errors when two species, A and B, are determined without correction for interference.

the operation is identical to determining both analytes simultaneously. The effect of interferences can thus be eliminated. One problem remains nevertheless and it has been discussed in the case of single-component analysis, namely, the matrix effect. This is a shortcoming of all external standard calibration procedures. As mentioned earlier, the standards used to estimate **K** must be identical in nature to the sample itself. We will return to this point later.

It is wrong to assume that the *best* results are obtained when the number of sensors is equal to the number of analytes. The effects of noise and uncertainties in $\mathbf{R}_s$ or $\mathbf{C}_s$ are determined by the nature of **K** [see Eq. (4.35)]. This is *not* unexpected since **K** reflects the overall sensitivity of the analytical system and the specificity (or selectivity) of each sensor. It has been shown that the effects of noise in **R** are minimized when the sensitivity of the analytical system increases [Eq. (4.28)]. This is also true for multicomponent systems. Keeping in mind that the sensitivity for a multicomponent system is given by the determinant of the square **K** matrix, consider the following example (units are arbitrary):

The amounts of two analytes are determined by monitoring three wavelengths, λ_1, λ_2, and λ_3. The system has the following **K** matrix (as determined by external standards):

	Sensor I	Sensor II	Sensor III
Analyte I	3.00	3.00	2.00
Analyte II	2.00	4.00	6.00

Assume that the "true" $\mathbf{C}_s$ is

$$\mathbf{C}_s = \begin{pmatrix} 1 \\ 2 \end{pmatrix}.$$

Then the "true" responses obtained with the sample are given by $\mathbf{R}_s^T = (7.00,\ 11.0,\ 14.0)$. Suppose that the experimental $\mathbf{R}_s$ obtained is subject to errors and is given by

$$\exp \mathbf{R}_s = \begin{pmatrix} 7.10 \\ 10.9 \\ 14.1 \end{pmatrix} \quad \begin{matrix} 1.4\% \text{ error in } R_1 \\ 0.9\% \text{ error in } R_2 \\ 0.7\% \text{ error in } R_3 \end{matrix}$$

If only two sensors are used to find $\mathbf{C}_s$, then

1. Using R_1 and R_2: $\mathbf{C}_s = \begin{pmatrix} 1.10 \\ 1.90 \end{pmatrix}$ $\begin{matrix} 10\% \text{ error; } \det(\mathbf{K}) = 6 \\ 5\% \text{ error} \end{matrix}$

2. Using R_2 and R_3: $\mathbf{C}_s = \begin{pmatrix} 0.90 \\ 2.05 \end{pmatrix}$ 10% error; det($\mathbf{K}$) = 10; 2.5% error.

3. Using R_1 and R_3: $\mathbf{C}_s = \begin{pmatrix} 1.03 \\ 2.01 \end{pmatrix}$ 3% error; det($\mathbf{K}$) = 14; 0.5% error.

It is clear that the accuracy of $\mathbf{C}_s$ depends on the choice of sensors. As det($\mathbf{K}$) increases the agreement between the calculated $\mathbf{C}_s$ and the true $\mathbf{C}_s$ gets better. This is indeed one disadvantage of having to choose only r sensors to determine r analytes. The analyst must choose the *best* r sensors [i.e., the sensors whose det($\mathbf{K}$) is highest]. A natural question at this point is: Can the analyst use more sensors than analytes? The answer is *yes*. As a matter of fact it is recommended where possible to have more sensors than analytes. In some cases, such as when one sensor's response is a linear combination of the other sensor responses, the number of sensors *must* be greater than the number of analytes. Note that if the number of sensors is greater than the number of analytes, $\mathbf{K}$ is not a square matrix and Equation (4.36) does not apply. In this case, the sensitivity is defined as

$$\text{Sensitivity} = [\det(\mathbf{K}^T\mathbf{K})]^{1/2}. \tag{4.37}$$

Also, Equation (4.35) represents an *overdetermined system* of equations. If all three wavelengths are used in the previous example to estimate $\mathbf{C}_s$, we obtain

$$\mathbf{C}_s = \begin{pmatrix} 0.998 \\ 2.010 \end{pmatrix} \quad \begin{matrix} 0.2\% \text{ error; } [\det(\mathbf{K}^T\mathbf{K})]^{1/2} = 18.2 \\ 0.5\% \text{ error} \end{matrix}$$

The above result is better than the results obtained with any combination of two wavelengths. The improvement in accuracy is a result of using more information (i.e., more measurements) to estimate $\mathbf{C}_s$. This leads to a signal-to-noise enhancement (9).

GENERALIZED STANDARD ADDITION METHOD (GSAM)

When matrix effects are suspect *and* specific sensors are not available, all calibration methods described above are inadequate. Until recently, the analyst had no recourse but to use matrix matched samples and multivariate calibration. Fortunately, now the method of standard additions can be generalized to include multicomponent analysis (10, 11). The analyst adds different standards to the sample and estimates C_s from the changes in the responses. The method is referred to as the *generalized standard addition method*, or GSAM. One advantage of GSAM, as in SAM, is that the analysis is performed in the sample matrix. Matrix effects are thus eliminated. Figure 4.11 shows a simple SAM for a case where two analytes give rise to a response on the sensor. The interference caused by the second analyte makes analysis of either analyte impossible.

It is therefore necessary to include all interfering analytes in the calibration/analysis scheme. Consider the following model to be the true relationship between response and concentration:

$$\mathbf{R}_0 = \mathbf{C}_0\mathbf{K}, \tag{4.38}$$

where $\mathbf{R}_0$ = matrix containing the responses of p sensors for n samples
$\mathbf{C}_0$ = matrix containing the initial concentration of r analytes for each sample
$\mathbf{K}$ = matrix of sensitivity coefficients (linear response constants)

Multiplying each response and concentration by the initial volume V_0 of the sample we obtain the so-called volume-corrected response equation:

$$\mathbf{Q}_0 = \mathbf{N}_0\mathbf{K}, \tag{4.39}$$

where $\mathbf{Q}_0$ = matrix of responses multiplied by the volume of the sample
$\mathbf{N}_0$ = matrix containing the initial amount (moles, grams, etc.) of each

Consider that one sample is to be analyzed for r analytes using p sensors ($p \geq r$) by making a series of n standard additions ($n \geq r$) to the sample and recording the sensor responses after each addition. The equation that must

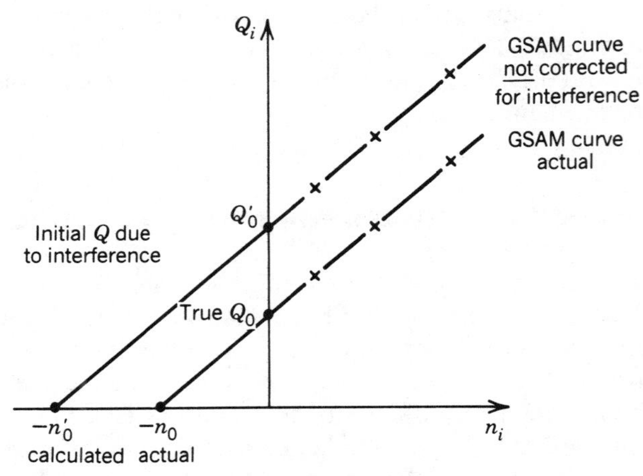

Figure 4.11 Effect of interference in an analysis. SAM corrects for matrix but not interference effects.

then be dealt with is

$$\mathbf{Q} = (\Delta\mathbf{N} + \mathbf{N}_0)\mathbf{K}, \tag{4.40}$$

where $\mathbf{Q}$ = matrix of volume-corrected responses, one row per addition $(n \times p)$

$\Delta\mathbf{N}$ = matrix of amounts added at each step $(n \times r)$

$\mathbf{N}_0$ = matrix of initial amounts of each analyte; all rows identical $(n \times r)$

$\mathbf{K}$ = linear response constants relating each sensor to each analyte $(r \times p)$

If the initial response for each sensor is subtracted from all corresponding column elements in $\mathbf{Q}$, a change in response matrix, $\Delta\mathbf{Q}$, results with

$$\Delta\mathbf{Q} = \Delta\mathbf{N}\,\mathbf{K}. \tag{4.41}$$

The reader should determine if the subtraction should be done before or after volume correction.

Since $\Delta\mathbf{Q}$ and $\Delta\mathbf{N}$ are known, $\mathbf{K}$ can be found using the generalized inverse method. The sensitivity coefficients matrix, $\mathbf{K}$, is given by

$$\mathbf{K} = (\Delta\mathbf{N}^T\Delta\mathbf{N})^{-1}\Delta\mathbf{N}^T\Delta\mathbf{Q} \tag{4.42}$$

and corrects the analysis for interference *and* matrix effects.

Having estimated $\mathbf{K}$, the vector of initial quantities, $\mathbf{n}_0$, can be obtained using the vector of volume-corrected initial responses, $\mathbf{q}_0$, by solving

$$\mathbf{q}_0 = \mathbf{K}^T\mathbf{n}_0. \tag{4.43}$$

The vector $\mathbf{n}_0$ is given by

$$\mathbf{n}_0 = (\mathbf{K}\mathbf{K}^T)^{-1}\mathbf{K}\mathbf{q}_0. \tag{4.44}$$

The effects of the errors in the measurements on the estimation of $\mathbf{K}$ and $\mathbf{n}_0$ are given according to ref. 10 as

$$\frac{\|\delta\mathbf{K}\|}{\|\mathbf{K}\|} \leq \mathrm{cond}(\Delta\mathbf{N})\frac{\|\delta\Delta\mathbf{Q}\|}{\|\Delta\mathbf{Q}\|} \tag{4.45}$$

and

$$\frac{\|\delta\mathbf{n}_0\|}{\|\mathbf{n}_0\|} \leq \mathrm{cond}(\mathbf{K})\left[\frac{\|\delta\mathbf{q}_0\|}{\|\mathbf{q}_0\|} + \frac{\|\delta\mathbf{K}\|}{\|\mathbf{K}\|}\right], \tag{4.46}$$

where, $\| \quad \|$ is the two norm, which for a vector is the length of the vector, the square root of the sum of the square of its elements, and for a matrix $\mathbf{X}$ is equal to the square root of the largest eigenvalue of $\mathbf{X}^T\mathbf{X}$. The condition number of a matrix $\mathbf{X}$ is given by $\text{cond}(\mathbf{X}) = [\text{cond}(\mathbf{X}^T\mathbf{X})]^{1/2}$ and is easily found by dividing the largest eigenvalue of $\mathbf{X}$ by its smallest eigenvalue. The following definitions apply to Equations (4.45) and (4.46):

$\delta\mathbf{K}$ = matrix of absolute deviations from the *true* elements of $\mathbf{K}$.
$\delta\Delta\mathbf{Q}$ = matrix of deviations from the *true* $\Delta\mathbf{Q}$ obtained in the absence of measurement error.
$\delta\mathbf{n}_0$ = vector of deviations from true analyte amounts.
$\delta\mathbf{q}_0$ = measurement errors in q_0.

Equations (4.45) and (4.46) have enormous significance to the analyst. Simply stated, Equation (4.45) says that the relative error in the estimated response constant is bounded by the relative error in the measurements and the manner in which the standard additions were made [the experimental design, cond($\Delta\mathbf{N}$)]. Equation (4.46) tells us that the relative overall error in the estimates of the analyte amounts is bounded by an equation that includes the relative errors in the linear response constants, the relative errors in the initial sensor responses and cond($\mathbf{K}$) which is a compact numerical description of the sensor/analyte relationship (the analytical measurement system). The reader should spend some time with these equations in order to understand their impact on chemical analysis. The good analyst will try to minimize every term in these two equations. Next, we will begin to see how this can be done.

EXPERIMENTAL DESIGNS IN GSAM

According to inequalities (4.45) and (4.46) the measurement errors are magnified by cond($\Delta\mathbf{N}$) and cond($\mathbf{K}$). An optimal experimental design is one where cond($\Delta\mathbf{N}$) and cond($\mathbf{K}$) are equal to 1. In the following sections, various experimental designs are investigated in order to minimize error amplification [find a low cond($\Delta\mathbf{N}$)]. The reader can verify that the identity matrix multiplied by a scalar yields the lowest condition number of unity. The analyst pays for unequal sensor sensitivities (nonequivalent), main diagonal elements, and interferences (nonzero off diagonal elements) by having to deal with cond($\mathbf{K}$) > 1.0 (error amplification).

Total Difference Calculations

Total difference calculations, TDC, simply define $\Delta\mathbf{Q}$ and $\Delta\mathbf{N}$ as the total change in volume-corrected responses and analyte amounts from the beginning of the experiment.

1. Case I: Assume that the amounts of two analytes are being estimated and that four additions are made such that the *total* $\mathbf{\Delta N}$ matrix is

$$\mathbf{\Delta N} = \begin{pmatrix} 1 & 0 \\ 1 & 1 \\ 2 & 1 \\ 2 & 2 \end{pmatrix}.$$

   We have

$$\mathbf{\Delta N}^T \mathbf{\Delta N} = \begin{pmatrix} 10 & 7 \\ 7 & 6 \end{pmatrix} \quad \text{and} \quad \text{cond}(\mathbf{\Delta N}) = 3.7.$$

   This is indeed a bad magnification of measurement error.

2. Case II: Now suppose that the four additions are made such that the total $\mathbf{\Delta N}$ matrix is

$$\mathbf{\Delta N} = \begin{pmatrix} 1 & 0 \\ 2 & 0 \\ 2 & 1 \\ 2 & 2 \end{pmatrix}.$$

   Now we have

$$(\mathbf{\Delta N}^T \mathbf{\Delta N}) = \begin{pmatrix} 13 & 6 \\ 6 & 5 \end{pmatrix} \quad \text{and} \quad \text{cond}(\Delta N) = 1.5.$$

   This method of making additions is superior to Case I.

3. Case III: Suppose again that the additions have proceeded such that

$$\mathbf{\Delta N} = \begin{pmatrix} 1 & 0 \\ 2 & 0 \\ 2 & 2 \\ 2 & 4 \end{pmatrix}.$$

   We now have

$$\mathbf{\Delta N}^T \mathbf{\Delta N} = \begin{pmatrix} 13 & 12 \\ 12 & 20 \end{pmatrix} \quad \text{and} \quad \text{cond}(\mathbf{\Delta N}) = 1.3.$$

   This is better than the above two cases.

The above three cases can be represented by the area covered by the experimental design in the *Information space* as shown in Figure 4.12.

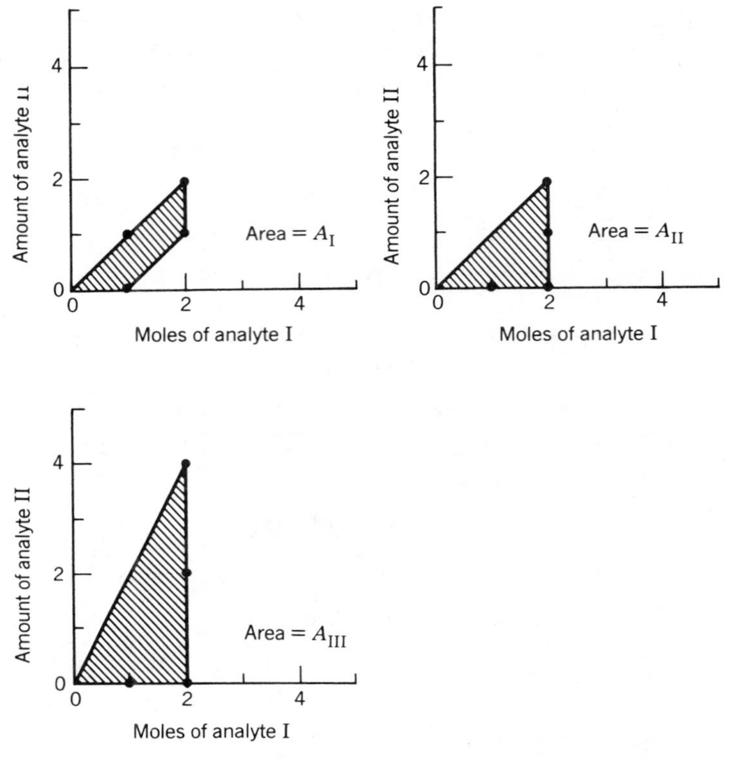

Figure 4.12 Experimental designs using the total difference calculation. Error magnification is inversely proportional to the area A.

The larger the area, the more "numerical leverage" available to minimize error amplification. Unfortunately, the TDC cannot be modified to get cond($\Delta\mathbf{N}$) = 1.0.

Incremental Difference Calculations

In this method the $\Delta\mathbf{N}$ matrix is calculated as the difference matrix after each incremented addition. In Case I $\Delta\mathbf{N}$ is now given as

$$\Delta\mathbf{N} = \begin{pmatrix} 1 & 0 \\ 0 & 1 \\ 1 & 0 \\ 0 & 1 \end{pmatrix}.$$

This results in

$$\Delta\mathbf{N}^T\Delta\mathbf{N} = \begin{pmatrix} 2 & 0 \\ 0 & 2 \end{pmatrix} \quad \text{and} \quad \text{cond}(\delta\mathbf{N}) = 1.0,$$

an optimal value.

In the total difference calculations $\mathbf{n}_0$ is found using the differences between the initial responses obtained after each standard addition. This requires that the initial responses be known very precisely. In the incremental difference calculations n_0 is calculated with minimal dependence on the initial responses. This can be advantageous when the initial responses are corrupted by much noise.

Another experimental design, Partition GSAM, has been suggested (10). In this method the sample is divided into a number of partitions equal to the number of analytes of interest. To every portion the addition of only one standard is made and the responses before and after the addition are obtained for every sensor. It is found that cond($\Delta\mathbf{N}$) for this design is always unity when the sensors have equal sensitivities.

As discussed earlier, using p sensors to determine r analytes such that $p > r$ is advantageous (9). The overdetermined system offers an improved S/N ratio. This is true in the method of external standards as well as with the GSAM.

The subject of multivariate calibration is expanding rapidly as analytical chemometricians develop mathematical solutions to the many problems associated with multicomponent analysis in all areas of spectroscopy and electrochemistry. This chapter should serve as an introduction to this topic. For more advanced study, the reader should consult the recent literature using the Fundamental Reviews (April during even-numbered years) in *Analytical Chemistry* as a guide. Recent Report section articles in the analytical literature also address this topic in more detail.

REFERENCES

1. N. R. Draper and H. Smith, *Applied Regression Analysis*, 2nd ed., Wiley, New York, 1981, Chap. 3.

   An excellent treatment of the analysis of residuals and the relationships between their behavior and the model used.

2. J. S. Garden, D. G. Mitchell, and W. N. Mills, "Nonconstant Variance Regression Techniques for Calibration-Curve-Based Analysis," *Anal. Chem.* **52**, 2310–2315 (1980).

   A discussion of weighted and unweighted least-squares regressions. Both methods are compared using hypothetical data. Confidence intervals for regression bands are given.

3. J. Mandel and F. Linnig, "Study of Accuracy in Chemical Analysis Using Linear Calibration Curves," *Anal. Chem.* **29**, 743–749 (1957).

   The joint confidence for β_0 and β_1 in linear models is demonstrated. The confidence band for a calibration curve is discussed and the equations of the boundaries are given.

4. P. D. Lark, B. R. Craven, and R. C. L. Bosworth, *The Handling of Chemical Data*, Pergamon Press, Oxford, U.K., 1968, pp. 177–180.

   A worked example of the estimation of the point of intersection of two lines.

5. A. Wald, "The Fitting of Straight Lines if Both Variables Are Subject to Error," *Ann. Math. Statis.* **11**, 284–300 (1940).

   A detailed discussion of least-squares regression and the estimation of β_0 and β_1 when the dependent and the independent variables are in error. Confidence intervals for β_0 and β_1 are also discussed. The method of grouping is demonstrated and rules of grouping the data are outlined.

6. J. A. Irvin and T. I. Quickenden, "Linear Least Squares Treatment When There Are Errors in Both X and Y," *J. Chem. Ed.* **60**, 711–712 (1983).

   A description of an iterative procedure for linear regression when both variables are subject to errors.

7. L. M. Schwartz, "Nonlinear Calibration," *Anal. Chem.* **49**, 2062–2068 (1977).

   Three methods to treat nonlinear data are presented. Confidence intervals for regression bands are given. The method of linear segments is discussed and its failures demonstrated. The three methods are demonstrated by an example.

8. J. P. Franke, R. A. de Zeeuw, and R. Hakkert, "Evaluation and Optimization of the Standard Addition Method for Absorption Spectrometry and Anodic Stripping Voltammetry," *Anal. Chem.* **50**, 1374–1380 (1978).

   A discussion of the propagation of error in SAM calculations. Confidence intervals for n_0 are derived for weighted least-squares estimation of n_0 and the case of constant coefficient of variation. Applications with simulated and real data are presented.

9. J. H. Kalivas, "Precision and Stability for the Generalized Standard Addition Method," *Anal. Chem.* **55**, 565–567 (1983).

   A discussion of overdetermined systems (i.e., when the number of sensors is greater than the number of analytes). A comparison is made between using 4, 18, and 36 sensors to estimate the amount of four analytes by GSAM.

10. C. Jochum, P. Jochum, and B. R. Kowalski, "Error Propagation and Optimal Performance in Multicomponent Analysis," *Anal. Chem.* **53**, 85–92 (1981).

    A detailed discussion of the theory of GSAM and error propagation in the calculations of the technique. Three experimental designs are presented with an example involving the determination of four analytes.

11. B. E. H. Saxberg and B. R. Kowalski, "The Generalized Standard Addition Method," *Anal. Chem.* **151**, 1038 (1979).

The generalized standard addition method (GSAM) is developed.

SUGGESTED READINGS

J. Agterdenbos, "Calibration in Quantitative Analysis—Part I. General Considerations," *Anal. Chim. Acta* **108**, 315–323 (1979).

J. Agterdenbos, F. J. M. J. Maessen, and J. Blake, "Calibration in Quantitative Analysis—Part 2. Confidence Regions for the Sample Content in the Case of Linear Calibration Relations," *Anal. Chim. Acta* **132**, 127–137 (1981).

A two part discussion on the principal models and assumptions involved in calibration. The effects of the number of calibration points on the precision of the values calculated from the model are presented.

M. Bader, "A Systematic Approach to Standard Addition Methods in Instrumental Analysis," *J. Chem. Ed.* **57**, 703–706 (1980).

A brief discussion of the merits of SAM is given. Several applications of SAM are treated.

Michael F. Delaney, "Chemometrics," *Anal. Chem.* **56**, 261R–277R (1984).

I. Frank and B. R. Kowalski, "Chemometrics," *Anal. Chem.* **54**, 232R–243R (1982).

B. R. Kowalski, "Chemometrics," *Anal. Chem.* **53**, 112R (1980).

These reviews include Sections on calibration.

D. L. Massart, A. Djikstra, and L. Kaufman, *Evaluation and Optimization of Laboratory Methods and Analytical Procedures*, Elsevier, Amsterdam, 1978, Chap. 17.

A discussion of multicomponent analysis. Calibration, analysis, and overdetermined systems are treated.

J. Sustek, "Method for the Choice of Optimal Analytical Positions in Spectrophotometric Analysis of Multicomponent Systems," *Anal. Chem.* **46**, 1676–1679 (1974).

Overdetermined systems are treated. The effects of the numbers of sensors and the uncertainties in sensitivity coefficients on the accuracy of the analysis are demonstrated.

CHAPTER

5

RESOLUTION OF ANALYTICAL SIGNALS

Resolution may be defined as the process by which a composite signal is reduced to simple forms. Generally, the signals acquired by analytical instruments are composite. The detector indicates a *total* signal, whose sources are, from the detector's point of view, indistinguishable. The signal may be generated by the analyte (or the process) of interest, as well as interfering components, closely related phenomena, and the background.

It is essential that the signal's component that is due to the analyte or the process of interest be filtered from the total signal. Examples of such a process have already been described. The effectiveness of the generalized standard addition method (GSAM) in resolving analytical signals affected by interferences and matrix effects has been covered in Chapter 4. The subtraction of background and baseline components has been discussed in Chapter 3. The use of derivative techniques to detect the presence of composite profiles and the application of Fourier transform deconvolution to remove the components of random noise have also been covered in Chapter 3. As Frank and Kowalski (1) have pointed out in their review, calibration, mathematical modeling, and spectral analysis are indeed all valid techniques for the resolution of signals.

In some areas of analytical chemistry, the term resolution is often used to indicate the degree of separation between signals. In chromatography, for example, the ability of the column to separate two components under certain chromatographic conditions is referred to as the chromatographic resolution, R, and is given as

$$R = \frac{2\Delta t}{W_1 + W_2}, \tag{5.1}$$

where Δt is the difference in retention times and W_i is the width of the ith peak at the base of the peak. Another example can be found in mass spectrometry where the instrumental resolution (or resolving power) is taken as the ability of the instrument to detect two signals of a small difference in molecular weights as two "separate" signals. The resolving power, R, is given as

$$R = \frac{M}{\Delta M}, \tag{5.2}$$

where M is the average M/Z of the two adjacent signals and ΔM is the difference between the two M/Z values. By convention, the two signals are considered separated if there is an overlap not exceeding 10% of their individual profiles. Some mass spectrometer manufacturers often allow the overlap to be as high as 50% of the individual profiles.

These two examples show useful, albeit arbitrary, working concepts of resolution. Nothing is necessarily being reduced to simple forms. This chapter demonstrates the various methods by which the analytical measurements are examined to determine whether they are elementary or composite. The techniques that can be used to reduce composite signals to simple forms will be discussed.

DETERMINING THE COMPLEXITY OF SIGNALS

Visual Inspection

Aided by sight, the human brain is a formidable pattern analyzer. Overlapping spectral bands or chromatographic peaks can often be detected by visual inspection of the instrumental output. A composite signal is indicated by the presence of shoulders and/or valleys in its profile. Lack of symmetry is not generally attributed to overlapping signals, as many physical processes often cause the signal to be distorted (e.g., tailing of peaks in chromatography). Visual inspection fails rather quickly, compared to other methods, as the separation of peaks decreases. The relative intensities of the individual bands also affect the detection of overlapping signals. The method is, nevertheless, a fast and powerful technique in evaluating the effects of noise in composite (or elementary) profiles. An experienced analyst can often distinguish a signal from noise spikes much more readily than a digital computer.

Differentiation of Signals

Consider a spectral band whose profile can be described by a Gaussian distribution. The spectral intensity as a function of frequency, $I(\nu)$, is given as

$$I(\nu) = \frac{1}{\sigma\sqrt{2\pi}} e^{-(\nu-\nu_0)^2/2\sigma^2}, \tag{5.3}$$

where ν_0 is the fundamental frequency or centroid of the distribution and σ is the bandwidth or standard deviation.

Differentiation of Equation (5.3) yields

$$I'(\nu) = -\left[\frac{1}{\sqrt{2\pi}}(\nu-\nu_0)\, e^{-(\nu-\nu_0)^2/2\sigma^2}\right]\sigma^{-3}. \tag{5.4}$$

The first derivative function is zero at $\nu = \nu_0$, positive at $\nu < \nu_0$, and negative at $\nu > \nu_0$. The second derivative of $I(\nu)$ is obtained by differentiating Equation (5.4). $I''(\nu)$ is given as

$$I''(\nu) = \left[\frac{1}{\sqrt{2\pi}} e^{-(\nu-\nu_0)^2/2\sigma^2}\right][(\nu-\nu_0)^2 - \sigma^2]\sigma^{-5}. \tag{5.5}$$

The second derivative is zero at $\nu = \nu_0 \mp \sigma$. These locations correspond to a maximum for $I'(\nu)$ at $\nu = \nu_0 - \sigma$ and a minimum for $I'(\nu)$ at $\nu = \nu_0 + \sigma$. In the range, $\nu_0 - \sigma < \nu < \nu_0 + \sigma$, $I''(\nu)$ is negative. The value of the second derivative is positive when $(\nu-\nu_0)^2 - \sigma^2 > 0$. This condition is satisfied when $\nu > \nu_0 + \sigma$ and when $\nu < \nu_0 - \sigma$. The second derivative function, therefore, has a central negative lobe and two positive side lobes. The side lobes are referred to as the *satellite pattern*. The central lobe is narrower than the original profile. Its width at the base is 2σ compared to approximately 6σ for the original profile. Equation (5.5) can be differentiated to obtain the third derivative function, $I'''(\nu)$:

$$I'''(\nu) = \left[\frac{1}{\sqrt{2\pi}}(\nu-\nu_0)\, e^{-(\nu-\nu_0)^2/2\sigma^2}\right][3\sigma^2 - (\nu-\nu_0)^2]\sigma^{-7}. \tag{5.6}$$

The third derivative is zero at $\nu = \nu_0$ and at $\nu = \nu_0 \mp \sigma\sqrt{3}$. These locations correspond to a minimum for $I''(\nu)$ (at $\nu = \nu_0$) and two maxima for $I''(\nu)$ (at $\nu = \nu_0 \mp \sigma\sqrt{3}$). The fourth derivative function $I''''(\nu)$ can be found by differentiating Equation (5.6) and is given as

$$I''''(\nu) = \left[\frac{1}{\sqrt{2\pi}} e^{-(\nu-\nu_0)^2/2\sigma^2}\right][3\sigma^4 + (\nu-\nu_0)^4 - 6\sigma^2(\nu-\nu_0)^2]\sigma^{-9}. \tag{5.7}$$

The fourth derivative function has a central positive lobe centered at $\nu = \nu_0$, and a satellite pattern that consists of two negative and two positive side lobes. The fourth derivative is equal to zero when the following condition is satisfied:

$$(\nu-\nu_0)^4 - 6\sigma^2(\nu-\nu_0)^2 + 3\sigma^4 = 0. \tag{5.8}$$

Equation (5.8) can be simplified by setting $(\nu-\nu_0)^2 = x$. Thus,

$$x^2 - 6\sigma^2 x + 3\sigma^4 = 0. \tag{5.9}$$

The roots of Equation (5.9) are given as

$$x = \sigma^2(3 \mp \sqrt{6}). \tag{5.10}$$

The fourth derivative is therefore equal to zero at

$$\left.\begin{aligned} \nu &= \nu_0 + \sigma(3+\sqrt{6})^{1/2}, \\ \nu &= \nu_0 + \sigma(3-\sqrt{6})^{1/2}, \\ \nu &= \nu_0 - \sigma(3-\sqrt{6})^{1/2}, \\ \nu &= \nu_0 - \sigma(3+\sqrt{6})^{1/2}. \end{aligned}\right\} \tag{5.11}$$

The width of the central positive lobe of the fourth derivative is approximately 1.48σ at the base $[I''''(\nu)=0$ at $\nu = \nu_0 \mp (3-\sqrt{6})^{1/2}\sigma]$. It is narrower than the central lobe of the second derivative. Note the proportionality of the derivative function to the standard deviation of the distribution, σ. The value of the nth derivative of a narrowband (small σ) is larger than that of a broadband (large σ). This enables the derivative function of a composite profile to be more sensitive to overlapping bands then the original profile itself (i.e., when the original profile does not show shoulders or valleys). The sensitivity increases as the order of the derivative function increases. Consequently, derivatives of order higher than one are preferred. In practice, second and fourth derivatives are popular functions in signal processing. The satellite patterns of side lobes make higher derivatives undesirable for resolution purposes. The pattern is often difficult to interpret.

The type of profile chosen for a particular study depends on the specific case being considered. Gaussian profiles have been popular in spectroscopy and chromatography. Spectroscopists often express Equation (5.3) as

$$\epsilon(\nu) = \epsilon_0 e^{-5.545(\nu-\nu_0)^2/2H^2}, \tag{5.12}$$

where $\epsilon(\nu)$ = absorption of frequency ν
ϵ_0 = molar absorption coefficient
ν_0 = mode of absorption band
H = width of the band at half the maximum height (or, full width at half maximum, FWHM).

The relationships between the derivative functions and H and ν_0 can be found in the same manner that Equations (5.4)–(5.11) are derived. Another profile that is very important in spectroscopy is the Lorentzian profile which is given as

$$\epsilon(\nu) = \frac{\epsilon_0 H^2}{H^2 + 4(\nu-\nu_0)^2}, \tag{5.13}$$

where $\epsilon(\nu)$, ϵ_0, ν_0, and H are as defined above. The conclusions derived from Equations (5.4)–(5.11) apply also to the profiles expressed in Equations (5.12)–(5.13). In practice, derivative functions are computed numerically. The fundamentals of the technique are discussed in Chapter 3.

Figure 5.1 shows a Lorentzian profile and its second derivative. Compared to the original profile the second derivative is narrower. This allows overlapping peaks that do not show shoulders to be detected.

Figure 5.2 shows two overlapping Lorentzian profiles and the second derivative of the composite profile. The original profile does not show any distortions. The second derivative, however, indicates the presence of a second independent signal.

The effectiveness of derivative functions in detecting the number of independent components in a composite profile depends on the separation

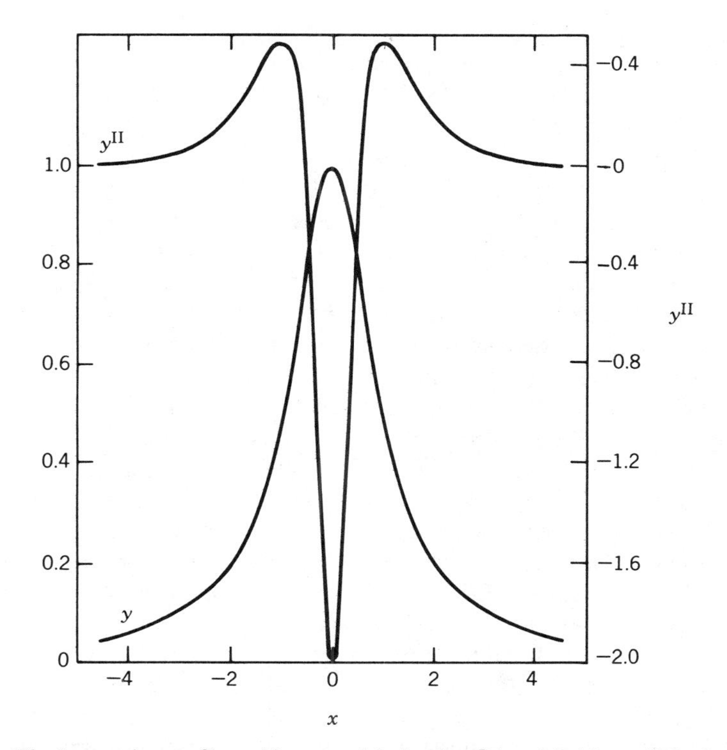

Figure 5.1 The Lorentzian profile and its second derivative. [From Maddams, "The Scope and Limitations of Curve Fitting," *Appl. Spectrosc.* **34**(3), p. 247 (1980), Fig. 3.]

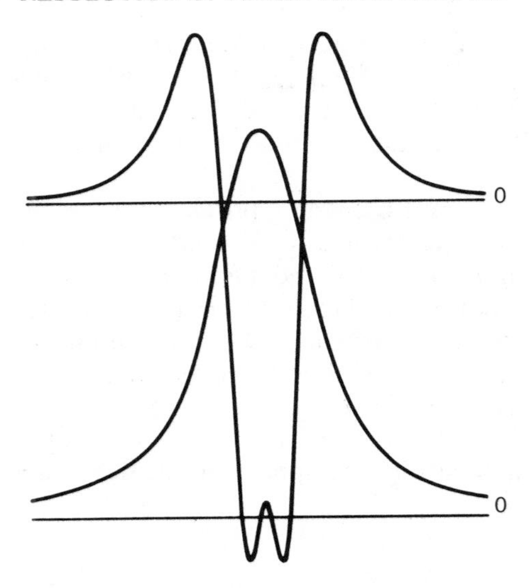

Figure 5.2 The composite peak and its second derivative, from two Lorentzian peaks separated by 40% of their half-widths. [From Maddams, "The Scope and Limitations of Curve Fitting," *Appl. Spectrosc.* **34**(3), p. 248 (1980).]

of the peaks (i.e., separation of the means of the distributions), the widths of the peaks, and the relative intensities of the overlapping bands. Generally, as the separation of the peak centers decreases and the bandwidths increase, the sensitivity of derivative functions decreases. Also, minor components are more difficult to detect. The effects of the factors that affect the sensitivity of derivative functions have been studied by Vandeginste and DeGalan (2). They have concluded that the most successful detection is obtained when the number of inflection points is equal to twice the number of individual peaks.

The level of noise in the measurements is an important factor in successful application of these methods. O'Haver and Green (3) have demonstrated the dependence of derivatives on the noise level. The level of noise increases as the order of the derivative function increases. The low level of noise in UV-Vis spectrophotometry allows for successful processing of second and higher derivatives. Talsky and co-workers (4) have reported the resolution of Congo Red spectra by means of a fourth derivative function. They have also found that while the UV-Vis spectra of trypsin and chymotrypsin are apparently similar, their fourth derivatives possess characteristic features that can be used for the identification of each component. When the experimental S/N ratio is low, derivative functions do not produce satisfactory results. The effects of noise can be reduced by smoothing the signal. The techniques used to smooth signals and increase the S/N ratio have been discussed in Chapter 3.

Factor Analysis

These two categories of signal manipulation are usually performed on *one* data vector (e.g., N spectral intensities, I_i, $i = 1, \ldots, N$). In many instances, several data vectors are acquired during the course of the experiment. In combined gas chromatography–mass spectrometry, for example, several mass spectra are acquired across the chromatographic profile. In kinetic studies, the spectra of the reactants and the products may be obtained over a certain period of time. The acquired spectra can be factor analyzed in order to estimate the number of independent components giving rise to the acquired data.

Factor and principal component analysis are powerful multivariate techniques that have found many applications in chemistry (5, 6). (See Chapter 6 for a detailed presentation.)

Geometric Approach to Factor Analysis

Assume that three data vectors (e.g., spectra) are obtained when four different channels (e.g., ion intensities of four M/Z) are being monitored during an experiment. The data vectors constitute a data matrix, $\mathbf{X}$, where x_{ij}, is the amplitude of the signal obtained in the ith channel when the jth data vector is acquired. Suppose that $\mathbf{X}$ is given as

$$\mathbf{X} = \begin{bmatrix} 10.40 & 19.74 & 19.42 \\ 47.34 & 81.12 & 36.63 \\ 100.00 & 97.04 & 0.40 \\ 3.46 & 100.00 & 100.00 \end{bmatrix}. \tag{5.14}$$

Since four different channels are monitored, the three data vectors may be considered in a four-dimensional space, where each independent axis represents a measurement channel. One purpose of factor analysis is to provide a representation of the data vectors in a space with a dimensionality, hopefully, less than four, while preserving the original information. Reducing the measurement dimensionality to the possibly smaller *intrinsic* dimensionality of the data allows for efficient manipulation of the data and a better understanding of the phenomenon under consideration. This can be achieved by observing the correlation between each pair of vectors. The correlation between any two unit-length vectors is the cosine of the angle between them. To calculate the correlation between the data vectors, the entries of each vector are first scaled such that their mean is zero and their variance is one. The new matrix is calculated according to

$$x'_{ij} = \frac{x_{ij} - m_j}{s_j}, \tag{5.15}$$

where m_j = the mean of the original entries of the ith vector
s_j = the standard deviation of the original entries of the ith vector.

The new transposed matrix is thus

$$\mathbf{X}'^T = \begin{bmatrix} -0.67624 & 0.15922 & 1.35023 & -0.83321 \\ -1.46272 & 0.17716 & 0.60302 & 0.68212 \\ -0.45580 & -0.05746 & -0.89603 & 1.40928 \end{bmatrix}. \tag{5.16}$$

The sample by sample correlation matrix, $\mathbf{C}$, is given as

$$\mathbf{C} = \left(\frac{1}{N-1}\right)\mathbf{X}'^T\mathbf{X}', \tag{5.17}$$

where N is the number of entries in each data vector (in this case four).
Equation (5.17) yields

$$\mathbf{C} = \begin{bmatrix} 1 & 0.42114 & -0.69514 \\ 0.42114 & 1 & 0.35920 \\ -0.69514 & 0.35920 & 1 \end{bmatrix}. \tag{5.18}$$

The angle between each two vectors can be calculated from $\mathbf{C}$. The angle matrix, θ, is given as

$$\theta_{ik} = \cos^{-1} C_{ik}. \tag{5.19}$$

Therefore,

$$\theta = \begin{bmatrix} 0 & 65.09 & 134.04 \\ 65.09 & 0 & 68.95 \\ 134.04 & 68.95 & 0 \end{bmatrix}. \tag{5.20}$$

The angle between the first vector and the second vector is 65.09°, and that between the second vector and the third vector is 68.95°. Similarly, the angle between the first vector and the third vector is 134.04° (65.09° + 68.95°). The three vectors can therefore be represented as being in a plane; they have an intrinsic dimensionality of two. This is schematically shown in Figure 5.3.

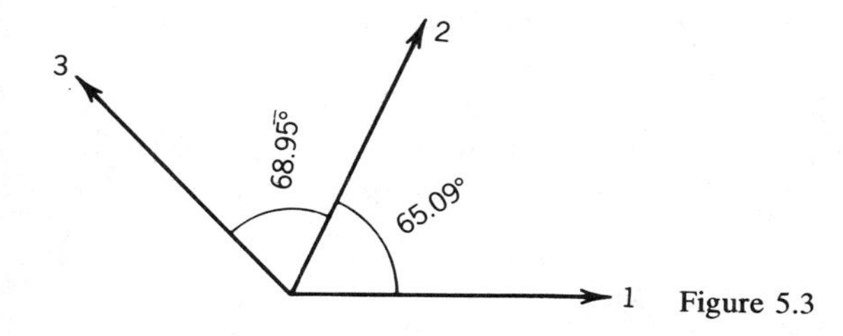

Figure 5.3

The representation shown in Figure 5.3 is *not* accompanied by any loss of information present in the original data. Each data vector can be represented by a pair of coordinates in the new two-dimensional space since three points always form a triangle which is planar. The data given in Equation (5.14) are, in fact, three different linear combinations of two independent mass spectra.

Consider the following 5×3 data matrix $\mathbf{X}$:

$$\mathbf{X} = \begin{bmatrix} 19.42 & 25.33 & 10.66 \\ 5.00 & 8.63 & 4.53 \\ 0.40 & 100.00 & 100.00 \\ 100.00 & 62.58 & 4.00 \\ 0.10 & 63.27 & 18.13 \end{bmatrix} . \tag{5.21}$$

After normalizing each data vector, as described earlier, the sample by sample correlation matrix is computed as

$$\mathbf{C} = \begin{bmatrix} 1 & 0.05695 & -0.39479 \\ 0.05695 & 1 & 0.77886 \\ -0.39479 & 0.77886 & 1 \end{bmatrix} . \tag{5.22}$$

The angles between the vectors are calculated as

$$\theta = \begin{bmatrix} 0 & 86.74 & 113.25 \\ 86.74 & 0 & 38.84 \\ 113.25 & 38.84 & 0 \end{bmatrix} . \tag{5.23}$$

The three vectors can *not* be placed in a plane with such angles between them. One of them has to be outside the plane of the other two. The data given in Equation (5.21) are three different combinations of three independent mass spectra.

As the number of data vectors and their entries increases, the above geometric interpretation becomes inconvenient. It is also difficult to represent data with intrinsic dimensionality greater than three geometrically. An algebraic approach is necessary (6).

Algebraic Approach to Factor Analysis

In order to represent vectors in an *n*-dimensional space, it is necessary to define the basis of the space and the projections of each data vector on each of the *n* orthogonal axes. This representation can be expressed in matrix form as

$$\mathbf{X} = \mathbf{C}\mathbf{V}^T, \tag{5.24}$$

where $\mathbf{X}$ = the data matrix whose rows are the experimental vectors
$\mathbf{V}$ = the matrix containing the orthogonal axes as its columns and
$\mathbf{C}$ = the matrix containing the projection coordinates of each data vector on the basis of the space as its rows

Factor analysis expresses the original data matrix in terms of linear combinations of orthogonal (independent) vectors. These orthogonal vectors define a space that, hopefully, contains the same information as the data matrix. One of the fundamental assumptions involved, when data vectors are factor analyzed in analytical chemistry is that the experimental vectors (e.g., mass spectra) are positive linear combinations of parent, independent vectors. Spectral data that obey Beer's Law constitute a popular example. If an $m \times n$ data matrix similar to that given by Equation (5.14) is normalized and factor analyzed, *n* orthogonal vectors are obtained. They are referred to as *eigenvectors*. Associated with each eigenvector is a descriptor of its importance referred to as the *eigenvalue*. The minimum number of eigenvectors (those with the largest eigenvalue) needed to account for the variations in the data is taken as the necessary number of independent vectors, *k*, needed in Equation (5.24). This number is also taken to be the number of independent vectors that make up the data vectors (i.e., the number of parent components). Factor analysis is said to remove the redundancy of the data when $k < n$; the dimensionality of the data matrix is reduced.

Because of experimental noise, the first *k* eigenvectors may not account for 100% of the data variance. In practice, slightly smaller values are found.

Factor analysis has been utilized for the study of composite profiles in many areas of chemistry (1). As the separation between signals decreases and the relative composition decreases (i.e., one of two components predominates) factor analysis may not be able to detect the presence of overlapping signals. The performance of factor analysis has been tested using data simulating coelution of components in combined gas chromatography-mass spectrometry (7). Gaussian profiles of equal standard deviation (equal widths) are used to simulate the coelution of two components. At a separation of 0.8 units of standard deviation, a second eigenvector is needed to account for 0.25% of the variance when the relative composition of the two components is 10:1. The data used in the study are noise free and should be considered as an upper limit for the effectiveness of factor analysis. In practice such a small percentage of the variance is often ignored because of random noise. When spectral data are used, the effectiveness of factor analysis is affected by the nature of the spectra. The more orthogonal the spectra (less signals in common) the more effective the factor analysis.

RESOLVING COMPOSITE SIGNALS

Deconvolution of Overlapping Signals

Deconvolution most commonly refers to resolving signals using the Fourier transform technique. The fundamentals of this method are discussed in Chapter 3. The removal of overlapping peaks is similar to the removal of noise components; the noise function is replaced by the function for one of the overlapping peaks. Kauppinen and co-workers (8) have demonstrated the effectiveness of such a strategy. Clearly, the shapes of the overlapping profiles have to be estimated a priori.

Resolving Signals By Mathematical Modeling (Curve Fitting)

Least-squares curve fitting is one of the most common procedures used to resolve overlapping peaks. The technique has been successfully utilized in chromatography, NMR, IR, Raman, and UV-Vis spectroscopy. Important physical parameters such as peak area, peak height, and peak width (or full width at half maximum) can be estimated from the model. The most common profiles used to model the signals are the Gaussian [Eq. (5.12)] and the Lorentzian [Eq. (5.13)] profiles. A combination of these two profiles (referred to as the *Voigt profile*) has also been used in many applications. In chromatography, a Gaussian profile is often used in combination with an exponentially decaying function to account for tailing peaks.

A composite profile can be resolved by modeling it as the sum (a linear combination is common) of two or more distributions using the least squares criterion in approximating the experimental profile. When such a technique is implemented, Occam's razor principle must be obeyed; the number of functions needed to account for the composite profile must be kept to a *minimum*.

The analyst must address three major points when curve fitting of composite profiles is considered. The first of these items is the number of profiles needed to reconstruct the original signal. An estimate can be obtained from the number of inflection points in the original profile, or from the derivative of the original profile. As Vandeginste and DeGalan (2) pointed out, it is imperative to have a reliable estimate of the number of individual components a priori. When applicable, factor analysis can be of valuable assistance in this regard. If the number of peaks in the profile exceeds twice the number of *inflection* points, errors are likely to occur.

The second item to be considered is the curve fitting technique itself. Because of the shapes of the profiles used, an *iterative*, nonlinear least-squares procedure must be used. Gans (9) has reviewed the most widely used procedures of nonlinear least-squares curve fitting. Most methods are affected by the models and the parameters.

The third concern is the initial parameters. Differentiation of the original profile enables the analyst to estimate adequate initial parameters. As shown earlier, the second derivative of a Gaussian profile is zero at $x = \mu \mp \sigma$. The peak width can thus be adequately estimated.

Suppose that the data are to be fitted to a Gaussian (or a Lorentzian) profile. The model can be expressed as

$$Y_i = f(\boldsymbol{\nu}_i, \mathbf{p}) + \epsilon_i \tag{5.25}$$

where Y_i is the ith experimental value, and $f(\boldsymbol{\nu}_i, \mathbf{p})$ is a function of the independent variables $\boldsymbol{\nu}_i$ and the parameters $\mathbf{p}$, and ϵ_i is the error of fitting the ith datum to the model.

The parameters $\mathbf{p}$ are selected such that

$$\chi^2(\mathbf{p}) = \sum_{i=1}^{n} \epsilon_i^2 = \text{minimum}. \tag{5.26}$$

The normal equations are thus given by

$$\sum_{i=1}^{n} [Y_i - f(\boldsymbol{\nu}_i, \mathbf{p})] \left[\frac{\partial f(\mathbf{v}_i, \mathbf{p})}{\partial p_k}\right] = 0, \tag{5.27}$$

where p_k is the kth parameter.

For nonlinear models, the solution of Equation (5.27) is difficult to obtain. Moreover, multiple solutions sometimes exist. Several iterative techniques have been developed to solve nonlinear least-squares problems (9–11). Some of these procedures will be outlined.

Taylor Series Linearization

This method is based on the Taylor expansion of the function about an initial "guess" for the values of the parameter, $\mathbf{p}^0$. The expansion is limited to the first derivative and yields a system of linear equations. At $\mathbf{p} \simeq \mathbf{p}^0$

$$f(\boldsymbol{\nu}_i, \mathbf{p}) = f(\boldsymbol{\nu}_i, \mathbf{p}^0) + \sum_{n=1}^{k} \left[\frac{\partial f(\mathbf{v}_i, \mathbf{p})}{\partial \mathbf{p}_n}\right]_{\mathbf{p}=\mathbf{p}^0} (\mathbf{p}_n - \mathbf{p}_n^0), \tag{5.28}$$

where k is the number of parameters.

Let

$$f(\boldsymbol{\nu}_i, \mathbf{p}^0) = f_i^0,$$

$$\mathbf{p}_n - \mathbf{p}_n^0 = \Delta \mathbf{p}_n^0,$$

and

$$\left[\frac{\partial f(\mathbf{v}_i, \mathbf{p})}{\partial p_n}\right]_{\mathbf{p}=\mathbf{p}^0} = D_{ni}^0,$$

where D_{ni}^0 is the value of the derivative evaluated at p^0 for the nth parameter at the ith data point. Equation (5.25) can now be written as

$$Y_i - f_i^0 = \sum_{n=1}^{k} [\Delta \mathbf{p}_n^0 D_{ni}^0] + \epsilon_i. \tag{5.29}$$

The change in parameter vector $\Delta \mathbf{p}^0$ can be computed according to

$$\Delta \mathbf{p}^0 = [(\mathbf{D}^0)^T \mathbf{D}^0]^{-1} (\mathbf{D}^0)^T (Y - f^0). \tag{5.30}$$

The vector $\Delta \mathbf{p}^0$ can be used to calculate a consecutive estimate for the parameters. This procedure is repeated until convergence occurs.

Consider the liquid chromatography resolution problem discussed in Chapter 2. The resolution is assumed to be determined by the ratio of the solvents, r, and the flow rate, f, and is given as

$$R = e^{-(r-1)^2} + e^{-(f-2)^2}. \tag{5.31}$$

Assume that the data shown in Table 5.1 are acquired and that the model used to approximate the behavior of resolution is

$$R = e^{-(r-\theta_1)^2} + e^{-(f-\theta_2)^2}. \tag{5.32}$$

Table 5.1 Resolution as a Function of r and f

r	f	R
0.5	0.7	0.96332
1.0	1.0	1.36788
1.5	0.6	0.91966
2.0	1.0	0.73576

Let $r=1$ and $f=1$ be chosen as initial estimates. (This example is for demonstration purposes. The choice of initial estimates can be very critical in practice. Also, the number of data points is related to the number of parameters being estimated. A minimum of 3–4 data points per parameter is typical in practice.) To calculate $\mathbf{D}^0$ the derivative of Equation (5.32) is evaluated:

$$\frac{dR}{d\theta_1} = 2(r-\theta_1)e^{-(r-\theta_1)^2} \tag{5.33}$$

and

$$\frac{dR}{d\theta_2} = 2(f-\theta_2)e^{-(f-\theta_2)^2}. \tag{5.34}$$

$\mathbf{R}^0$ is calculated by substituting the values of r and f in Table 5.1 in Equation (5.32) after setting $\theta_1 = 1$ and $\theta_2 = 1$. The following table can be constructed:

r	f	$\mathbf{R}$ (exp)	$\mathbf{R}^0$ [Eq. (5.32)]	$\mathbf{R}-\mathbf{R}^0$	$\mathbf{D}^0$	
0.5	0.7	0.96332	1.69273	−0.72941	−0.77880	−0.54836
1.0	1.0	1.36788	2.00000	−0.63212	0	0
1.5	0.6	0.91966	1.63094	−0.71128	0.77880	−0.68172
2.0	1.0	0.73576	1.36788	−0.63212	0.73576	0

$$\chi^2 = 1.83711.$$

The first column in $\mathbf{D}^0$ is computed by substituting the corresponding values of r in Equation (5.33). Similarly, the second column in $\mathbf{D}^0$ is calculated by substituting the corresponding values of f in Equation (5.34). The vector of parameter increments, $\Delta\mathbf{p}$, is obtained according to Equation (5.30).

$$\Delta\mathbf{p} = (\mathbf{D}^{0T}\mathbf{D}^0)^{-1}\mathbf{D}^{0T}(\mathbf{R}-\mathbf{R}^0). \tag{5.35}$$

Solving Equation (5.35) for $\Delta\mathbf{p}$ gives

$$\Delta\mathbf{p} = \begin{pmatrix} -0.19014 \\ 1.13073 \end{pmatrix}.$$

The next estimates are thus $\theta_1 = 0.80986$ and $\theta_2 = 2.13023$.

This procedure is repeated and the following table can be constructed:

r	f	$\mathbf{R}$	$\mathbf{R}^0$	$\mathbf{R}-\mathbf{R}^0$	$\mathbf{D}^0$	
0.5	0.7	0.96332	1.03776	−0.07444	−0.56299	−0.36988
1.0	1.0	1.36788	1.24325	0.12463	0.36678	−0.63012
1.5	0.6	0.91966	0.71725	0.20241	0.85727	−0.29433
2.0	1.0	0.73576	0.52133	0.21443	0.57740	−0.63012

$$\chi^2 = 0.10802.$$

Solving for $\Delta\mathbf{p}$ gives

$$\Delta\mathbf{p} = \begin{pmatrix} 0.20623 \\ -0.11194 \end{pmatrix}.$$

The new estimates are $\theta_1 = 1.01609$ and $\theta_2 = 2.01829$. These estimates yield $\chi^2 = 6.7 \times 10^{-4}$ and

$$\Delta\mathbf{p} = \begin{pmatrix} -0.01601 \\ -0.01834 \end{pmatrix}.$$

The following iteration therefore uses $\theta_1 = 1.00008$ and $\theta_2 = 1.99995$. (The true values are [Eq. (5.31)] $\theta_1 = 1$ and $\theta_2 = 2$.) The process can be repeated until the desired accuracy is obtained.

Grid Search

In this method, $\chi^2(\mathbf{p})$ is minimized with respect to each parameter separately. Clearly, this is only valid when the parameters affect $\chi^2(\mathbf{p})$ independently. The value of one parameter, $\mathbf{p}_i$, is changed by an amount $\Delta\mathbf{p}_i$ such that $\chi^2(\mathbf{p})$ decreases. The same parameter is then changed by the same amount, $\Delta\mathbf{p}_i$, until $\chi^2(\mathbf{p})$ increases. The last three values of $\chi^2(\mathbf{p})$ are then fitted to a parabola and the minimum of the parabola is determined. Let the behavior of $\chi^2(\mathbf{p})$ obtained with the last three values of $\mathbf{p}_i$ be modeled as

$$\chi^2(\mathbf{p}) = a_0 + a_1 p_i + a_2 p_i^2, \tag{5.36}$$

then $\chi^2(\mathbf{p})$ is a minimum at

$$p_i = -\frac{a_1}{2a_2}. \tag{5.37}$$

Equation (5.37) is obtained by setting the derivative of Equation (5.36) to zero. If the last three values of p_i (p_1, p_2, p_3 where $p_1 < p_2 < p_3$) are coded as $-1, 0, +1$, and they correspond to $\chi^2(1)$, $\chi^2(2)$, and $\chi^3(3)$, respectively, then the minimum of the model is at

$$p_i' = -\frac{\Delta p_i}{2}\left[\frac{\chi^2(3) - \chi^2(1)}{\chi^2(3) - 2\chi^2(2) + \chi^2(1)}\right], \tag{5.38}$$

where p_i' is the *coded* value of p_i.

Method of Steepest Descent

This is one method in a class of procedures known as the gradient techniques. In these techniques, the χ^2 surface is examined and the parameters $\mathbf{p}$ are changed concurrently in the direction of maximum change in χ^2. The gradient of the surface, $\nabla\chi^2$, can be estimated with respect to each parameter numerically according to

$$[\nabla\chi^2]_i = \frac{\partial\chi^2}{\partial p_i} \simeq \frac{\chi^2(p_i + \delta p_i) - \chi^2(p_i)}{\delta p_i}, \tag{5.39}$$

where δp_i is a small increment in p_i that is, of course, smaller than the step size, Δp_i. Gradient methods are iterative procedures that proceed such that the initial estimate of parameters, $\mathbf{p}_i$, is replaced by $\mathbf{p}_{i+1}$ according to

$$\mathbf{p}_{i+1} = \mathbf{p}_i + a_i\boldsymbol{\nu}_i, \tag{5.40}$$

where $\boldsymbol{\nu}_i$ is a vector in the direction of the proposed ith step and a_i is scalar, chosen such that $\chi^2(\mathbf{p}_{i+1}) < \chi^2(\mathbf{p}_i)$.

Equation (5.40) is the general model of *all* gradient techniques. Individual methods differ in the process by which a_i, the step size, and $\boldsymbol{\nu}_i$, the step direction, are computed. The method of steepest descent is the simplest of the gradient methods. The iterative procedure proceeds according to:

$$\mathbf{p}_{i+1} = \mathbf{p}_i - a_i[\nabla\chi^2 |_{\mathbf{p}_i}], \tag{5.41}$$

where $\nabla\chi^2 |_{\mathbf{p}_i}$ is the gradient of χ^2 evaluated at $\mathbf{p}_i$.

Newton Method

Another useful method is referred to as the Newton–Raphson method. (For convenience, the derivation of the iteration equation will be given for the case of one parameter, p_i). The behavior of $\chi^2(p_{i+1})$ can be approximated by the Taylor expansion of $\chi^2(p_i)$ about p_i. Curtailing the expansion to the second derivative term yields

$$\chi^2(p_{i+1}) = \chi^2(p_i) + \left[\frac{d\chi^2}{dp_i}\bigg|_{p_i}\right](p_{i+1} - p_i) + \frac{1}{2}\left[\frac{d^2\chi^2}{dp_i^2}\bigg|_{p_i}\right](p_{i+1} - p_i)^2. \tag{5.42}$$

The terms in brackets are the first and second derivatives, respectively, of χ^2 evaluated at p_i.

Equation (5.42) shows that $\chi^2(p_{i+1})$ is quadratic in p_{i+1}. It is reasonable to assume that the function is curved and possesses a minimum. To find p_{i+1} at which $\chi^2(p_{i+1})$ is a minimum, Equation (5.42) is differentiated with respect to p_{i+1}, and the value of the derivative is set to equal zero.

$$\frac{d\chi^2(p_{i+1})}{dp_{i+1}} = 0 = \left.\frac{d\chi^2}{dp_i}\right|_{p_i} + \left[\left.\frac{d^2\chi^2}{dp_i^2}\right|_{p_i}\right](p_{i+1} - p_i). \tag{5.43}$$

Equation (5.43) yields the following iterative formula:

$$p_{i+1} = p_i - \left[\left.\frac{d^2\chi^2}{dp_i^2}\right|_{p_i}\right]^{-1}\left[\left.\frac{d\chi^2}{dp_i}\right|_{p_i}\right], \tag{5.44}$$

where

$$\left[\left.\frac{d^2\chi^2}{dp_i^2}\right|_{p_i}\right] \neq 0.$$

If $\nabla\chi^2$ is defined as the gradient of χ^2, $d\chi^2/d\mathbf{p}_i$, and $\mathbf{H}$ is defined as the matrix of second derivatives with element $H_{ij} = \partial^2\chi^2/\partial p_i \partial p_j$ ($\mathbf{H}$ is also known as the Hessian matrix), then Equation (5.44) can be generalized for the case of more than one parameter, and written as

$$\mathbf{p}_{i+1} = \mathbf{p}_i - [\mathbf{H} \mid p_i]^{-1}[\nabla\chi^2 \mid p_i], \tag{5.45}$$

where $\mathbf{H}$ is nonsingular.

The restriction on $\mathbf{H}$ to be nonsingular, in addition to the need to compute second derivatives, makes the Newton method impractical in many situations. Several improvements have been introduced. These modifications aim at either overcoming the problem of a singular $\mathbf{H}$ or eliminating the need for second derivatives. The techniques include the Marquardt method, the Gauss method, and the variable metric procedure of Fletcher and Powell (11). The Fletcher–Powell optimization procedure can be summarized as follows (12):

1. Choose $\mathbf{p}_i$ and set $\mathbf{A}_i$, an $n \times n$ matrix, equal to the identity matrix, $\mathbf{I}$. n is the number of parameters.
2. Compute $\chi^2(p_i)$ and the gradient $\mathbf{G}_i = \nabla\chi^2 \mid p_i$.
3. Find m at which $\chi^2(\mathbf{p}_i + m\mathbf{A}_i\mathbf{G}_i)$ is a maximum, and define it as m^*. (For minimization problems the function is multiplied by -1.)
4. Set $\mathbf{p}_{i+1} = \mathbf{p}_i + m^*\mathbf{A}_i\mathbf{G}_i$.
5. Calculate $\chi^2(\mathbf{p}_{i+1})$. If $\chi^2(\mathbf{p}_{i+1}) - \chi^2(\mathbf{p}_i)$ is within an acceptable tolerance, the procedure is stopped. If not:

6. Find $\mathbf{Y} = \nabla\chi^2|\mathbf{p}_i - \nabla\chi^2|\mathbf{p}_{i+1} = \mathbf{G}_i - \mathbf{G}_{i+1}$.
7. Set $\mathbf{A}_{i+1} = \mathbf{A}_i + \left[\frac{1}{\mathbf{B}_i^T\mathbf{Y}}\right]\mathbf{B}_i\mathbf{B}_i^T - \left[\frac{1}{\mathbf{Y}\mathbf{A}_i\mathbf{Y}}\right]\mathbf{A}_i\mathbf{Y}\mathbf{Y}^T\mathbf{A}_i$, where $\mathbf{B}_i = m^*\mathbf{A}_i\mathbf{G}_i$.
8. Set $i = i + 1$ and go to step 3.

Resolving Signals Using Multiple Regression and Optimization Techniques

When the components of a composite signal are known, or can be adequately estimated, resolution of the signal can be achieved in a straightforward manner. The composite signal is assumed to be a linear combination of the signals of the individual components (or of a suitable transformation of the components' signals). Examples of such a situation can be found in spectroscopic systems that adhere to Beer's Law. The relationship between the acquired signals and its components (or parent signals) can be written as

$$R_i = \beta_{1i}P_1 + \beta_{2i}P_2 + \cdots + \beta_{ni}P_n + \epsilon_i, \tag{5.46}$$

where R_i = ith composite signal obtained
β_{ji} = coefficient (or molar fraction) of the jth parent component that contributes to R_i
P_i = signal obtained with the pure jth component and
ϵ_i = error term

Equation (5.46) can be generalized and written in matrix form as

$$\mathbf{R} = \mathbf{P}\boldsymbol{\beta} + \epsilon, \tag{5.47}$$

where $\mathbf{R}$ = data matrix whose columns are the acquired signal vectors (e.g., mass spectra)
$\mathbf{P}$ = matrix containing the signals of the parent components in its columns
$\boldsymbol{\beta}$ = coefficient matrix. Each column in β holds the coefficients of parent components needed to account for the corresponding data vector.
ϵ = error matrix

If $\mathbf{R}$ and $\mathbf{P}$ are known, $\boldsymbol{\beta}$ can be estimated as $\mathbf{b}$ using the least-squares criterion. The generalized inverse solution of Equation (5.47) is given as

$$\mathbf{b} = (\mathbf{P}^T\mathbf{P})^{-1}\mathbf{P}^T\mathbf{R}, \tag{5.48}$$

when $(\mathbf{P}^T\mathbf{P})^{-1}$ exists. Weighted least squares is recommended when the experimental errors associated with elements of $\mathbf{R}$ differ markedly. The total signal R_i can then be partitioned into the parent components.

The model expressed in Equation (5.47) can be examined by analysis of variance (13). A basic ANOVA table is shown in Table 5.2, where p is the number of parameters in the model, and n is the number of entries in $\mathbf{R}_i$ (e.g., the number of wavelengths in an acquired spectrum).

The square of the correlation, r^2, between the experimental, R_i, and the modeled, $\hat{R}_i$, values is a measure of the adequacy of the model. r^2 is in the range [0, 1] and can be computed according to

$$r^2 = \frac{\sum_{i=1}^{n} (\hat{R}_i - \bar{R})^2}{\sum_{i=1}^{n} (R_i - \bar{R})^2}, \tag{5.49}$$

where $\bar{R} = \sum R_i/n$. If the model is accurate, the residual mean square is an acceptable estimate of the variance, σ^2:

$$\sigma^2 \simeq s^2 = \frac{\mathbf{R}^T\mathbf{R} - \mathbf{b}^T\mathbf{P}^T\mathbf{R}}{n-p}. \tag{5.50}$$

The uncertainty in the parameters is given as

$$\mathbf{V}(\mathbf{b}) = (\mathbf{P}^T\mathbf{P})^{-1}\sigma^2 \simeq (\mathbf{P}^T\mathbf{P})^{-1}s^2, \tag{5.51}$$

where $(\mathbf{P}^T\mathbf{P})^{-1}$ exists. $\mathbf{V}(\mathbf{b})$ is the variance–covariance matrix. The diagonal elements of $\mathbf{V}(\mathbf{b})$ are the variances, and the off-diagonal elements are the covariances. The above procedure can be applied to one data vector, as well as to the case when several data vectors are acquired. An example of the latter is combined gas chromatography–mass spectrometry, where several mass spectra can be acquired during the coelution of unseparated components. Sharaf and Kowalski (14) have used this procedure to resolve the chromatograms of overlapping components using the mixture mass spectra. Christian and co-workers (15) utilized multiple regression to resolve fluorescence spectra. Tunicliff and Wadsworth (16) applied this procedure to resolve mixture mass spectra. In

Table 5.2 ANOVA Table for Model 5.47

Source of Variation	Degrees of Freedom, ν	Sum of Squares, SS	Mean Square, SS/ν
Due to Regression	p	$\mathbf{b}^T\mathbf{P}^T\mathbf{R}$	(MS) regression
Residual	$n-p$	$\mathbf{R}^T\mathbf{R}-\mathbf{b}^T\mathbf{P}^T\mathbf{R}$	(MS) residual

their study, they selected one parent component at a time and continually updated the model to account for the variance.

There are two major drawbacks to the least squares multiple regression strategy. The first is that when the parent spectra are similar, the results are often inaccurate. One way to avoid this problem is to allow the system to be excessively overdetermined. This is limited by the sampling frequency resolution of the acquisition system itself. The second problem is that the procedure is sensitive to random noise. Negative coefficients can thus be obtained. Since the solution of Equation (5.47) has to be physically meaningful, negative coefficients are rejected. The analyst may simply consider any negative coefficient to be equal to zero. This, however, may be suitable only in special cases.

To overcome the problems of least squares multiple regression, several strategies have been developed. The objective of these strategies is to restrict the least squares solution to only positive coefficients. The sum of the squares of the deviations in Equation (5.47) is minimized, while *all* coefficients are greater than or equal to zero. This restriction, however, precludes a simple solution like that given in Equation (5.48). The problem of estimating $\boldsymbol{\beta}$ is thus transformed into an optimization problem, which can be solved by an iterative technique. Leggett (17) has demonstrated the utility of non-negative least squares and simplex optimization in resolving multicomponent spectra. He has compared both methods with multiple regression. Unacceptable coefficients were obtained when multiple regression was used. Optimization criteria other than least squares have also been used. Fausett and Weber (18) resolved mixture mass spectra by minimizing the absolute deviations and the maximum absolute deviation, as well as the sum of the squares of the deviation. When a sample contains components that are responsible for some of the measured signal, but were not included in the $\mathbf{P}$ matrix in Equation (5.47), erroneous results are to be expected. It should be understood that negative coefficients and low values of the goodness of fit are useful indicators of this condition known as "the background problem."

These methods can be successfully implemented when *all* the parent components are known. Warner and co-workers (15) have shown that optimization methods can be used for partial resolution. If only one component is known to be present, its contribution sometimes can be filtered by simplex optimization.

Method of Rank Annihilation

Rank annihilation is based on factor analysis. Consider a data matrix $\mathbf{X}$ that is composed of m data vectors. If each of the data vectors is a linear combination of n independent components, then, under favorable conditions, factor analysis will show that **X** has a rank of n (i.e., the data matrix **X** defines an n-dimensional space, or n eigenvectors are needed to account for the variance in the data). If the contribution of one component is completely subtracted from each data vector, a new data matrix $\mathbf{X}^*$ will result that will have a rank equal to $n - 1$. The reduction in the rank of the data matrix may be used as an indication of the successful removal of information pertaining to the component being subtracted. This technique has been demonstrated by Christian and co-workers (19, 20) and applied to fluorescence data. The fluorescence spectrum of one component is subtracted from the spectra of a multicomponent sample. In the absence of experimental noise, all of the variance associated with the component is completely subtracted. In practice, however, one eigenvalue reaches a minimum. This procedure can also be used for partial resolution. If only k components are known to be present in an n-component mixture ($k < n$), the contributions of the k known components can be estimated. An obvious requirement is that the spectra of the components of interest must be known a priori. See Suggested Readings for more on this topic.

Biller–Biemann Technique

Biller and Biemann (21) have developed a simple, but widely used method for the analysis of unresolved peaks in combined gas chromatography–mass spectrometry. The technique can be applied to any chromatography/spectrometry system. It is based on scanning the acquired spectra and determining the intensities of individual key signals (e.g., specific mass-to-charge ratios). Suppose that two components are partially separated and that each of them has a specific M/Z signal. These specific signals are not subject to interferences from any other component and can be used to trace the analyte of interest. The plot of the intensities of the signals versus time, known as a mass chromatogram, will indicate the presence of a composite signal, since the mass chromatograms of the specific signals maximize at two different times. As such, the technique provides only qualitative information. It also indicates the time (i.e., the scan number) at which one component predominates. If the overlap is not severe, good estimates of the parent spectra can be obtained. Dromey and co-workers (22) considered the specific mass chromatograms as the elution profile of the individual analytes. By modeling the mass chromatograms, the contributions of the background, as well as the interfering analyte(s), can be removed from the original data, thus resolving the chromatographic profile and the mass spectra. Successful application requires that the specific signals are known a priori to the analyst.

Resolution Using Eigenvectors Scores Space

Several workers have used factor analysis to estimate the spectra of the parent components in mixtures (23, 24). The procedures that were developed require that specific signals are present, and that they are known to the analyst a priori. Lawton and Sylvester (25) have shown that the information obtained from factor analysis of UV-Vis data can be used to estimate the spectra of the independent components. If the spectra of mixtures of two components are factor analyzed, two major eigenvectors, $\mathbf{V}_1$ and $\mathbf{V}_2$, are obtained. All experimental spectra are linear combinations of these vectors. If $\mathbf{S}_i$ denotes the ith experimental spectrum, then

$$\mathbf{S}_i = \alpha_i \mathbf{V}_i + \beta_i \mathbf{V}_2, \tag{5.52}$$

where α_i and β_i are coordinates in the factor scores space representing $\mathbf{S}_i$. The coefficients α_i and β_i are also known as the scores of $\mathbf{S}_i$. The parent spectra, $\mathbf{p}_1$ and $\mathbf{p}_2$, are also linear combinations of $\mathbf{V}_1$ and $\mathbf{V}_2$:

$$\mathbf{p}_1 = \alpha_1^0 \mathbf{V}_1 + \beta_1^0 \mathbf{V}_2 \quad \text{and} \quad \mathbf{p}_2 = \alpha_2^0 \mathbf{V}_1 + \beta_2^0 \mathbf{V}_2. \tag{5.53}$$

Estimates of $\mathbf{p}_1$ and $\mathbf{p}_2$ can be computed once the region in the factor scores space that contains α_1^0, β_1^0, α_2^0, and β_2^0 is restricted. The procedure of Lawton and Sylvester (25, 26) is the most general approach available to analyze binary mixtures. The case where specific signals are present is, indeed, a special case in the above general solution. Qualitative resolution is achieved when the solution is restricted to only scores that yield physically meaningful spectra (e.g., non-negative elements in $\mathbf{p}_1$ and $\mathbf{p}_2$). It is possible to obtain a band for the parent spectra. It has been shown that specific signals are not required for adequate estimation of parent spectra (7). Sharaf and Kowalski (14) have shown that, for binary mixtures, the factors' scores space contains information pertaining to the fractional composition of the parent component in each mixture spectrum. This has made it possible to resolve the total signal quantitatively. The method of multivariate curve resolution (14, 26) was shown to be more accurate than multiple regression when experimental noise is present. The method is based on extracting relevant information from the scores' space and achieves qualitative and quantitative resolution of binary mixtures without any simplifying assumptions. It is also independent of the shapes of the elution profiles. This makes it the most general technique available to tackle binary mixtures. Since extensions of this resolution method are in great demand, the student should consult the recent literature for further advances. (If the data on p. 40 (Chapter 2) are factor analyzed, a plot of the scores will indicate that the third method is different from the other four.)

Other Resolution Techniques

There are many strategies available to the analyst for resolution of overlapping signals (1, 27). Abramson (28) has introduced reverse searching of spectral libraries. In this approach, the spectra of reference compounds are "searched for" in the spectrum of a suspected mixture. McLafferty and co-workers (29) have used this method in the identification of unknown mixtures, relying on a mass spectral library. After a best match is obtained, spectral stripping is used to subtract the suspected amount of the best match from the original spectrum. The residual, **R**, is given as

$$\mathbf{R} = \mathbf{E} - \alpha\mathbf{B}, \tag{5.54}$$

where $\mathbf{R}$ = residual spectrum
$\mathbf{E}$ = experimental spectrum of a suspected mixture
$\mathbf{B}$ = best match for $\mathbf{E}$ found by reverse search of mass spectral library and
α = estimated fraction of $\mathbf{B}$ in $\mathbf{E}$

The value of α in Equation (5.54) is determined empirically by testing several mixtures (29). The residual, **R**, can consequently be identified by reverse search. In theory, this process can be repeated until all components in the mixture are identified. Consecutive spectral subtraction, however, can not be performed indefinitely in practice. Differences between library spectra and experimental spectra of the same compounds, as well as the nonorthogonality of the spectra, introduce errors as early as the second stage in the process. Equation (5.54) often yields negative entries in **R** which have to be set to zero (29).

Subtractive techniques have been used in electrochemistry and spectroscopy. The signal of a suitable blank (the contributions of the background and interfering processes) is subtracted from the analytical signal of interest. Preparing a suitable blank, however, may be a difficult task. In electrochemistry, the supposedly identical working solutions and electrodes may be different in crucial surface characteristics. Wang and Dewald (30) have demonstrated a subtractive technique that requires only one working electrode in combined flow injection analysis (FIA)—anodic stripping voltammetry (ASV). In order to remove the contribution of HCOONa from the signal of HCOONa–DCOONa mixture, Kakihana and Okamoto (31) placed HCOONa in the reference cell of a double-beam infrared spectrometer. They were thus able to obtain the signal of DCOONa.

REFERENCES

1. Ildiko E. Frank and Bruce R. Kowalski, "Chemometrics," *Anal. Chem.* **54**, 232R–243R (1982).

   This review of the developments in this field, include mathematical modeling, calibration, resolution, factor analysis, and pattern recognition.

2. B. G. M. Vandeginste and L. DeGalan, "Critical Evaluation of Curve Fitting in Infrared Spectrometry," *Anal. Chem.* **47**, 2124–2132 (1975).

   This discusses the effects of peak separation, peak band widths, and relative compositions on the detection and resolution of overlapped bands by derivative functions.

3. T. C. O'Haver and G. L. Green, "Numerical Error Analysis of Derivative Spectrometry for the Qualitative Analysis of Mixtures," *Anal. Chem.* **48**, 312–318 (1976).

   The utility of derivative functions for quantitative analysis is examined. The effects of random noise, peak separation, and relative composition are evaluated.

4. Gerhard Talsky, Lotar Mayring, and Haus Kreuzer, "Higher-Order Derivative Spectrophotometry for the Fine Resolution of UV/VIS Spectra," *Angew. Chem. Int. Ed. Engl.* **17**, 532–533 (1978).

   A fourth derivative function is used to resolve the spectrum of Congo Red. A fourth derivative function is also used to distinguish between trypsin and chymotrypsin.

5. E. R. Malinowski and D. G. Howery, *Factor Analysis in Chemistry*, Wiley, New York, 1980.

   An overview of the fundamentals of factor analysis.

6. R. J. Rummel, *Applied Factor Analysis*, Northwestern University Press, Evanston, Ill., 1970.

   A comprehensive treatment of the theory and practice of factor analysis.

7. M. A. Sharaf and B. R. Kowalski, "Extraction of Individual Mass Spectra from Gas Chromatography-Mass Spectrometry Data of Unseparated Mixtures," *Anal. Chem.* **53**, 518–522 (1981).

   The eigenvectors of mixture spectra data matrices are used to calculate spectral bands for the parent components in binary mixtures.

8. Jyrki K. Kauppinen, Douglas J. Moffatt, Henry H. Mantsch, and David G. Cameron, "Fourier Transforms in the Computation of Self-Deconvoluted and First-Order Derivative Spectra of Overlapped Band Contours," *Anal. Chem.* **53**, 1454–1457 (1981).

   Spectral bands are deconvoluted using Lorentzian profiles as intrinsic line shapes.

9. P. Gans, "Numerical Methods for Data-Fitting Problems," *Coord. Chem. Rev.* **19**, 99–124 (1976).

An excellent review of the various techniques that are available for the solution of nonlinear least-squares problems. The techniques include the grid method, simplex optimization, Hooke and Jeeves technique, and gradient procedures.

10. Philip R. Bevington, *Data Reduction and Error Analysis for the Physical Sciences*, McGraw-Hill, New York, 1969, Chap. 11.

    A discussion of the methods of nonlinear least-squares curve fitting. Four procedures are used to solve one problem and their performances are compared.

11. Yonathan Bard, *Non-linear Parameter Estimation*, Academic Press, New York, 1974.

    A detailed mathematical treatment of nonlinear curve fitting.

12. Richard Bronson, *Operations Research*, McGraw-Hill, New York, 1982, Chap. 11.

    A treatment of nonlinear programming. Some of the popular methods, such as Newton–Raphson and Fletcher–Powell, are demonstrated.

13. Norman Draper and Harry Smith, *Applied Regression Analysis*, 2nd ed., Wiley, New York, 1981, Chap. 2.

    A discussion of the linear algebraic approach to linear regression. Analysis of variance used to estimate the adequacy of models.

14. M. A. Sharaf and B. R. Kowalski, "Quantitative Resolution of Fused Chromatographic Peaks in Gas Chromatography–Mass Spectrometry," *Anal. Chem.* **54**, 1291–1296 (1982).

    The scores obtained from factor analysis of mixture spectra are used to calculate mole fractions of the components. This results in the quantitative resolution of mixtures.

15. I. M. Warner, E. R. Davidson and G. D. Christian, "Quantitative Analysis of Multicomponent Fluorescence Data by the Methods of Least Squares and Non-negative Least Sum of Errors," *Anal. Chem.* **49**, 2155–2159 (1977).

    Fluorescence spectra are resolved using multiple regression (when all components are known) and linear programming (when only one of the components is known). Simplex optimization and multiple regression are compared.

16. D. D. Tunnicliff and P. A. Wadsworth, "A Step-wise Regression Program for Quantitative Interpretation of Mass Spectra," *Anal. Chem.* **37**, 1082–1085 (1965).

    Spectra are chosen from a reference library such that their linear combination approximate a mixture spectrum. The selection of the reference spectra proceeds in a stepwise fashion (i.e., one spectrum is chosen at a time).

17. D. J. Leggett, "Numerical Analysis of Multicomponent Spectra," *Anal. Chem.* **49**, 276–281 (1977).

    Analysis of multicomponent samples by non-negative least-squares and simplex optimization. The former method is compared with multiple regression. Negative concentrations are obtained when multiple regression is used.

18. Donald W. Fausett and James H. Weber, "Mass Spectral Pattern Recognition via Techniques of Mathematical Programming," *Anal. Chem.* **50**, 722–731 (1978).

    A subset of a reference library is chosen such that a linear combination of the members of the subset approximates an unknown mixture spectrum. Criteria other than least squares are used.

19. C. N. Ho, G. D. Christian, and E. R. Davidson, "Application of the Method of Rank Annihilation to Quantitative Analyses of Multicomponent Fluorescence Data from the Video Fluorometer," *Anal. Chem.* **50**, 1108–1113 (1978).

    The spectrum of a component is removed from a matrix of mixture spectra. Successful removal is indicated by the reduction of the rank of the matrix containing the mixture spectra.

20. C. N. Ho, G. D. Christian, and E. R. Davidson, "Simultaneous Multicomponent Rank Annihilation and Applications to Multicomponent Fluorescent Data Acquired by the Video Fluorometer," *Anal. Chem.* **53**, 92–98 (1981).

    This is a generalized rank annihilation technique. A Fletcher–Powell optimization technique is used to minimize the eigenvalues.

21. J. E. Biller and K. Biemann, "Reconstructed Mass Spectra—A Novel Approach for the Utilization of Gas Chromatograph–Mass Spectrometer Data," *Anal. Lett.* **7**, 515–528 (1974).

    Mass chromatograms are used to detect unresolved components. Specific masses are then used to reconstruct the mass spectrum of each component.

22. R. G. Dromey, Mark J. Stefik, Thomas C. Rindfleisch, and Alan M. Duffield, "Extraction of Mass Spectra Free of Background and Neighboring Component Contributions from Gas Chromatography–Mass Spectrometry Data," *Anal. Chem.* **48**, 1368–1375 (1976).

    The mass chromatograms obtained by the Biller–Biemann method are modeled. The mathematical model is used to filter the signal from the background and interfering components.

23. Fritz J. Knorr and J. H. Futrell, "Separation of Mass Spectra of Mixtures by Factor Analysis," *Anal. Chem.* **51**, 1236–1241 (1979).

    A data matrix containing mixture mass spectra is factor analyzed. If the parent components possess specific signals, then the factor loadings can be used to compute the parent spectra.

24. John M. Halket, "Numerical Analysis of Chromatographic–Spectrometric Data," *J. Chromatogr.* **186**, 443–455 (1979).

    Factor analysis is used to analyze mixture spectra obtained in combined high-performance liquid chromatography (HPLC)–ultraviolet spectrometry

(UV). GC–MS data are also used. The spectra of the parent components are extracted mathematically.

25. William H. Lawton and Edward A. Sylvester, "Self-Modeling Curve Resolution," *Technometrics* **13**, 617–633 (1971).

    UV-Vis spectra of mixtures are factor analyzed. The eigenvectors are used to estimate the spectra of the parent components. The conditions needed to obtain solution bands are discussed.

26. Muhammad Abdallah Sharaf, PhD. Dissertation, University of Washington, Seattle, 1982.

    A detailed treatment of the method of multivariate curve resolution. The technique is used for the qualitative and quantitative resolution of mixtures using GC–MS data.

27. Bruce R. Kowalski, "Chemometrics," *Anal. Chem.* **52**, 112R–122R (1980).

    A review of the developments in resolution and other applications of chemometrics from 1976 to 1979.

28. Fred P. Abramson, "Automated Identification of Mass Spectra by the Reverse Search," *Anal. Chem.* **47**, 45–49 (1975).

    A discussion of forward and reverse search strategies. A reverse search program is developed and its performance is evaluated.

29. Barbara L. Atwater (Fell), Rengachari Venkataraghavan, and Fred W. McLafferty, "Matching of Mixture Mass Spectra by Subtraction of Reference Spectra," *Anal. Chem.* **51**, 1945–1949 (1979).

    A reference library is searched for a best match for a mixture spectrum. The best match is consequently subtracted and the residual is analyzed.

30. Joseph Wang and Howard D. Dewald, "Subtractive Anodic Stripping Voltammetry with Flow Injection Analysis," *Anal. Chem.* **56**, 156–159 (1984).

    The background signal is subtracted from the signal of interest in combined flow injection analysis–anodic stripping voltammetry (FIA–ASV). One electrode is used in order to minimize the effects of unmatched working electrodes.

31. Masato Kakihana and Makoto Okamoto, "Application of Spectral Subtraction Technique to Solid Samples of HCOONa and CH_3COONa with Small Amounts of D-Variants," *Appl. Spectrosc.* **38**, 66–67 (1984).

    The spectra of HCOONa and DCOONa are resolved by placing the former in the reference cell of a double-beam spectrometer.

SUGGESTED READINGS

Barry E. Barker and Malcolm Fox, "Computer Resolution of Overlapping Electronic Absorption Bands," *Chem. Soc. Rev.* **9**, 143–184 (1980).

A review of the methods of band resolution and their applications. The techniques include differential methods such as PEAK and LOGDIFF and least-squares procedures.

Dana M. Barry and Louis Meites, "Titrimetric Applications of Multiparametric Curve-Fitting," *Anal. Chim. Acta* **68**, 435–445 (1974).

The data obtained during a titration are modeled according to the theoretical titration equation. Accurate estimation of end points is thus possible.

Thomas A. Brubaker and Kelly R. O'Keefe, "Nonlinear Parameter Estimation," *Anal. Chem.* **51**, 1385A–1388A (1979).

A discussion of nonlinear curve fitting emphasizing gradient methods. Examples from kinetics and spectroscopy are given.

Joan Grimault, Hotensia Iturriaga, and Xavier Tomas, "The Resolution of Chromatograms with Overlapping Peaks by Means of Different Statistical Functions," *Anal. Chim. Acta* **139**, 155–166 (1982).

The performances of gamma, log-normal, and Weibull distributions are examined.

Peter Jochum and Erich Schrott, "Deconvolution of Multicomponent Ultraviolet-Visible Spectra," *Anal. Chim. Acta* **157**, 211–226 (1984).

A non-negative least-squares procedure is developed to resolve UV-Vis spectral data.

A. Lorber, "Features of Quantifying Chemical Composition from Two-dimensional Data Array by the Rank Annihilation Factor Analysis Method," *Anal. Chem.* **57**, 2395 (1985).

Jeffrey T. Lundeen and Richard S. Juvet, Jr., "Quantitative Resolution of Severely Overlapping Chromatographic Peaks," *Anal. Chem.* **53**, 1369–1372 (1981).

Chromatographic signals are modeled as polynomials in concentrations. Chromatographic peaks are resolved by solving nonlinear equations simultaneously.

W. F. Maddams, "The Scope and Limitations of Curve Fitting," *Appl. Spectrosc.* **34**, 245–267 (1980).

Curve fitting of spectral data is discussed in terms of the number of bands present, peak separation, peak shape, and peak bandwidth.

J. R. Morrey, "On Determining Spectral Peak Positions From Composite Spectra with a Digital Computer," *Anal. Chem.* **40**, 905–914 (1968).

Derivative functions are used to examine the complexity of overlapping signals.

David W. Osten and Bruce R. Kowalski, "Multivariate Curve Resolution in Liquid Chromatography," *Anal. Chem.* **56**, 991–995 (1984).

HPLC–UV data are analyzed by the multivariate curve resolution technique. The performance of the method is compared with the perpendicular drop method for quantitative analysis.

David W. Osten and Bruce R. Kowalski, "Background Detection and Correction in Multicomponent Analysis," *Anal. Chem.* **57**, 908 (1985).

J. W. Perram, "Interpretation of Spectra," *J. Chem. Phys.* **49**, 4245–4246 (1968). A Gaussian peak is decomposed into two Gaussian peaks!!

E. Sanchez and B. R. Kowalski, "Generalized Rank Annihilation Factor Analysis," *Anal. Chem.*, accepted for publication in *Anal. Chem.*, January, 1986.

CHAPTER

6

EXPLORATORY DATA ANALYSIS

In this chapter we will turn our attention to the part of the analytical process of Figure 6.1 in which chemical sense is made of data once they are obtained. It is worthwhile at this point to review our progress in this scheme of things. The goal of the analytical process is to learn something about the properties or behavior of a system. The starting point, and the first place at which bias is introduced, is to take samples to represent the system. Techniques from sampling theory discussed in Chapter 1 are used to ensure as representative a sampling scheme as possible. Then, we probe the samples in a variety of ways, making measurements, that is, recording signals of one form or another such as voltages and currents (Chapter 3), using optimization techniques presented in Chapter 2 and again in Chapter 7 to maximize or enhance the responses. These measurements are in turn related to chemical information via the resolution and calibration steps discussed in Chapters 4 and 5. All of these techniques discussed so far help to provide a solid foundation from which the next step is taken: to extract knowledge from the information about the system.

This is a relatively new endeavor for many chemists concerned with making measurements. Traditionally, analytical chemists have concerned themselves with the steps of making measurements and relating them to chemical information using very straightforward calibration methods. Data interpretation and even instrument optimization have been the domains of statisticians, engineers, and managers. Such a division of labor was no doubt necessary before the advent of highly computerized laboratory instrumentation when obtaining a single datum was time-consuming and costly. However, the microprocessor revolution has enabled chemists to acquire and store great quantities of data easily and cheaply; the bottleneck has become data processing and interpretation. Instead of merely generating data, analytical chemists are becoming chemical problem solvers and are using multivariate data analysis methods to uncover the meaning of the chemical information they produce.

Finally, it is important to bear in mind that bias is inescapably introduced all along the way in the analytical process of Figure 6.1. The

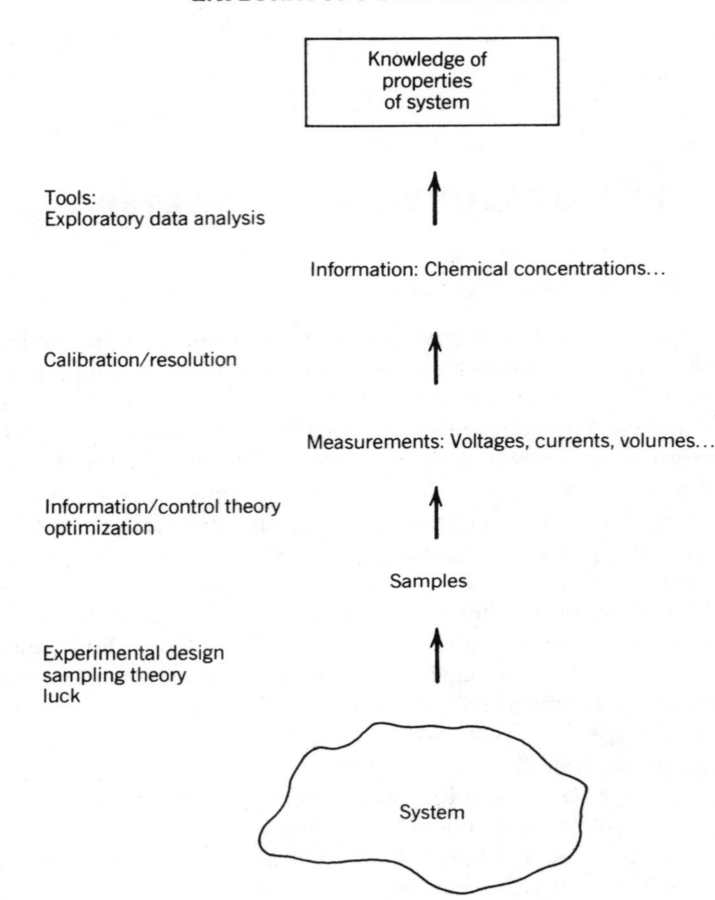

Figure 6.1 The analytical process.

ultimate success of multivariate data analysis thus relies heavily on the integrity of each of these steps.

MULTIVARIATE LEVERAGE

The broad and well-developed field of *univariate statistics* concerns itself with one-dimensional distributions of points. In other words, given a set of samples, one type of measurement is made on each sample and these values can be plotted as a frequency histogram or on a number line and characterized in a variety of ways as discussed in Chapter 1. As such, it is useful for *confirmatory* but not *exploratory* purposes, since no information is gained about how different types of measurements are related. At the risk of understatement, nature is a complex web of interrelated factors and processes, and so it is only reasonable that to understand it, we have to study more than one thing at a time.

The values of a measurement, m_1, for a set of samples are plotted in Figure 6.2. In this plot, the axis represents all possible values of measurement m_1 and thus each point positions each sample along the axis according to the value of the measurement that sample has.

As additional measurements are made on the set of samples, we can begin to study the *covariance* of the measurements, and the potential to extract more useful knowledge from the data is greatly enhanced. For example, if the results of a second measurement, m_2, were available, the

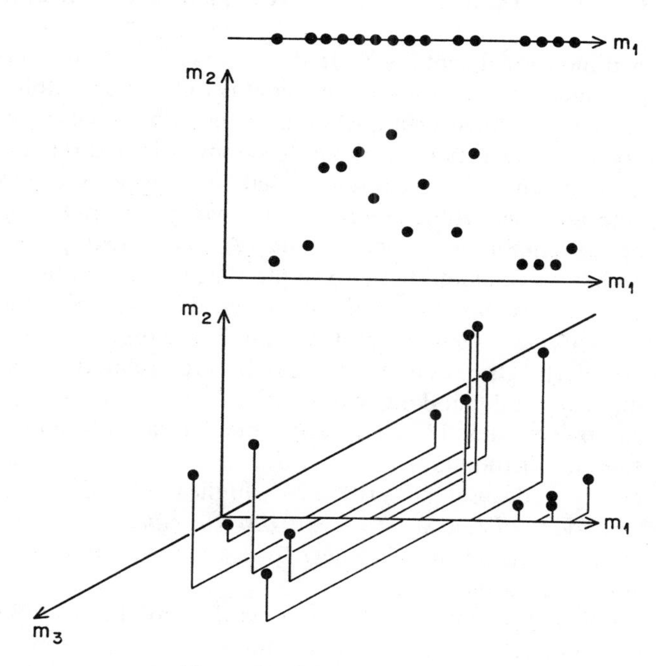

Figure 6.2 Multivariate leverage.

samples in Figure 6.2 would be plotted as points in the (m_1, m_2) plane. Techniques of analyzing such plots are quite familiar, ranging from visual inspection for outliers and groups to regression analysis or the Fourier transform. In this particular case, the two-dimensional plot seems to suggest the existence of two distinct groups of samples. If a third measurement were made, we could attempt a projection of a 3-D plot on 2-D paper, or perhaps use computer graphics to project and rotate the axes to get a good look at the *data structure*.

If three dimensions are good, then four or more dimensions ought to be better—except that we cannot *see* what we are doing. The operations we do routinely in 2- and 3-space can be extended to any number, n, of dimensions with the help of the computer. For the sacrifice of working in hyperspace we gain the ability to study the covariance of measurements, the key to multivariate data analysis as we shall see in the following sections.

CATEGORY VERSUS CONTINUOUS PROPERTY DATA

Given an n-dimensional data set, that is, a set of samples with n measurements made on each sample, our goal is to learn something about a property or behavior, answer a question, or test a hypothesis about the system. The type of property or question to be investigated influences the selection of data analysis techniques that are appropriate to use. Generally, data analysis problems are one of two types: involving either *category data* or *continuous property data*. A *continuous property* is a dependent variable associated with a sample that has a continuous range of possible values. A familiar example of a data analysis technique appropriate for this case would be multivariate linear regression where the continuous property (the dependent variable) is related to a set of measurements (the independent variables). A method of continuous property data analysis called Partial Least Squares Path Modeling will be covered at the end of the chapter.

A *category* is a designated group of samples. Categories that are entirely independent of one another are *discrete categories*, while those with some dependence on one another, such as "low, middle, high" are called *continuous categories*.

Some examples of samples, the categories they might be divided into and the criterion of interest are the following:

Samples	Categories	Classification Criterion
Oxygen-containing organic molecules	Alcohol, nonalcohol	Have —OH functional group?
Oxygen-containing organic molecules	Alcohol, ester, aldehyde ketone, ether	Type of oxygen functional group?
Orange juices	Florida, California, Brazil	Source of orange?
Orange juices	Frozen, fresh, other	Condition?
Electronic parts	Good, faulty	Performance of part?

Note that samples can be a wide range of objects and they may be classified in any way depending on the property of interest. The same set of samples may be grouped in more than one way.

In category-type data analysis, the property of interest about the system relates to the category membership of the data, which is obviously not amenable to a regression approach as used for continuous properties. Pattern recognition, a collection of techniques designed to assist a human in analyzing n-dimensional data, is especially designed for category data analysis problems.

PATTERN RECOGNITION: THE APPROACH

As a branch of artificial intelligence, pattern recognition began as a collection of computer tools to solve classification-type problems in a variety of areas. *Artificial intelligence* is the field that is concerned with using computers to mimic or model human intelligence (see Suggested Readings).

To compete with the human's ability to recognize patterns in only two or three dimensions is a fairly difficult task. The human being is a formidable pattern recognizer in this domain. Consider for the moment how easily and quickly we recognize objects and distinguish people, and imagine instructing a computer to reproduce that ability. An interesting study of how we recognize faces and how a computer might be instructed to do so can be found in reference (9) of the Suggested Readings.

In terms of large tables of numbers with many samples and many measurements, however, the human's ability to "see" patterns in the data is very poor. Even though the potential to learn about a system increases with an increasing number of samples and measurements, our ability to get at the information diminishes without the help of the computer.

A classic example and one of the earliest uses of pattern recognition is the recognition of handwritten and printed alphanumeric characters. Speech recognition, speaker recognition, fingerprint identification, radar and sonar signal processing, electrocardiogram analysis, weather forecasting, particle tracking, and stock market analysis are only some of the many applications of pattern recognition techniques. Regardless of the type of application, there is a central goal and strategy common to them all that can be stated in the following way.

Given a collection of objects characterized by a set of measurements made on each object, the goal is to find and/or predict a property of the objects that is not directly measurable itself (or for whatever reason, we do not choose to measure the property itself) but that is thought to be indirectly related to the measurements via some unknown or undetermined relationship.

It was not until the late 1960s that this approach was actually applied to chemical problems (1–3). It is instructive to interpret this approach in a chemical context and to define certain terms used in chemical applications. By *collection of objects* we usually mean many *samples* taken to represent a system. They could be anything—chemical compounds, water samples, protein molecules, or people. A *variable* is a facet or an aspect of a sample that can be measured, that is, it represents an axis in n-dimensional space. A *measurement* is the experimentally determined value for a variable used to characterize the samples. A variable could be the concentration of iron in a sample, refractive index, heat of fusion, absorbance at 254 nanometers, and so on. A measurement would therefore be the experimentally obtained numerical value of iron concentration, for example.

Properties that are not directly measurable might be the quality of a product, source of a material, or the environmental impact of a pollutant. It is also desirable to predict a property at an early advantageous time even though it is actually measurable, such as predicting the performance of a mechanical part via nondestructive analyses. Other examples would be predicting the consumer acceptance of a product before it is put on the market, the biological activity of an organic molecule before it is synthesized, or the presence of disease in a human being early and without invasive procedures.

Often one of the most challenging aspects of solving such a chemical problem—especially for the novice—is to express the problem in terms of the general pattern recognition framework discussed above. For this reason a variety of examples on pattern recognition applications will be presented throughout this chapter.

In addition to the goal or properties of interest in a study, there are statistical considerations that determine the type of techniques to be used. The result of a classification problem, be it "yes" or "no," "Class A" or "Class B," "Florida orange" or "California orange," can only be associated with a probability of being correct when we have a wealth of statistical information about the system to begin with. *Parametric* methods depicted in the schematic diagram in Figure 6.3 of pattern recognition depend on knowing or estimating the probability density functions of the

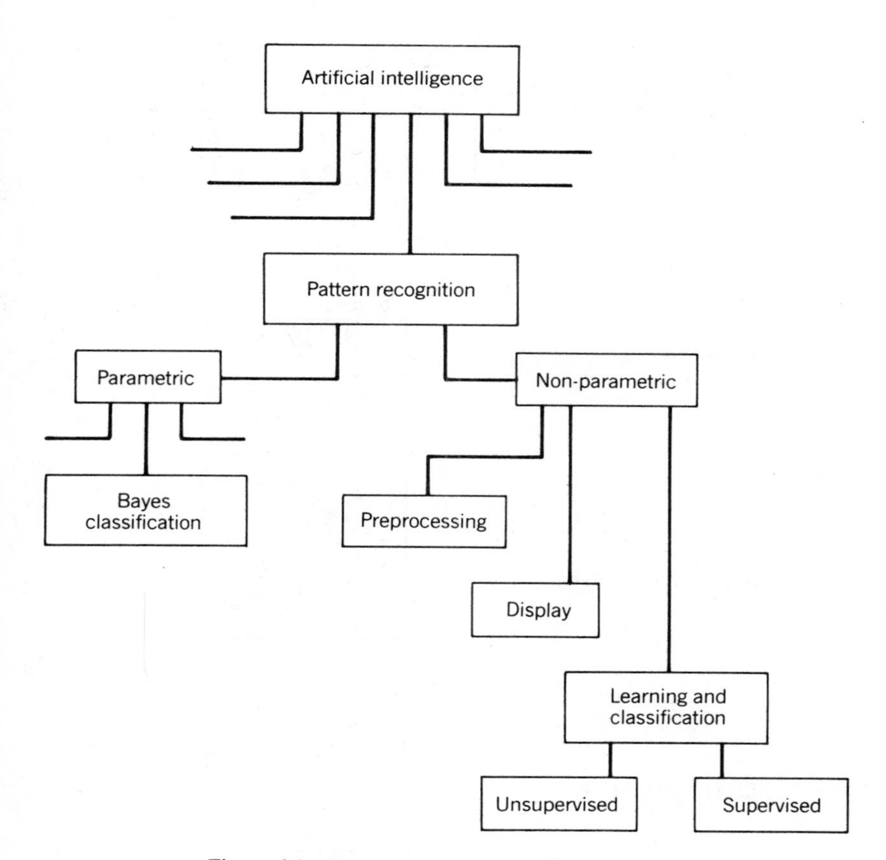

Figure 6.3 Branches of pattern recognition.

categories and thus, require an extremely large data base. Because in most practical applications this information is not available, *non-parametric* techniques are used that make no assumptions about the underlying statistical distribution of the data. However, no probabilistic information *in* means none *out* either—classification results are not accompanied by any confidence estimates.

The next branch in the pattern recognition scheme of Figure 6.3 occurs to address the issue of the form or representation of the data, which must be conformable with the pattern recognition algorithms. *Preprocessing techniques* are designed to transform the data into the most informative representation in the context of the goal of the study.

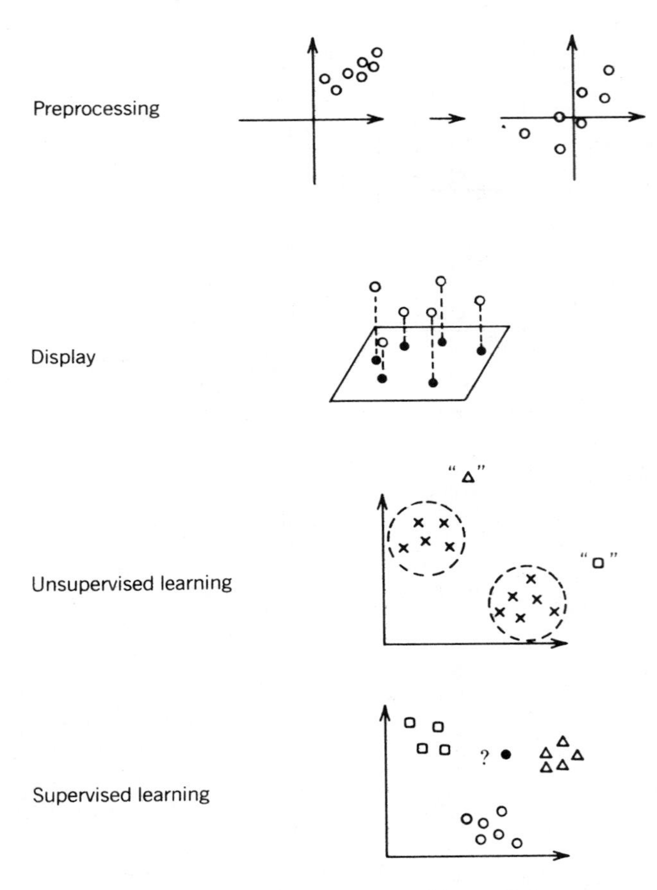

Figure 6.4 Types of pattern recognition techniques.

The data analyst may then choose to *display* the data in two, or with computer graphics, possibly three, dimensions for human inspection but still preserving the maximum amount of information about the structure of the data base as it exists in the n-dimensional space.

Unsupervised learning refers to methods that make no a priori assumptions about category-membership of the samples, but rather assist the analyst in uncovering intrinsic clusters or other patterns in the data.

Conversely, in *supervised learning* the computer "learns" to optimally classify the samples based on advance knowledge about their category membership. The goal of such a study is to develop a classification rule or algorithm in order to either simply test the validity of the hypothesis, or beyond that, to be able to classify an unknown.

These pattern recognition methods are schematically depicted in Figure 6.4.

Finally, the success of a pattern recognition study depends on a number of assumptions. Obviously, we must have multiple samples from a system with multiple measurements consistently made on each sample. For some techniques the system should be well *overdetermined*, that is, the ratio of number of samples to number of measurements should be at least three. Furthermore, these techniques operate under the assumption that *the*

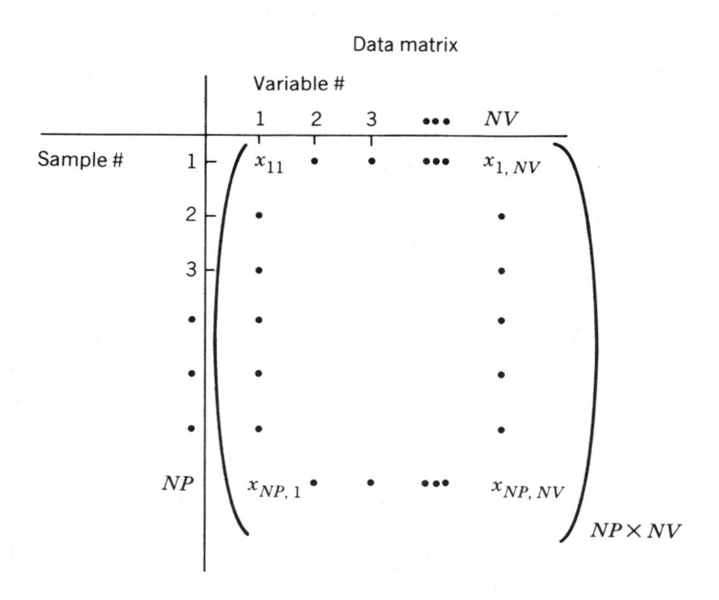

Figure 6.5 Construction of the data matrix.

nearness of points in n-space faithfully reflects the similarity of the properties of the samples. Last but not least, the data can be arranged in a data matrix, with one row per sample and the entries of each row being the measurements made on the sample, as shown in Figure 6.5. The information needed to answer our questions must indeed be *implicitly contained* in the data matrix, and the data representation must be conformable with the pattern recognition algorithms we use.

PREPROCESSING TECHNIQUES

Preprocessing refers to *any* transformation of the original data. After any transformation, a variable is thenceforth referred to as a *feature* to distinguish it from the original variable. This convention is in keeping with our intuitive notion of a feature as simply a characteristic of an object. Examples of preprocessing techniques range from simple scalar multiplication to logarithmic transformation to eigenanalysis—in other words, anything done to the original data to enhance the information representation. The type of preprocessing performed on a data set depends on the context of the problem: the goals/questions involved, any physical or chemical factors, even common sense. It also depends on whether the data are *single-source*, coming from the same instrument, or *multi-source*, coming from a variety of instruments. Spectroscopy is an example of the single-source case since measurements are obtained on a sample through digitizing a spectrum. In this case, transformations such as autocorrelation and Fourier may be useful. In the absence of any other information, a standard protocol is recommended and will be discussed below.

Missing Data

The best thing that can be said about missing data is—*don't have any*! A measurement designates an axis in n-space. In order to plot points all on the same graph and compare them, they must have a "coordinate" on each axis. Note that "zero" is *not* a suitable value to fill in for a missing datum. If data are missing, they must be filled in by appropriate techniques before any pattern recognition algorithms may be applied.

In the *mean fill* technique, a missing datum in the data set is given the mean value of the variable for the category to which the data vector belongs. A missing value in the test data would be filled with the mean of that variable over the entire set.

It is also possible to use a principal component model (see Eigenvector

Rotation) of a category to estimate missing data values. The *principal component fill* builds more correlation into the data, however.

The most conservative method, producing results that are probably *worse* than reality, is the *random fill* method. Data are filled in by random selection from the measurements for the appropriate category.

Regardless of the method used, filling in missing data will alter the data structure and is a limitation to the success of pattern recognition algorithms. The mean and principal component fills make categories look better than they are, and the random fill makes them look worse. Any solution to the missing data problem is merely finding a way to hurt oneself the least. Also, the nature and purpose of the study will certainly influence the choice of a suitable technique.

Redundant/Constant Variables

The data set should be checked for constant and redundant variables before attempting pattern recognition since variance is required in the data. Usually redundant variables are detected by examining the correlation matrix [see Eq. (5.17)] for "high" coefficients of correlation (e.g., 1.0).

Translation

The purpose of *translation* is to change the position of the data set with respect to the axes. Usually, it is desirable to have the origin coincide with the mean of the data set. Thus, to *mean-center* the data, let x_{ik} be the datum associated with the kth measurement on the ith sample; thus, this value will be located in the ith row and kth column of the data matrix. From this value is subtracted the mean over all the samples of the kth measurement, x_k:

$$x'_{ik} = x_{ik} - \bar{x}_k \tag{6.1}$$

and

$$\bar{x}_k = \frac{1}{NP} \sum_{i=1}^{NP} x_{ik}, \tag{6.2}$$

where NP represents the total number of samples.

This procedure is performed on all of the values of the data matrix $\mathbf{X}$ to produce a new data matrix $\mathbf{X}'$ whose variables are now referred to as *features*. An example of the effect of mean-centering on a data matrix is shown in Figure 6.6.

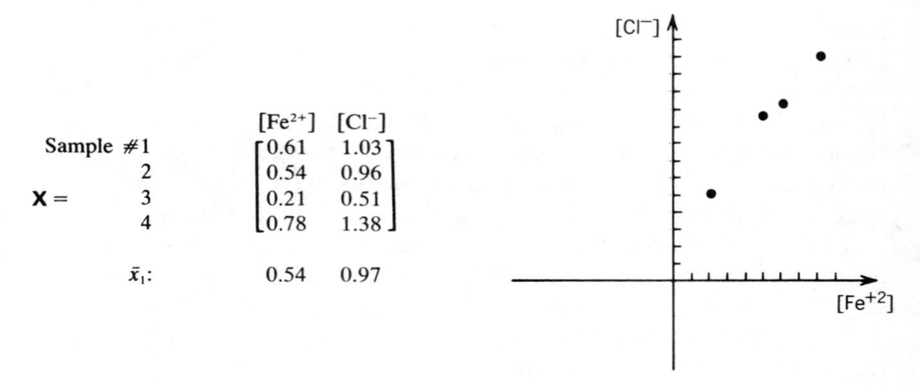

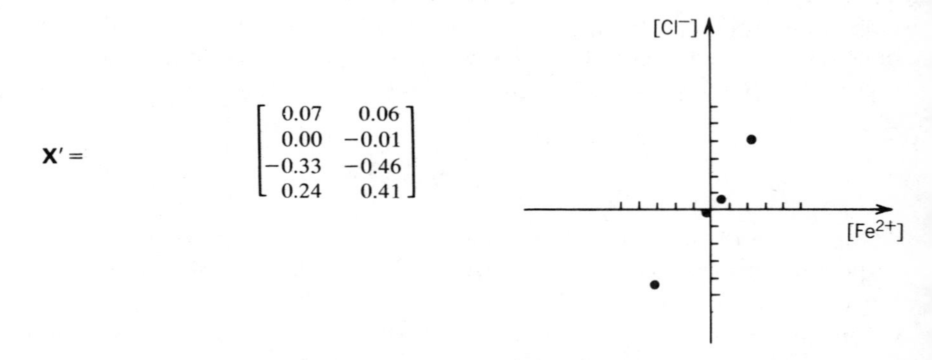

Figure 6.6 Mean-centering a data matrix.

Normalization

Normalization makes the lengths of all the data vectors in the data set the same, that is, the sum of the squares of the elements of a data vector is the same for all the samples in the entire data set. Let c_i be the sum obtained for an unnormalized sample i:

$$\sum_{k=1}^{NV} x_{ik}^2 = c_i. \tag{6.3}$$

To normalize the data vectors to the constant N, each element of the data vector would be multiplied by $N/\sqrt{c_i}$.

Familiar examples would be normalizing vectors to unit length or normalizing the area under a set of curves to unit area. It is thus a procedure performed over a single data vector, as opposed to translation, which is done within each variable over all the samples. Normalization effectively removes the variance in a data set due to arbitrary differences in magnitudes of a suite of measurements. It is applied as a preprocessing step when such an effect in the data set is not meaningful and would obscure the important variance.

Scaling

The notion of scaling is intuitive to anyone who has ever plotted some points on a piece of graph paper. Typically, our criteria for scaling are to fill the page and to show about equal magnitude of variation in the points on each axis so it "looks right." However, in pattern recognition, a more rigorous approach to scaling is needed.

In the absence of any a priori information the data should be scaled so

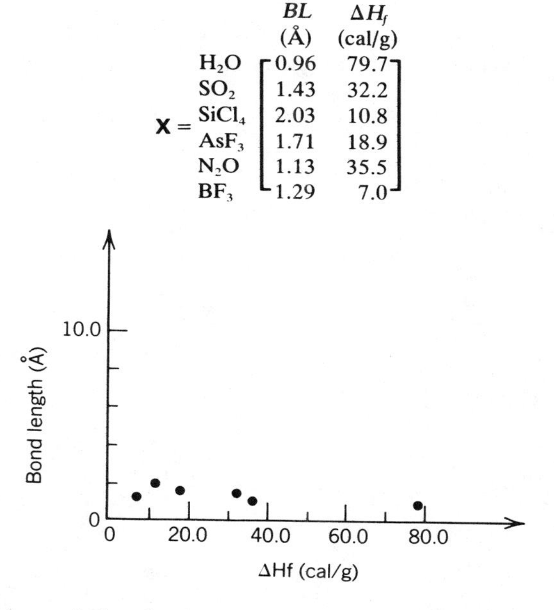

Figure 6.7 Arbitrary choice of units creates problems for unscaled data. In a plot of bond length versus heat of fusion for various molecules, the variance in bond length appears to be insignificant in comparison to the large variance in heat of fusion.

as to put all the variables on an equal footing in terms of their variance. Later, under Feature Weighting, we may choose to *weight* axes according to their ability to discriminate between categories.

As an example of the problems encountered with unscaled data, consider a plot of bond length versus heat of fusion for a set of molecules as shown in Figure 6.7. In terms of absolute numerical magnitude, the variation in bond length is very small in comparison to the variation in the heat of fusion. Due to our arbitrary choice of units, the latter variance would obscure the information contained in the bond length variable, which could be significant even though small.

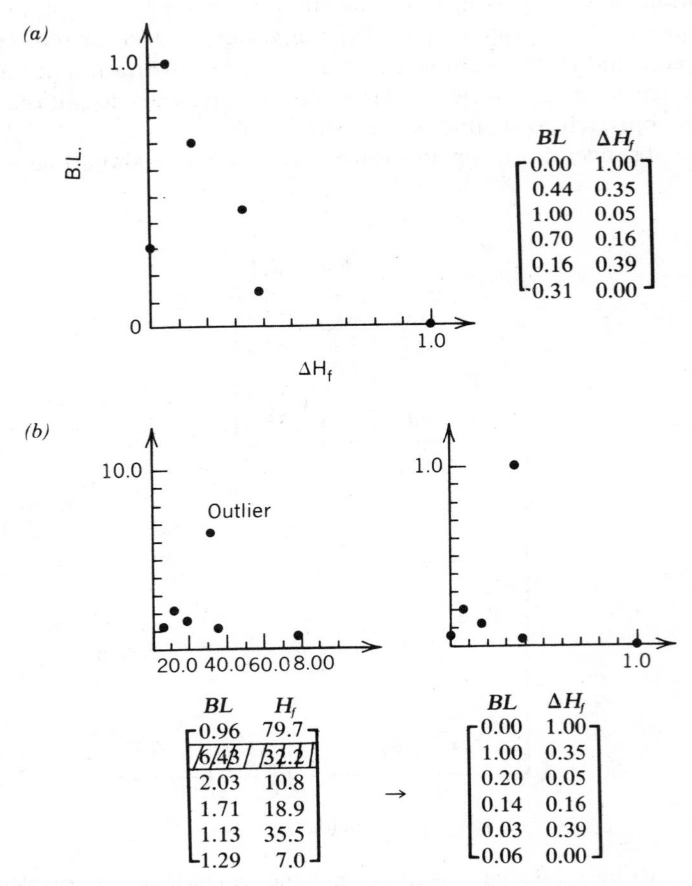

Figure 6.8 Range scaling: (*a*) Effect on data on Figure 6.7. (*b*) Results of range scaling if an outlier had been present in the data of Figure 6.7.

Range scaling is perhaps the most familiar form of scaling. It is done by placing the minimum value of each variable at the origin and dividing by the range of the data, $x_k(\max) - x_k(\min)$, so that the maximum value of each new feature is 1.0:

$$x'_{ik} = \frac{x_{ik} - x_k(\min)}{x_k(\max) - x_k(\min)}, \qquad 0.0 \leq x'_{ik} \leq 1.0. \tag{6.4}$$

This method is sensitive to the presence of outliers, however, as depicted in Figure 6.8.

Autoscaling

Autoscaling (2) removes any inadvertent weighting that arises due to arbitrary units, but is not as sensitive to outliers as range scaling. Two versions of autoscaling are in common practice in the chemometrics literature. Both involve mean-centering but differ in their scaling factor. The net effect is the same in both cases.

Autoscaling to unit variance refers to mean-centering followed by dividing by the standard deviation, s_k, on a variable by variable basis:

$$x'_{ik} = \frac{x_{ik} - \bar{x}_k}{s_k} \tag{6.5}$$

and

$$s_k = \left[\frac{1}{NP - 1} \sum_{i=1}^{NP} (x_{ik} - \bar{x}_k)^2\right]^{1/2}. \tag{6.6}$$

The reader may demonstrate that the variance of an autoscaled feature is equal to 1.0.

The other version of autoscaling in common use is

$$x'_{ik} = \frac{x_{ik} - \bar{x}_k}{\left[\sum_{i=1}^{NP} (x_{ik} - \bar{x}_k)^2\right]^{1/2}}, \tag{6.7}$$

which results in a variance for every feature of $1/(NP - 1)$; there are $NV/(NP - 1)$ units of variance in the entire autoscaled data set.

Figure 6.9 illustrates the effect of autoscaling on the bond-length—ΔH_{fusion} data for the case where an outlier is present. The choice of a scaling method depends on the context of the problem. In the absence of any information that would preclude its use, however, autoscaling is recommended for most applications.

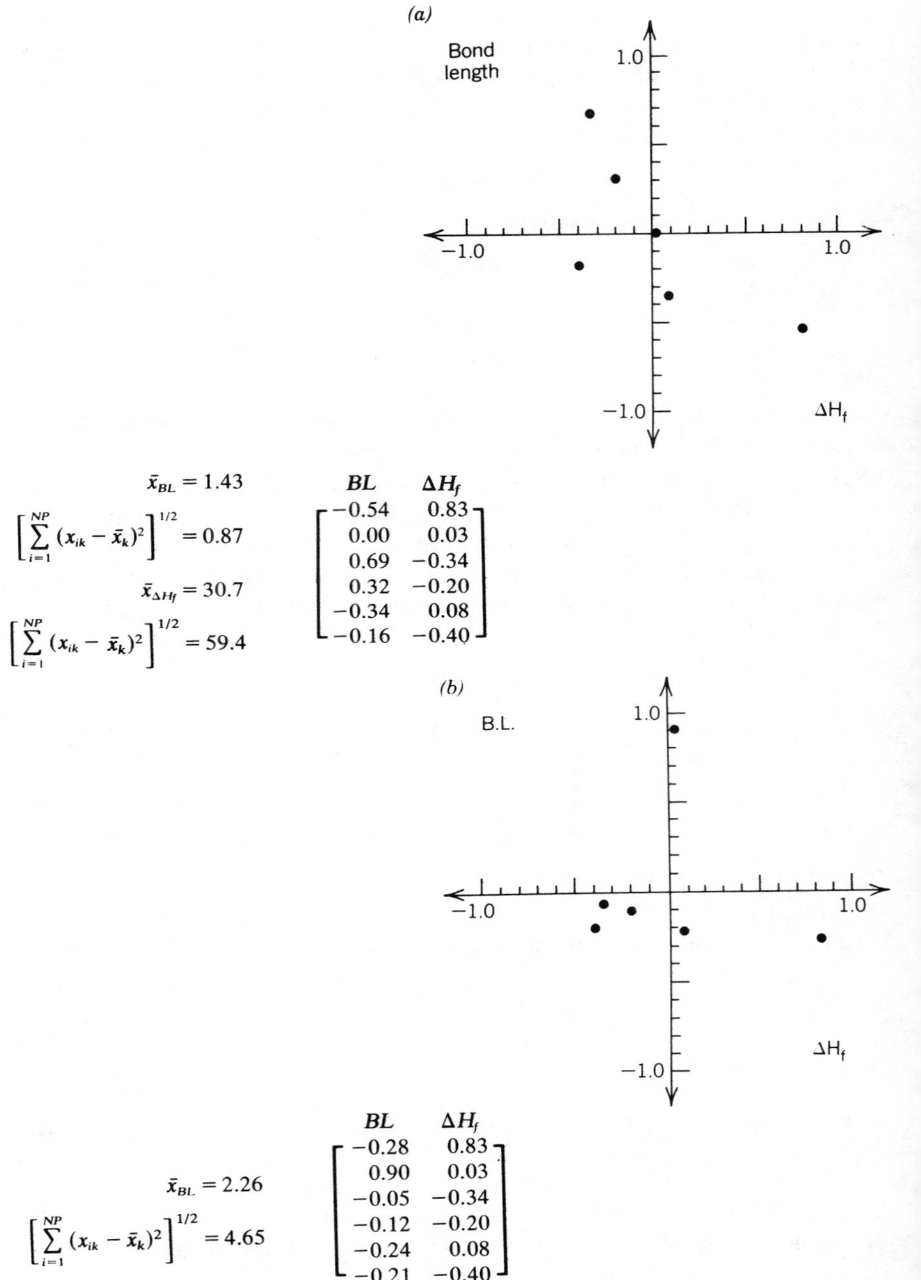

Figure 6.9 Graph of the autoscaled data of Figure 6.7: (*a*) No outlier; (*b*) outlier present.

Feature Weighting

The purpose of *feature weighting* is two-fold: (1) to provide a measure of the discriminating ability of a variable in terms of category separation, and (2) to improve classification results by "stretching," for example, scaling each axis according to its overall feature weight, w_k.

Given the frequency histogram with respect to feature x_k of two categories, I and II, in Figure 6.10, we would like to derive an expression for how well this feature separates the two curves. A very discriminating feature will yield widely separated distributions; conversely, the distributions will be very nearly superimposed for a poor one.

The *variance weight* for these categories, $w_k(\mathrm{I, II})$, is calculated as the ratio of the intercategory variances to the sum of the intracategory variances:

$$w_k(\mathrm{I, II}) = 2 \times \frac{(1/N_\mathrm{I}) \sum x_\mathrm{I}^2 + (1/N_\mathrm{II}) \sum x_\mathrm{II}^2 - (2/N_\mathrm{I} N_\mathrm{II}) \sum x_\mathrm{I} \sum x_\mathrm{II}}{(1/N_\mathrm{I}) \sum (x_\mathrm{I} - \bar{x}_\mathrm{I})^2 + (1/N_\mathrm{II}) \sum (x_\mathrm{II} - \bar{x}_\mathrm{II})^2}. \quad (6.8)$$

Such a value can be computed for all pairs $NJ = \frac{1}{2} NV(NV - 1)$ of features. It is desirable to compute a single, overall rating of each feature's discriminating ability, w_k:

$$w_k = \left[\prod_{j=1}^{NJ} w_k(j) \right]^{1/NJ}. \quad (6.9)$$

In this way, w_k is always greater than or equal to 1.0 units of variance even if one or more features have no discriminating ability.

Alternately, the *Fisher weight* replaces the numerator in Equation (6.8)

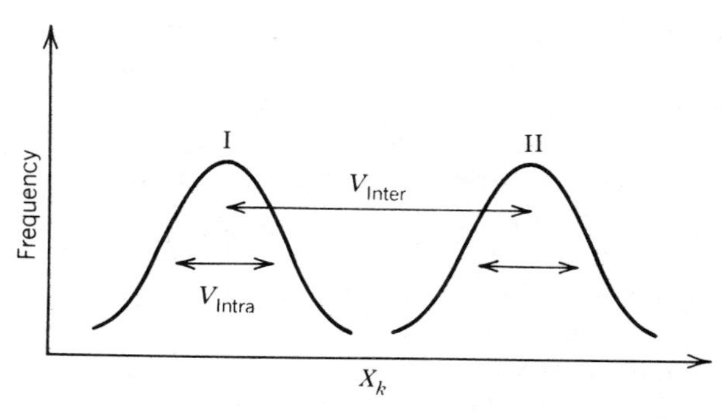

Figure 6.10 Frequency histogram of data in two categories for variable x_k.

with the difference of the category means:

$$w_k(\mathrm{I}, \mathrm{II}) = \frac{|\bar{x}_{\mathrm{I}} - \bar{x}_{\mathrm{II}}|}{(1/N_{\mathrm{I}}) \sum (x_{\mathrm{I}} - \bar{x}_{\mathrm{I}})^2 + (1/N_{\mathrm{II}}) \sum (x_{\mathrm{II}} - \bar{x}_{\mathrm{II}})^2}. \tag{6.10}$$

Since the Fisher weight may actually go to zero for a nondiscriminating feature, the overall weight is calculated as

$$w_k = \frac{1}{NJ} \sum_{j=1}^{NJ} w_k(j). \tag{6.11}$$

Feature weighting is of particular utility to the analytical chemist in revealing whether or not a given list of measurements actually contains the information necessary to solve a problem, and, if so, the minimum subset of the list that will do the job. For example, in forensic chemistry, the ability to quickly determine which measurements are necessary and sufficient to answer a question about a sample may save analysis cost, precious time, and may even "crack the case."

In an application, either the Fisher or Variance weights would be calculated and the axis scaled by these values to produce a new set of features x'_{ik}:

$$x'_{ik} = w_k x_{ik}. \tag{6.12}$$

The values of the weights depend on the classification objectives of the problem. In fact, it is not uncommon to first pose one question about the data, calculate the weights and proceed with the classification analysis, and then to pose an entirely different question on the same data and repeat the procedure.

The ability to quickly solve a classification problem is important in forensic analysis. Let us imagine that a scrap of white paper was found at the scene of a crime; detectives need to know what kind of paper it is in order to track down the suspect. The forensic chemist obtains 119 sheets of paper representing 40 different grades of paper from nine manufacturers (4). Thirteen elements are measured by neutron activation analysis and then the data are autoscaled. The chemist calculates the Variance and Fisher weights to find the most discriminating variables with respect to *paper grade*; these are shown in Table 6.1(a). These weights are exceptionally good and indicate that there is indeed sufficient information to distinguish between paper grades based on these chemical measurements, and moreover, perhaps several of the measurements made could be omitted from the analysis. We will see in Feature Selection in Classification how this can be done. To distinguish between manufacturers, the category assignments of paper grade are now removed and

Table 6.1 Variance and Fisher Weights in the Paper Study

	(a) Paper Grade		(b) Manufacturer	
Element	Variance	Fisher	Variance	Fisher
Na	7.39	6.62	1.49	0.048
Al	66.95	22240.	3.03	0.650
Cl	9.96	137.1	1.67	0.085
Ca	13.24	11.08	1.98	0.141
Ti	17.94	12.78	1.66	0.092
Cr	8.67	41.53	1.75	0.106
Mn	13.01	15.43	2.37	0.182
Zn	4.87	18.24	1.99	0.163
Sb	10.19	31.68	1.92	0.138
Ta	2.06	1.71	1.25	0.013

Source: From *Anal. Chem.* **47**, 528 (1975).

replaced with the *manufacturer category*. The Variance and Fisher weights are recalculated and appear in Table 6.1(b). These measurements are still fairly useful to discriminate between manufacturers although not as dramatically as between paper grades. At any rate, using this training set it would be possible then to predict the grade and manufacturer of the scrap of paper left at the scene of this crime and, furthermore, to pare down the number of variables used in future cases.

Rotation

In general, a set of coordinate axes may be rotated through an angle θ to change the relative orientation of a set of points to the axes. Mathematically this is done by multiplying a transformation matrix $\mathbf{R}^T$ times the original data, $\mathbf{A}$, in column matrix form, to obtain the new column matrix of coordinates $\mathbf{B}$:

$$\mathbf{B} = \mathbf{R}^T \mathbf{A}. \tag{6.13}$$

For example, in two space, the transformation matrix for rotation through the angle θ is obtained from

$$\mathbf{R}^T = \begin{pmatrix} \cos\theta & \sin\theta \\ -\sin\theta & \cos\theta \end{pmatrix}. \tag{6.14}$$

In Figure 6.11 for example, the data points are held fixed and the coordinates with respect to the original, solid axes are given in the matrix

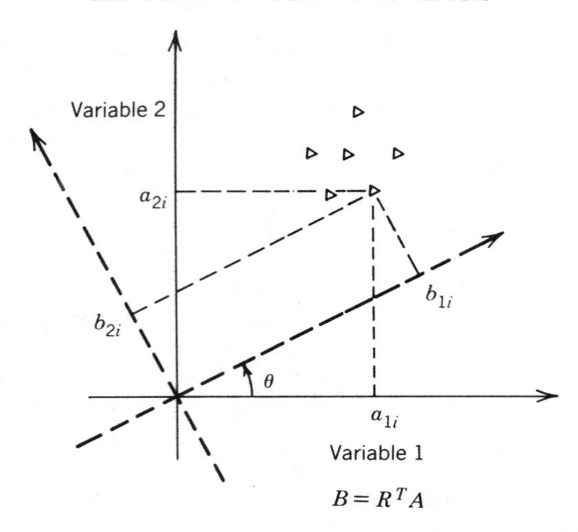

Figure 6.11 Rotation of coordinate axis. Coordinates a_{ji} refer to sample i in the original reference system, b_{ji} to the new rotated system.

$\mathbf{A}$. Premultiplying $\mathbf{A}$ by $\mathbf{R}^T$ has the effect of rotating the solid axes counterclockwise to the position of the dashed axes. Since the points remain fixed, they now have coordinates $\mathbf{B} = \mathbf{R}^T\mathbf{A}$ in the new system.

Eigenvector Rotation

As a preprocessing step, it is extremely useful to rotate all the axes involved in an n-dimensional data set so that the first new axis corresponds to the direction of greatest variance in the data, and each successive axis represents the maximum residual variance. Without changing the data structure, we want to find orthogonal axes to represent the directions of maximum variance. This will be true when the correlation matrix of the data, in terms of the new rotated axes, is a diagonal matrix.

Our data matrices have been defined as row matrices and thus, we must take the transpose of Equation (6.13). Letting $\mathbf{Y} = \mathbf{B}^T$ and $\mathbf{X} = \mathbf{A}^T$ we obtain

$$\mathbf{Y} = \mathbf{X}\mathbf{R} \tag{6.15}$$

and

$$\mathbf{Y}^T = \mathbf{R}^T\mathbf{X}^T. \tag{6.16}$$

For autoscaled data, the correlation matrix, $\mathbf{C}$, is given by

$$\mathbf{C} = \frac{1}{N-1}(\mathbf{X}^T\mathbf{X}) \tag{6.17}$$

where N is the number of columns in $\mathbf{X}$. Now, we wish to find a transformation matrix that, when applied to $\mathbf{X}$, yields a new set of coordinates $\mathbf{Y}$ for which

$$\mathbf{Y}^T\mathbf{Y} = \mathbf{R}^T\mathbf{X}^T\mathbf{X}\mathbf{R} = \mathbf{R}^T\mathbf{C}\mathbf{R} = \Lambda, \tag{6.18}$$

where Λ is a diagonal matrix. In other words, this is an eigenvector problem, in which we would like to find the vectors in $\mathbf{R}$ that, when applied to the system are converted to scalar multiples of themselves:

$$\mathbf{C}\mathbf{R} = \lambda\mathbf{R}, \tag{6.19}$$

$$\mathbf{C}\mathbf{R} - \lambda\mathbf{R} = 0, \tag{6.20}$$

and

$$(\mathbf{C} - \lambda\mathbf{I})\mathbf{R} = 0. \tag{6.21}$$

Here λ is a scalar variable, whose solutions are the diagonal elements of Λ. This problem has a nontrivial solution when

$$|\mathbf{C} - \lambda\mathbf{I}| = 0. \tag{6.22}$$

Equation (6.22) is solved for the roots of λ which are the eigenvalues, that is, the variances associated with each new axis. Once the solutions for λ are known, they can be substituted back into Equation (6.21) to solve for the vectors (columns) of $\mathbf{R}$.

As an example, consider the data matrix $\mathbf{A}$ whose points are plotted in Figure 6.12:

$$\mathbf{A} = \begin{pmatrix} 2 & 1 \\ 3 & 2 \\ 4 & 3 \end{pmatrix}. \tag{6.23}$$

The data are first autoscaled to yield $\mathbf{X}$ whose plot is shown in Figure 6.12.

$$\mathbf{X} = \begin{pmatrix} -1/\sqrt{2} & -1/\sqrt{2} \\ 0 & 0 \\ 1/\sqrt{2} & 1/\sqrt{2} \end{pmatrix}. \tag{6.24}$$

The correlation matrix is then

$$\mathbf{C} = \tfrac{1}{2}\mathbf{X}^T\mathbf{X} = \begin{pmatrix} 1 & 1 \\ 1 & 1 \end{pmatrix}, \tag{6.25}$$

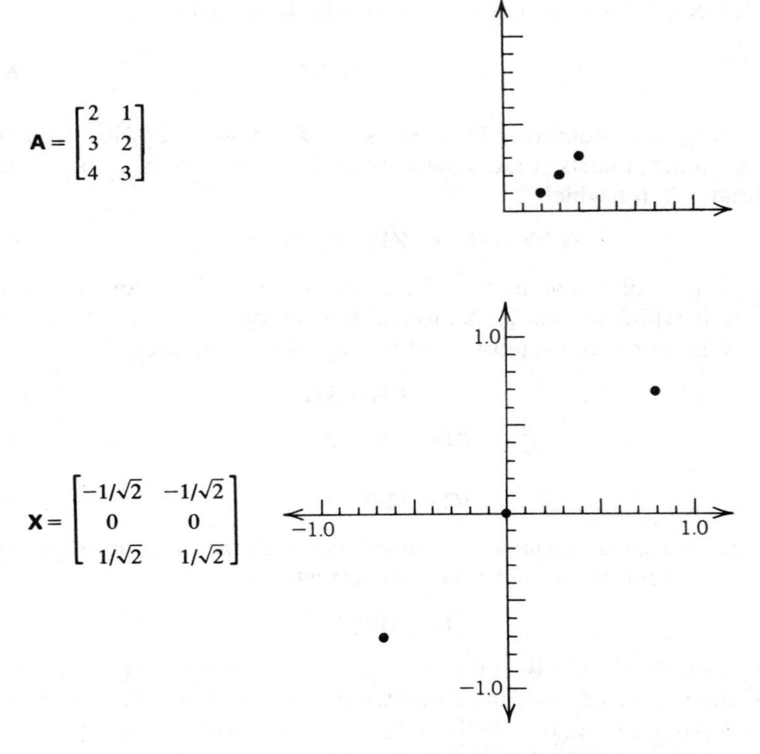

Figure 6.12 Original and autoscaled data before eigenvector rotation.

and we must solve the following for the roots of λ:

$$|\mathbf{C} - \lambda\mathbf{I}| = \begin{vmatrix} 1-\lambda & 1 \\ 1 & 1-\lambda \end{vmatrix} = 0, \tag{6.26}$$

$$(1-\lambda)^2 - 1 = 0, \tag{6.27}$$

$$\lambda(\lambda - 2) = 0, \tag{6.28}$$

so

$$\lambda_1 = 2 \quad \text{and} \quad \lambda_2 = 0. \tag{6.29}$$

The eigenvectors associated with these eigenvalues are determined by successively substituting the values of λ back into Equation (6.21). For $\lambda_1 = 2$, we obtain

$$\begin{pmatrix} 1-2 & 1 \\ 1 & 1-2 \end{pmatrix}\begin{pmatrix} v_{11} \\ v_{21} \end{pmatrix} = \begin{pmatrix} 0 \\ 0 \end{pmatrix}, \tag{6.30}$$

where v_{ij} is the ith element of the jth eigenvector. Equation (6.30) yields two equations in two unknowns:

$$-v_{11} + v_{21} = 0, \tag{6.31}$$

and

$$v_{11} - v_{21} = 0, \tag{6.32}$$

which both reduce to

$$v_{11} = v_{21}. \tag{6.33}$$

Applying the constraint of normalization to unity leads to the solution for v_1:

$$v_1 = \begin{pmatrix} 1/\sqrt{2} \\ 1/\sqrt{2} \end{pmatrix}. \tag{6.34}$$

Eigenvector v_2 is obtained analogously, noting that the *phase* (relative sign) of the elements v_{12} and v_{22} is arbitrary:

$$v_2 = \begin{pmatrix} 1/\sqrt{2} \\ -1/\sqrt{2} \end{pmatrix}. \tag{6.35}$$

Thus the entire transformation matrix is

$$\mathbf{R} = \begin{pmatrix} 1/\sqrt{2} & 1/\sqrt{2} \\ 1/\sqrt{2} & -1/\sqrt{2} \end{pmatrix}. \tag{6.36}$$

The reader is invited to fill in all the steps of these calculations as an exercise.

Armed with this transformation matrix, let us return to the objective of eigenvector rotation, as a preprocessing step. We would now like to rotate the original axes, obtain the coordinates of the data in the new system and plot the data. From Equation (6.15),

$$\mathbf{Y} = \mathbf{XR}. \tag{6.15}$$

$$\mathbf{Y} = \begin{pmatrix} -1/\sqrt{2} & -1/\sqrt{2} \\ 0 & 0 \\ 1/\sqrt{2} & 1/\sqrt{2} \end{pmatrix} \begin{pmatrix} 1/\sqrt{2} & 1/\sqrt{2} \\ 1/\sqrt{2} & -1/\sqrt{2} \end{pmatrix}. \tag{6.37}$$

$$\mathbf{Y} = \begin{pmatrix} -1 & 0 \\ 0 & 0 \\ 1 & 0 \end{pmatrix}. \tag{6.38}$$

The plot of $\mathbf{Y}$ is shown in Figure 6.13. The coordinate axes are now the

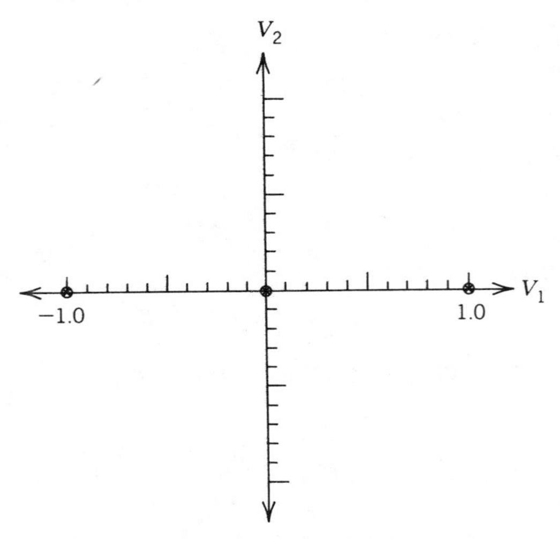

Figure 6.13 Data of Figure 6.11 preprocessed by eigenvector rotation.

eigenvectors, which are linear combinations of the original variables. The coordinates Y are referred to as *scores* and the elements of the eigenvectors—eigenvector coefficients—are called the *loadings*. One might think of the eigenvector coefficient as indicating how much a variable is "loaded into" an eigenvector, for example, the magnitude of the contribution of a variable in comprising the eigenvector.

Perhaps the most startling and remarkable result of making this eigenvector plot is the fact that axis v_1 now contains 100% of the information (variance) in the data. The *intrinsic dimensionality* of this data set is found to be *one*. In a chemical application, this would hopefully lead to some chemical understanding about the underlying factor controlling the system. It also provides a means of *data reduction*, that is, the discarding of factors that do not contain significant information about the data. Instead of two variables, we really need only one factor to characterize the behavior of the system in the example. Note that since the sum of these represents the total variance in the data set, the percent variance remaining after dimensionality reduction, % Var, is given by

$$\%\ \mathrm{Var} = \frac{\sum_{i=1}^{NC} \lambda_i}{\sum_{i=1}^{NV} \lambda_i} \times 100, \tag{6.39}$$

where NC is the number of components retained.

Furthermore, this preprocessing method can result in a signal/noise enhancement. Ideally, the chemical information will be contained in the first one or more eigenvectors having the largest variances, and the noise contained in the eigenvectors of smallest eigenvalue may be discarded.

Eigenvector plots are extremely useful to display n-dimensional data, because they provide a way to preserve the maximum amount of information in a two-dimensional projection. They are thus an aid to the human pattern recognizer in n dimensions. Eigenanalysis and subsequent inspection of the eigenvector plots is one of the first and foremost procedures that should be done when tackling a data set. Eigenanalysis is also useful in modeling of the data. For example, each category may be modeled by one or more of its eigenvectors and a classification algorithm can be developed using this type of model. Such a procedure forms the basis of SIMCA, a classification method discussed later in this chapter.

Principal component analysis, the statistician's tool, refers to the diagonalization of the covariance matrix, [COV], which is obtained from the mean-centered data:

$$x'_{ik} = x_{ik} - \bar{x}_k \tag{6.40}$$

$$[\mathrm{COV}] = \frac{1}{NV - 1} \mathbf{X}'^T \mathbf{X}'. \tag{6.41}$$

It is important to note how to get back and forth computationally between scores and data. Given the score y_i, the original datum x_{ik} can be obtained as

$$x_{ik} = \bar{x}_k + \sum_{m=1}^{NV} y_{im} v_{km}. \tag{6.42}$$

Conversely, a score can be computed as

$$y_{ik} = \sum_{m=1}^{NV} (x_{im} - \bar{x}_k) v_{mk}. \tag{6.43}$$

The power of eigenanalysis is illustrated by an example from clinical chemistry. The differential diagnosis of certain liver diseases has been primarily based on the concentration of one or two blood enzyme concentrations and has traditionally been quite problematic. Figure 6.14 shows a plot of two blood enzyme concentrations for 55 patients with liver disorders diagnosed as type A and B, or with unknown diagnosis, X (5).

It is obvious from this plot that there is not sufficient information in these two measurements to distinguish liver disorders A and B. Six additional blood enzymes, shown in Table 6.2, were measured and an eigenanalysis of the eight-dimensional data set lead to the eigenvector plot

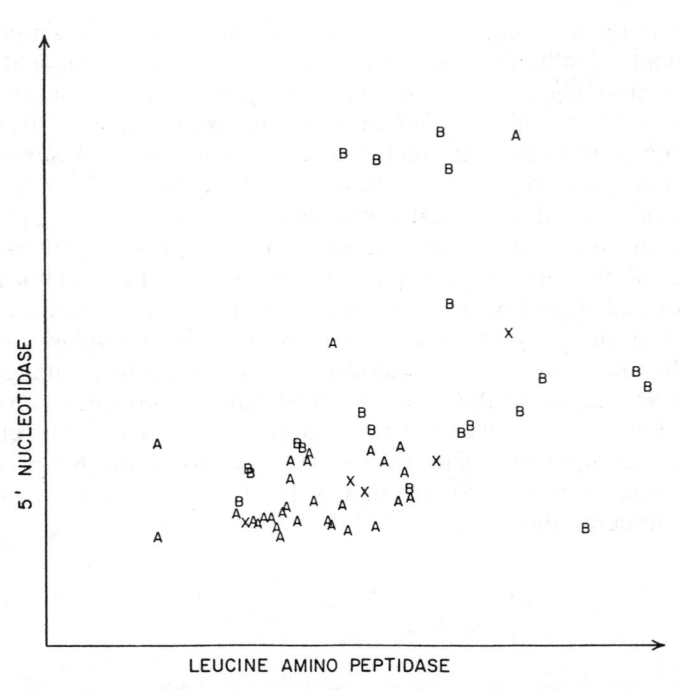

Figure 6.14 Plot of two blood enzyme concentrations for 55 patients. A = Patient with liver disorder A; B = liver disorder B; X = unknown diagnosis. [From Kowalski, "Measurement Analysis by Pattern Recognition," *Anal. Chem.* **47**, p. 1153A (1975), Fig. 1.]

Table 6.2 Blood Constituents Measured in Liver Disorder Patients, and Their Variance Weights with Respect to *A* and *B* Category Separation

Variable	Variance Weight
Leucine aminopeptidase	1.7
5′-Nucleotidase	1.5
Glutamate oxaloacetate transaminase	1.6
Glutamate pyruvate transaminase	1.8
Ornithine carbamoyl transferase	1.3
Isocitrate dehydrogenase	1.2
Alkaline phosphatase	3.5

Source: From *Anal. Chem.* **47**, 1152A (1975).

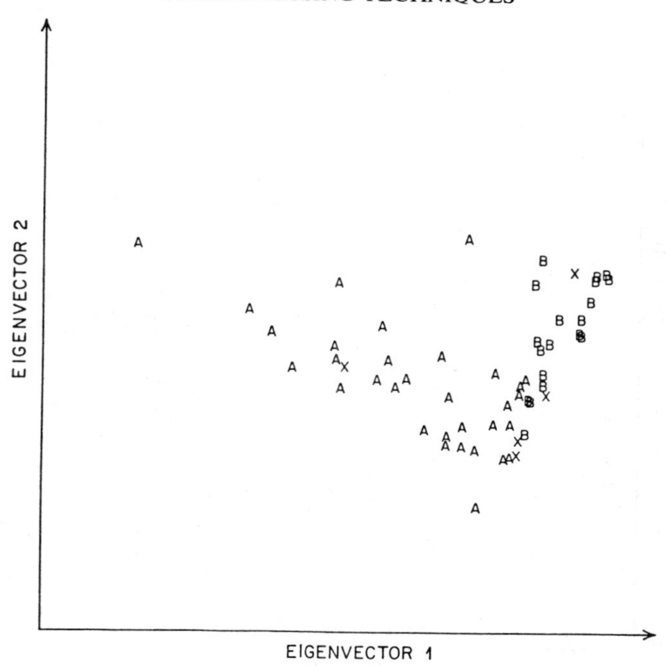

Figure 6.15 Results of eigenvector rotation of eight-dimensional data set on liver disorder patients; 65% of the total information in the data set is retained in this plot. [From Kowalski, "Measurement Analysis by Pattern Recognition," *Anal. Chem.* **47**, p. 1154A (1975), Fig. 2.]

in Figure 6.15. Now the separation of *A*'s and *B*'s is greatly improved, even though 65% of the total variance in the eight-dimensional data is retained in this plot. No single one of the eight enzymes is sufficient to solve the problem, yet together in a linear combination they can. An even greater improvement was achieved by weighting the data with the variance weights of Table 6.2 before eigenanalysis (Fig. 6.16).

The data now possess a radial structure, and it was found that the farther out on the "*A*" spoke or the "*B*" spoke, the more severe the disease; "normal" patients were located at the hub of the wheel. In fact, when more data were included in the analysis for patients with a third liver disease, another "spoke," or "arm" of the "starfish," so to speak, was observed when computer graphics were used to simulate a three-dimensional plot of the first three eigenvectors. Thus, it was possible to identify both the type and extent of disease of these patients.

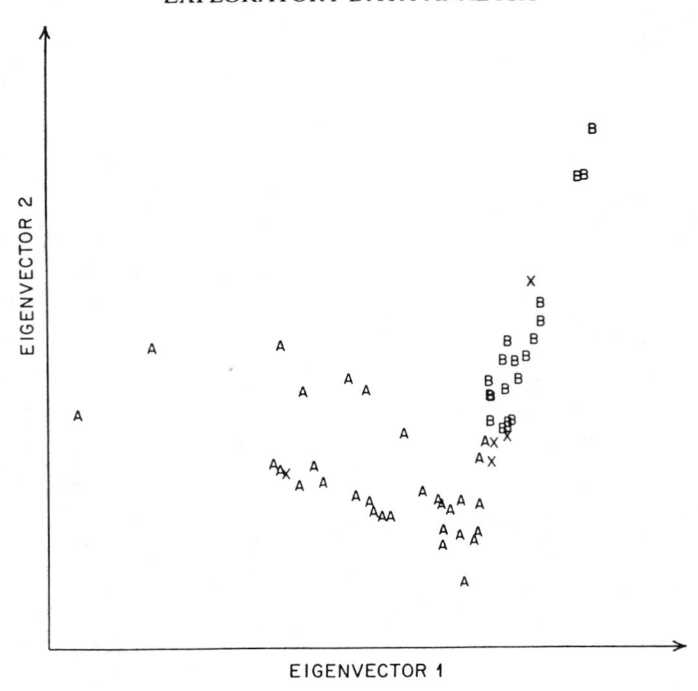

Figure 6.16 Eigenvector rotation of the variance weighted data on liver disorder patients. [From Kowalski, "Measurement Analysis by Pattern Recognition," *Anal. Chem.* **47**, p. 1155A (1975), Fig. 3.]

Varimax Rotation

In view of our goal to obtain knowledge about the properties of a system, it is desirable to try to interpret the chemical significance of the eigenvectors (principal components) by analyzing their loadings. As an example, a group of chemists and atmospheric scientists were interested in learning about the rainwater chemistry in a Pacific Northwest region of the United States (6). Rainwater samples were collected at selected sites in Figure 6.17 under appropriate meteorological conditions and analyzed for the 16 trace metals and ionic species shown in Table 6.3. Eigenanalysis of the autoscaled data yielded the eigenvectors in that table, ordered by decreasing magnitude of the eigenvalues. The coefficients, or loadings, indicate the relative contribution of the features to each eigenvector. For example, the sodium loading in the first eigenvector is 0.351 and thus effectively contains 12.3% of the information in that eigenvector. The largest

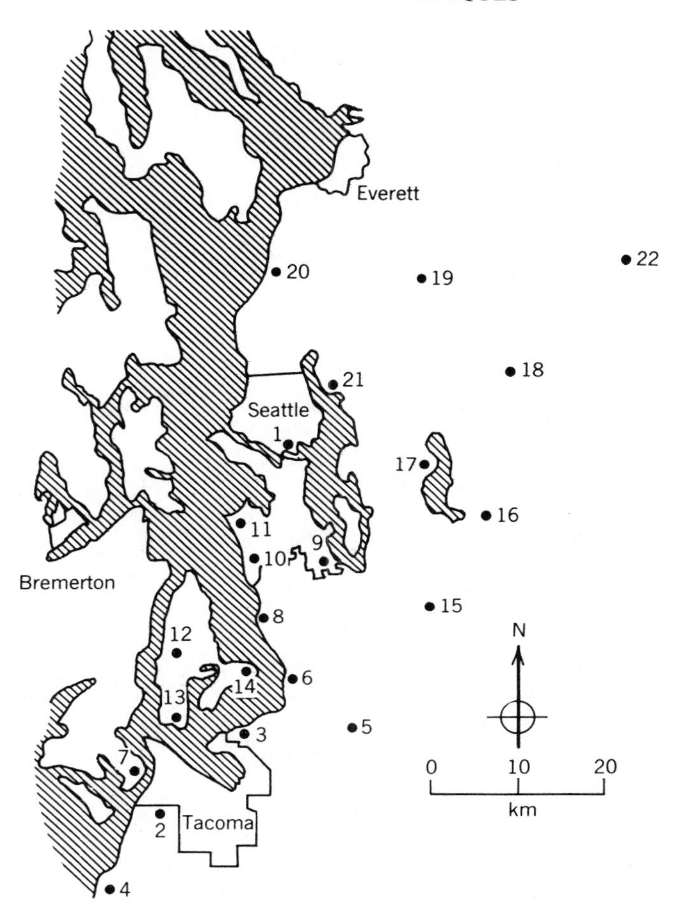

Figure 6.17 Pacific northwest region showing rainwater sampling sites in a rain chemistry study. [From Knudson et al., in *Chemometrics: Theory and Application*, B. R. Kowalski, Ed., ACS Symposium Series 52, ACS, Washington, D.C., 1977, p. 93, Fig. 1.]

contribution (positive or negative) to the first eigenvector are from elements typically associated with sea salts. That this vector spans 42.3% of variance in the data suggests that total dissolved ion concentration is important in characterizing the behavior of this system. The second eigenvector spans 15.2% of the variance, and consists primarily of arsenic, copper, antimony, and cadmium. At this point the human pattern recognition talents come into play: what is the significance of these features

Table 6.3 Eigenvalues and Eigenvectors of the Rainwater Data

Eigenvalue	Variance	H^+	NH_4^+	Na	K	Ca	Mg	Zn	Cu	Pb	SO_4^{2-}	Mn	Cd	As	Sb	Cl^-	NO_3^-
6.76900	42.3	.194	.198	.351	.258	.310	.311	.238	.238	.225	.162	.205	.239	.204	.287	.321	.161
2.42600	15.2	−.162	.253	.092	.165	.197	.227	.003	−.473	.208	.106	.164	−.255	−.508	−.348	.163	.034
1.85900	11.6	.464	−.375	.130	.132	−.153	.145	−.277	−.050	.105	.080	−.263	−.226	−.042	−.099	−.002	.580
1.30900	8.2	.261	−.035	−.102	.347	.150	−.143	−.016	−.130	−.328	−.406	.407	.416	−.149	−.151	−.205	.196
1.15000	7.2	.080	.366	.212	−.133	−.053	.196	−.082	−.023	−.277	−.622	−.375	−.061	.071	−.074	.360	−.011
.67660	4.2	−.205	−.010	.089	.430	.019	.137	−.707	−.004	−.298	.227	−.021	.058	.100	.131	.080	−.263
.46680	2.9	−.293	−.386	.166	.330	−.275	.297	.458	.233	−.279	.014	−.134	.048	−.116	−.283	.040	−.085
.44740	2.8	.362	.129	.064	−.181	.192	.054	.174	.038	−.652	.353	.145	−.400	−.030	.004	−.081	−.078
.35650	2.2	.142	−.263	.003	.253	.570	−.189	.162	−.129	.066	−.116	−.504	−.067	−.159	.213	−.075	−.304
.22010	1.4	−.133	.468	−.298	.389	−.062	−.364	.143	.146	−.107	.231	−.345	−.017	.014	−.053	.033	.373
.16440	1.0	.240	.156	−.146	.396	−.117	.113	.047	.101	.272	−.213	.150	−.443	.393	−.220	−.251	−.312
.08173	.5	.509	.136	−.172	−.047	−.232	.080	−.019	.006	.108	.281	−.204	.447	−.193	−.284	.148	−.397
.03505	.2	−.013	.160	−.250	.034	−.192	.541	.145	−.366	−.064	.004	−.160	.116	−.018	.464	−.402	.075
.02757	.2	−.145	.129	.194	−.198	.388	.204	−.106	.095	.023	.107	−.230	.235	.253	−.476	−.500	.114
.00609	.1	−.006	.084	−.246	−.060	.135	.246	−.195	.671	.084	−.157	.032	−.112	−.533	.109	−.140	.023
.00249	0.0	.070	.236	.676	.083	−.320	−.283	−.002	.048	.080	−.006	−.028	.001	−.286	.173	−.402	−.103

Source: Reference 6.

covarying? Is there a geographical reason, an atmospheric process, or a geological source? In fact, there is a copper smelter located at the southwestern corner of this region, upwind during the typical southwesterly onshore flow that brings rain to the area. The third eigenvector is predominantly H^+ and NO_3^- and accounts for 11.6% of the variance in the data set.

The task of interpreting the significance of eigenvectors is not always straightforward. The technique of *Varimax rotation* (8) has been developed as an aid to this process. Often, but not always, will it make the eigenvectors chemically more interpretable.

Varimax rotation perturbs the eigenvectors so as to maximize the variance within each vector. As a result, in each vector the number of variables with intermediate loadings is decreased, and the number with either very large (absolute magnitude) or very small loadings is increased. In other words, "the rich get richer and the poor get poorer."

To see how the method works, let us define a quantity called the *communality*, h_k^2, to be the amount of variance in feature k that is spanned by NC eigenvectors, where $NC \leq NV$:

$$h_k^2 = \sum_{m=1}^{NC} v_{mk}^2. \tag{6.44}$$

The Varimax rotation is the orthogonal transformation that maximizes the following function:

$$\sum_{\ell=1}^{NC} \left[NV \sum_{j=1}^{NV} \left[\frac{(v_{j\ell}\sqrt{\lambda_j})^2}{h_j^2} \right]^2 - \left[\sum_{j=1}^{NV} \frac{(v_{j\ell}\sqrt{\lambda_j})^2}{h_j^2} \right]^2 \right]. \tag{6.45}$$

Varimax rotation is a pairwise iterative process on the original loadings, in which, beginning with the first two vectors, then the first and third, and so on, the vectors are "tweaked" (rotated) until the criterion of Equation (6.45) is satisfied. A graphic way of visualizing the effect of Varimax rotation on the loadings is seen in Figure 6.18 where the terms of h_k^2 are plotted versus the eigenvector index (first eigenvector, second, third, etc.). Before Varimax rotation, plot (a) is typically observed in which a gradual decrease in magnitude of v_{kj}^2 occurs. Varimax rotation, plot (b), accentuates the higher loadings and minimizes the smaller. Concurrently, the loadings within any given eigenvector will typically undergo the same type transformation so that, hopefully, they will be easier to interpret, that is, the features primarily involved in each eigenvector will be more easily distinguished from the minor ones.

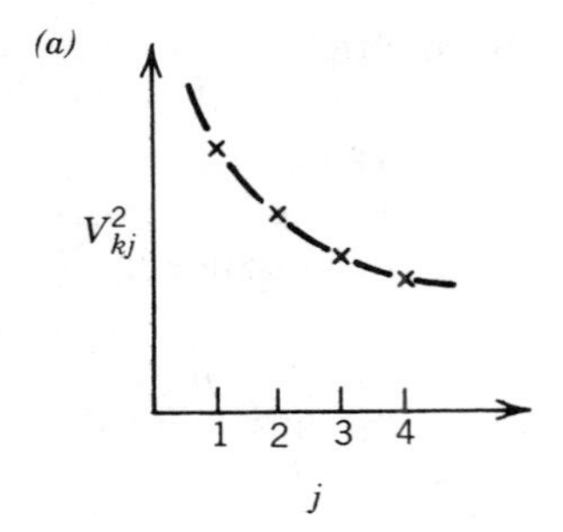

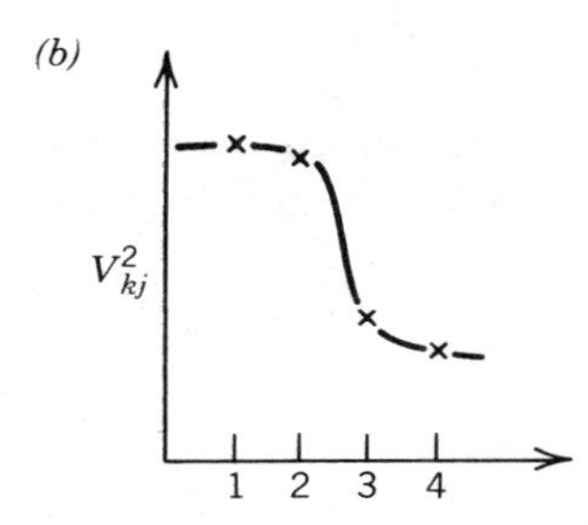

Figure 6.18 Effect of Varimax rotation on the eigenvector loadings.

In the rainwater example, the Varimax rotation of the eigenvectors did indeed clear up the picture. As shown in Table 6.4 some rearrangement of the vectors has occurred in terms of variance per vector as well as their order. Now the first two factors are about equal in information content. The eigenvectors and *varivectors* (Varimax rotated eigenvectors) are more easily compared by examining histograms, or bar graphs, of their loadings in Figure 6.19. It can be seen at a glance that the information in the varivectors is compressed into fewer features with a much more dramatic cutoff between large and small loadings. The first varivector clearly represents the smelter factor and the second, the sea factor. The third, comprised almost entirely of hydrogen and nitrate ions, may now be attributed to automobile and other urban sources. Thus, the Varimax rotation has facilitated the interpretation of the underlying factors that determine rainwater chemistry in this region. The reader may well ask why we stopped at these three factors, and how in general to decide the number of statistically significant vectors. This issue will be discussed later.

Table 6.4 Varimax Rotated Eigenvectors of Rainwater Example

Varivalue	Variance	H^+	NH_4^+	Na	K	Ca	Mg	Zn	Cu	Pb	SO_4^{2-}	Mn	Cd	As	Sb	Cl^-	NO_3^-
3.26100	20.4	−.173	−.011	−.163	−.033	−.064	−.045	−.135	−.518	−.027	−.064	−.018	−.323	−.545	−.479	−.119	.004
3.13000	19.6	.096	.237	.479	.218	.241	.512	.130	.088	.186	.097	.068	.046	.053	.124	.477	.130
1.81700	11.4	.637	−.124	.194	.174	.070	.140	−.006	.077	.093	.011	.021	.043	.070	.051	.073	.676
1.43100	8.9	.032	.185	.091	.232	.294	.105	.181	.009	.067	.092	.776	.389	−.008	.079	.004	.012
1.13800	7.1	.040	.056	−.097	−.101	−.120	−.146	−.048	−.092	−.252	−.883	−.129	.192	.015	−.160	−.048	−.059
1.05900	6.6	−.012	−.145	−.135	−.024	−.240	−.104	−.865	−.224	−.136	−.044	−.173	−.158	−.004	−.067	−.120	.009
1.01500	6.3	−.073	.833	.093	.025	.297	.097	.135	−.048	.067	−.038	.161	.053	.035	.082	.355	−.068
.98730	6.2	.074	.019	.160	.830	.289	.184	.027	.057	.060	.075	.177	.291	−.020	.035	.103	.142
.98550	6.2	.036	−.061	−.130	−.068	−.163	−.152	−.134	.035	−.879	−.213	−.068	−.015	−.019	−.132	−.189	−.166
.63030	3.9	.229	.139	.216	.145	.790	.000	.113	−.095	.087	.059	.117	.125	−.046	.379	.116	−.122
.30770	1.9	.072	−.019	−.028	−.118	−.100	.072	−.068	−.099	−.010	.063	−.071	−.933	.095	−.161	−.163	−.094
.14090	.9	.779	−.004	−.064	.001	.072	.117	.002	−.002	−.017	−.011	.007	−.036	.096	−.049	−.071	−.593
.04224	.3	.020	−.006	.152	−.003	.018	−.065	−.007	.363	−.014	−.012	−.012	−.024	.221	−.880	−.123	−.021
.03987	.2	.036	.029	−.080	−.003	−.010	.619	−.004	−.174	−.004	−.001	.010	−.025	.156	−.055	−.741	−.032
.01009	.1	−.006	−.011	−.090	.006	−.006	.011	.003	.762	−.005	.009	.001	.012	−.634	−.002	.097	.011
.00381	0.0	−.011	.011	.845	.002	.008	−.478	.002	−.028	.002	.001	.004	.001	.057	−.027	−.228	.007

Source: Reference 6.

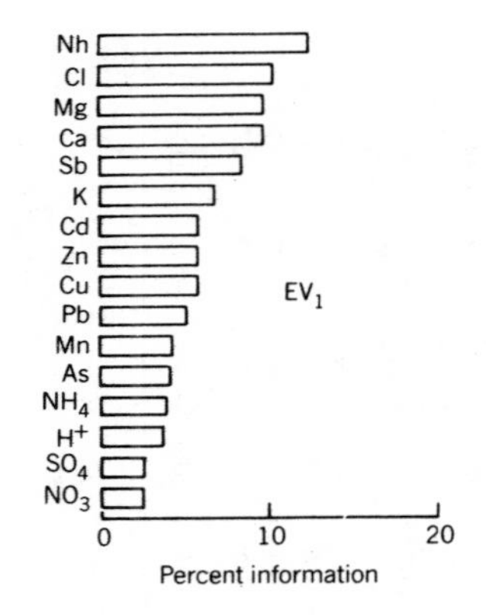

Nh
Cl
Mg
Ca
Sb
K
Cd
Zn
Cu
Pb
Mn
As
NH_4
H^+
SO_4
NO_3
EV_1
0
10
20
Percent information

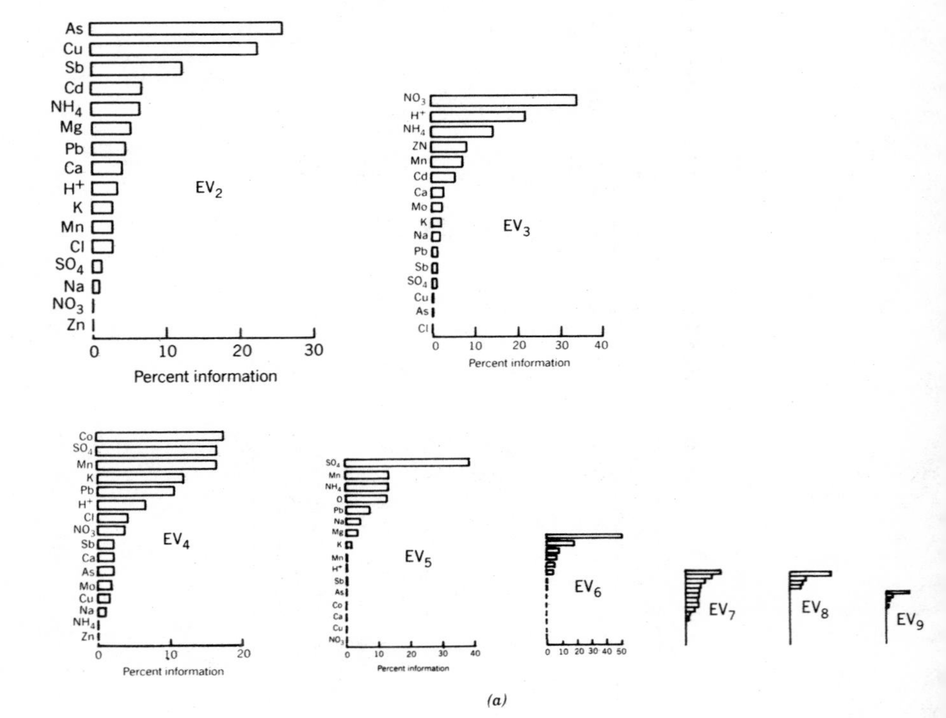

As
Cu
Sb
Cd
NH_4
Mg
Pb
Ca
H^+
K
Mn
Cl
SO_4
Na
NO_3
Zn
EV_2
0
10
20
30
Percent information
NO_3
H^+
NH_4
ZN
Mn
Cd
Ca
Mo
K
Na
Pb
Sb
SO_4
Cu
As
Cl
EV_3
0
10
20
30
40
Percent information
Co
SO_4
Mn
K
Pb
H^+
Cl
NO_3
Sb
Ca
As
Mo
Cu
Na
NH_4
Zn
EV_4
0
10
20
Percent information
SO_4
Mn
NH_4
O
Pb
Na
Mg
K
Mn
H^+
Sb
As
Co
Ca
Cu
NO_3
EV_5
0
10
20
30
40
Percent information
EV_6
0
10
20
30
40
50
EV_7
EV_8
EV_9
(a)

Figure 6.19a.

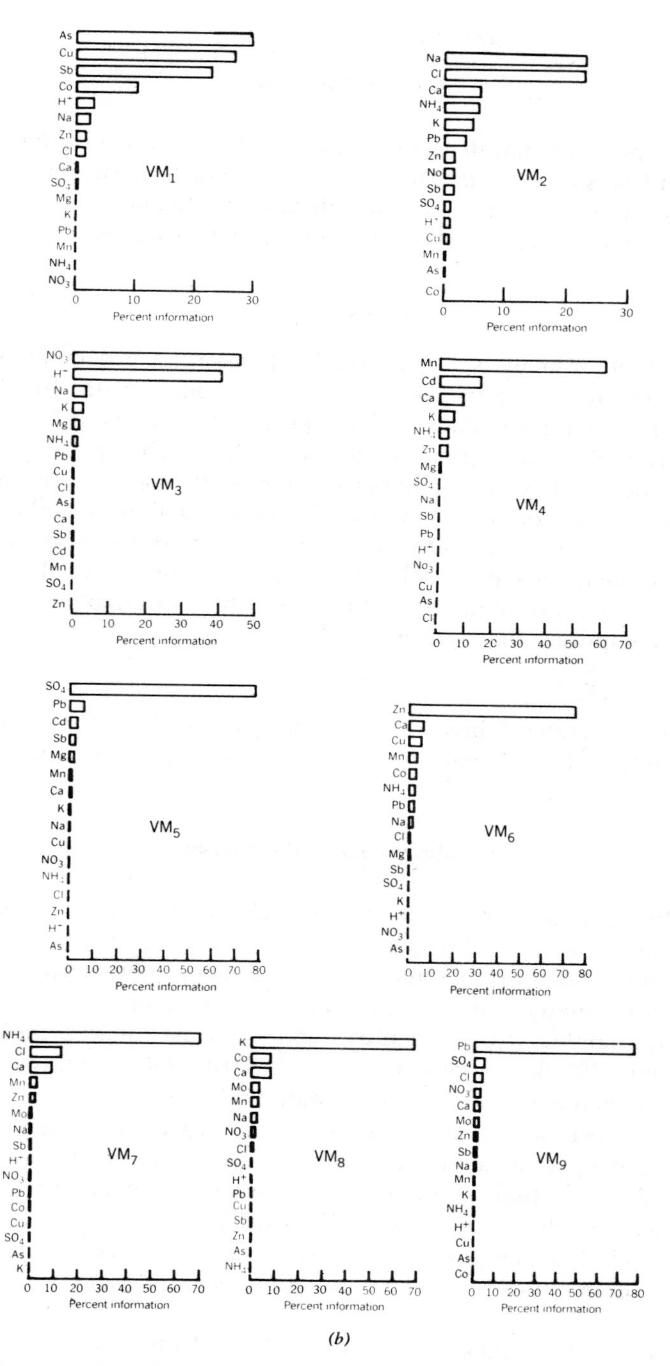

Figure 6.19 Histogram of the eigenvector and varivector loadings in the rain chemistry study. [From Knudson et al., in *Chemometrics: Theory and Application*, B. R. Kowalski, Ed., ACS Symposium Series 52, ACS, Washington, D.C., 1977, pp. 104–105, Figs. 13 and 14.]

Factor Analysis

Factor analysis (7, 8) has its roots in psychology. Psychologists back in the 1930s and 1940s attempted to understand individuals' performance, X, on intelligence tests in terms of some intrinsic intelligence factors, L, and some individual scores on those factors, S, with some residual random error, E:

$$X = SL + E. \tag{6.46}$$

Thus, in factor analysis, an observed data matrix $\mathbf{X}$ is decomposed into the product of two matrices plus some residual error. In chemical applications, we are often interested in principal component factor analysis—that is, finding the intrinsic factors L by diagonalizing the correlation or covariance matrix. The number of significant eigenvectors is determined by an appropriate means (see Cross Validation) and the error is discarded. As seen in the last section, Varimax rotation can then be applied to improve the chemical interpretability of the factors. Once the Varimax rotated factors L are obtained, the data may be plotted in the new factor space by computing the coordinates Y:

$$Y = XL^T(LL^T)^{-1}, \tag{6.47}$$

where the generalized inverse has been used in Equation (6.47). The reader is referred to the many texts on this subject (see, e.g., refs. 7 and 8).

Nonlinear Factor Analysis

The methods of principal component analysis and factor analysis presented thus far assume a linear relationship among the variables. Nature, on the other hand, is often nonlinear, as the reader knows only too well. A number of techniques outside of the chemical domain have been devised to treat the nonlinear case such as nonlinear least-squares, multidimensional scaling (9), and parametric mapping (10). Only recently have such techniques been applied to chemical data (11).

To see how the nonlinear projection methods work, imagine a set of data points lying along a curved surface in three dimensions as depicted in Figure 6.20 (11). If these data were analyzed by linear principal component analysis, three factors would be extracted even through there are only two underlying nonlinear independent variables. These can be seen as 1) the spread of points along the z direction and, 2) the curve in the (x, y) plane.

To perform a nonlinear projection analysis of these data, the distances

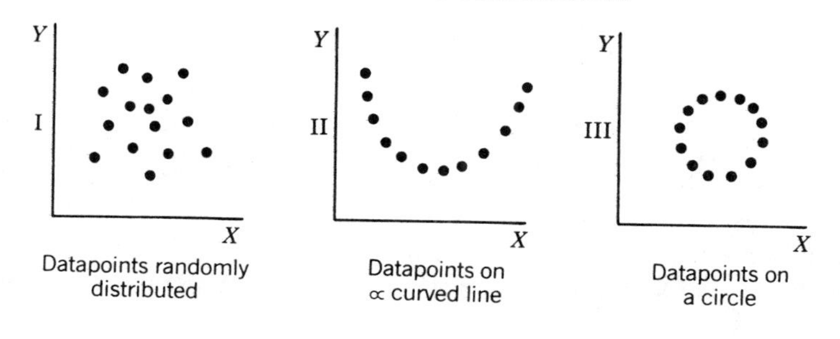

Dataset	Principal Component Analysis	Multidimensional Scaling	Parametric Mapping
	Dimensionality found by		
I	2	2	2
II	2	1	1
III	2	2	1

Figure 6.20 Results of linear and nonlinear methods for analyzing the underlying structure of some data sets. [From C. Jochum and B. R. Kowalski, *Anal. Chim. Acta*, **133** (1981) 583.]

d^*_{ij} between all m data vectors in n-space are calculated, where $i, j = 1, \ldots, m$. The points are subsequently arranged in an r-dimensional space where $r < n$, in such a manner as to minimize the stress, S:

$$S = \left[\frac{\sum_{i,j} (d^*_{ij} - d_{ij})^2}{\sum_{i,j} d^{*2}_{ij}} \right]^{1/2} . \tag{6.48}$$

Here, d_{ij} refers to the recalculated distances in r-space. A small value of S indicates a good fit or a faithful representation of the data vectors in r-space compared with their configuration in n space. The methods of multidimensional scaling and parametric mapping differ principally in their measure of this goodness of fit and may yield different results depending on the nature of the data structure. For example, Figure 6.20 contrasts the results obtained from these methods for three different simulated data structures.

DISPLAY TECHNIQUES

Introduction

Methods to somehow display n-dimensional data in two dimensions for the scientist to inspect are extremely important for two reasons, (1) they allow the human pattern recognizer to analyze the data and (2) they provide a way to direct and supervise the application of other pattern recognition algorithms (3).

Feature by feature plots are generally not an acceptable way of scrutinizing n-dimensional data. Given a small data set, by pattern recognition standards, of only ten variables, there would be $\frac{1}{2}(NV)(NV-1) = \frac{1}{2}(10)(9) = 45$ plots to analyze. Now, figuring just ten minutes to examine and think about each plot, a total of 450 minutes or an entire work day would be required to see this ten-dimensional data. And even then, the more complex relationships involving three or more features would be missed. Display techniques geared for the multivariate approach are obviously needed.

Linear Methods

Linear methods of display are often called *projections* (3). The axes used to plot the data are linear combinations of the original features. A simple variable by variable plot is a trivial case where the two coordinate axes have a trivial relationship with just two of the original variables. It is possible to rotate objects or data in hyperspace through any angle in any dimension and then project it on to a plane for inspection. Two-dimensional rotation matrices, $R_{ij}(\alpha)$, are used to rotate points in n-space by, angle α in the i, j plane. A total of $\frac{1}{2}n(n-1)$ two-dimensional rotations are therefore needed to effect a full rotation, α, in n-space. For example, in four dimensions the final rotation matrix would be,

$$\mathbf{R} = \mathbf{R}_{12}(\alpha_1)\mathbf{R}_{13}(\alpha_2)\mathbf{R}_{14}(\alpha_3)\mathbf{R}_{23}(\alpha_4)\mathbf{R}_{24}(\alpha_5)\mathbf{R}_{34}(\alpha_6). \tag{6.49}$$

Now each new axis is a linear combination of the original variables. This type of display is most useful in combination with the interactive graphics terminal so that the rotations can be made in real time. If not, usually there is insufficient information a priori to know what rotation to apply.

For this reason, the method of eigenvector projection is most commonly used, since this special rotation finds the axes, or linear combinations of the original variables, that align with the "directions" of greatest variance in the data. In this method, the scores of the data, Y, are obtained from Equation (6.13) are are plotted in eigenvector space.

Usually, the first two or three (or the statistically significant number) are used so that the total number of plots is very small.

The amount of variance, or information, retained in an eigenvector plot is given by

$$\% \text{ Information} = \frac{\lambda_i + \lambda_j}{\sum_{k=1}^{NV} \lambda_k} \times 100, \qquad (6.50)$$

where λ_i and λ_j are the eigenvalues corresponding to the two eigenvectors used in the plot. Figure 6.13 is an example of an eigenvector plot, as is Figure 6.15.

Nonlinear Methods

Nonlinear display methods are referred to as *mappings* and produce displays whose coordinate systems are nonlinear combinations of the original variables. A variety of such techniques have been developed.

One example is *nonlinear mapping* (NLM) (12) which tries to find the best two-dimensional representation of the data that optimally preserves the n-space interpoint distances. The starting point for the NLM map can be the eigenvector plot described in the preceding section. The n-space interpoint distances, d^*_{ij}, are calculated as

$$d^*_{ij} = \left[\sum_{k=1}^{NV} (x_{ik} - x_{jk})^2 \right]^{1/2}, \qquad (6.51)$$

and all of the two-space interpoint distances, d_{ij}, are given by

$$d_{ij} = [(y_{i1} - y_{j1})^2 + (y_{i2} - y_{j2})^2]^{1/2}, \qquad (6.52)$$

where the y's are the scores, i, j refer to the points, and $1, 2$ refer to the coordinates. The goal of NLM is to iteratively change the coordinates y for each point so as to minimize the following error function, E:

$$E(\rho) = \frac{1}{\sum_{i>j} (d^*_{ij})^\rho} \sum_{i>j} \frac{(d^*_{ij} - d_{ij})^2}{(d^*_{ij})^\rho}. \qquad (6.53)$$

Thus, the two-space interpoint distances are made to be as close as possible to the n-space distances. The exponent, ρ, is included to allow weighting of small or large distances (see Fig. 6.21).

As an intuitive way to understand NLM, imagine a set of point masses interconnected by springs so that in three-space the total energy is zero. We desire to squash this three-dimensional object into two dimensions so that the strain is minimized in the springs. Note that a linear

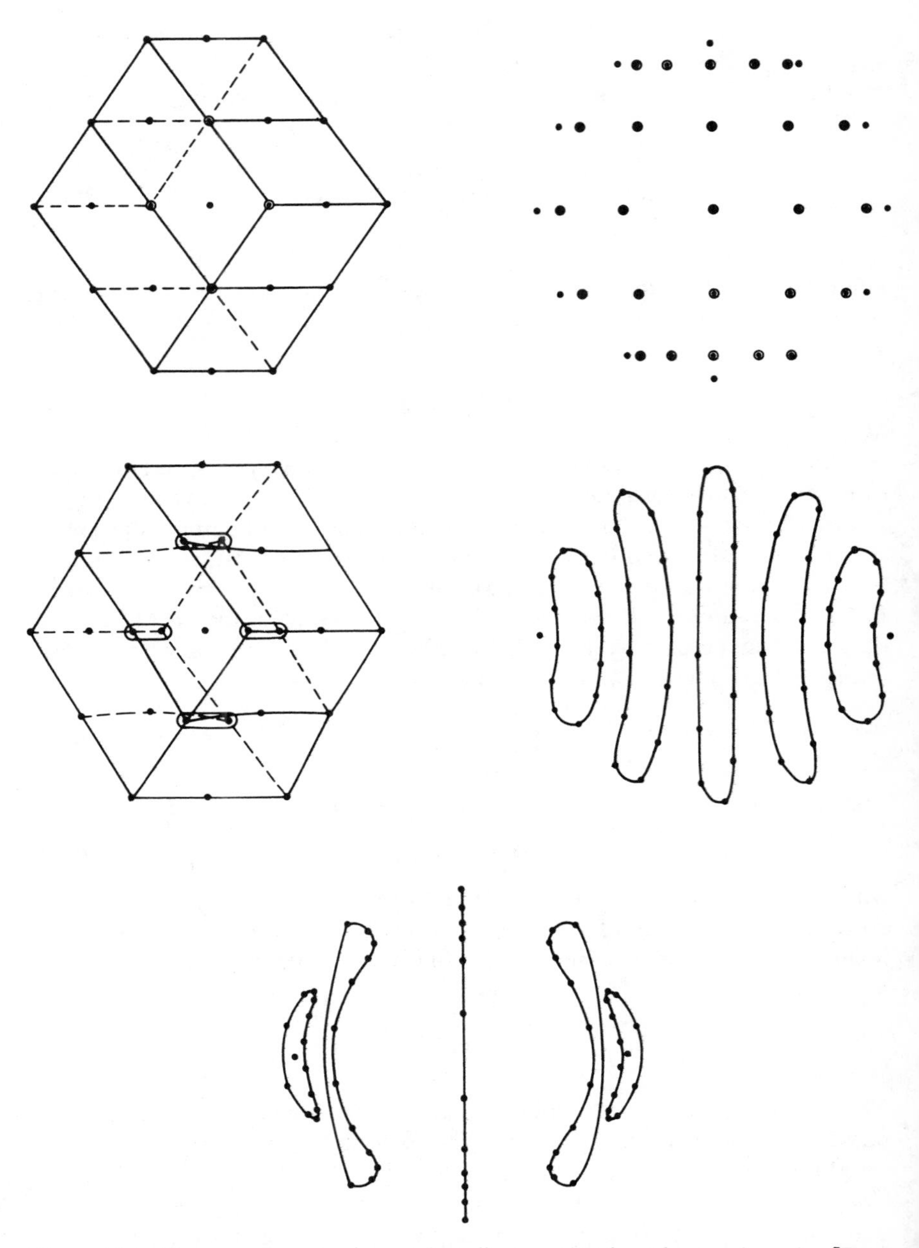

Figure 6.21 Eigenvector projections and nonlinear maps of a sphere and a cube. [From Kowalski and Bender, "Linear and Nonlinear Methods for Displaying Chemical Data," *J. Am. Chem. Soc.* **95**, pp. 690–692 (1973), Figs. 9-10, 9-13, 9-14, and 9-15.]

method would not allow the points to move during projection and would produce a high energy two-space configuration, especially when points would overlap. NLM allows the points to move around until they find the lowest energy configuration in two-space. Thus, NLM is an iterative process where E is minimized as a function of the Y coordinates using an appropriate gradient method.

Figure 6.21 compares the eigenvector projections of a sphere and a cube with their nonlinear maps (3). Note first of all that the circles show overlapping points in the eigenvector projections. Secondly, data sets such as these containing axes of symmetry do not give unique eigenvector projections. In the nonlinear map of the cube, the ellipses enclose the points that were overlapped in the eigenvector projections. The two nonlinear maps of the sphere illustrate the effect of the weighting factor, ρ. With $\rho = 2$, small and large distances are equally weighted and thus the same effort is expended to preserve distances of all magnitudes. However, when $\rho = -2$, the large distances are preserved at the expense of small ones.

UNSUPERVISED LEARNING

The goal of unsupervised learning is to find a property of a collection of objects via measurements made on the objects. Without making a priori assumptions about the data, we would like to uncover intrinsic structure or underlying behavior of a data set. Unsupervised learning is also useful to check the validity of a data set and to detect outliers and strange or faulty results.

Most unsupervised learning techniques are based on the idea of finding *clusters* of points in the data, hence the term *cluster analysis*. Points are grouped together based on their nearness or similarity into clusters and one of our initial assumptions (from Pattern Recognition: The Approach) is invoked—we assume that the nearness of points in n-space reflects the similarity of their properties.

The general approach of *hierarchical cluster analysis* is depicted in Figure 6.22. Given a set of samples representing the system, we may choose one of two routes. Typically, measurements are made on the samples and used to calculate interpoint distances. Similarity values, S_{ij}, are calculated as

$$S_{ij} = 1 - \frac{d_{ij}}{d_{ij}(\max)}, \tag{6.54}$$

where d_{ij} is the distance between points i and j, and $d_{ij}(\max)$ is the largest

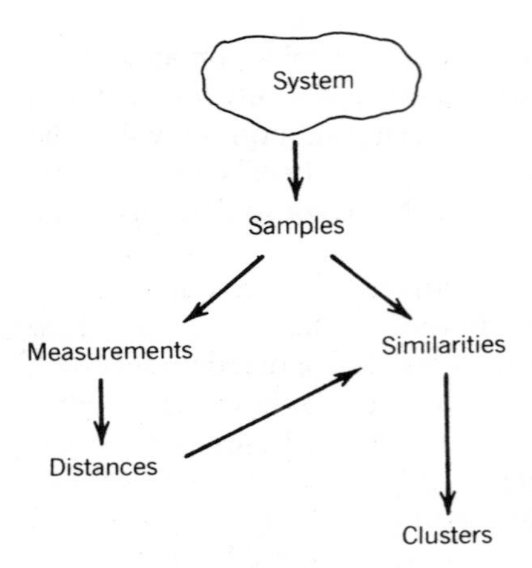

Figure 6.22 Cluster analysis approach.

distance in the distance matrix. In this way, the two most distant points in the data set have $S = 0.0$ and identical points would have the maximum similarity of 1.0. The other route is to obtain similarities directly from the samples, in which case one may proceed directly. This will be discussed later.

In hierarchical cluster analysis, each point is initially assumed to be a lone cluster. The similarity matrix is scanned for the largest value, whose corresponding points are clustered together and treated as a single point. The respective rows and columns of the old points are deleted and the new "cluster" is added as the average of the two points. This process is repeated until all the points have formed one single cluster. The result of this procedure is a diagram called a *connection dendrogram* which can be analyzed for clusters using a variety of criteria.

The above description represents an overview of *hierarchical Q-mode clustering*, however, there are a number of specific options to be selected depending on the context of the problem.

First of all, the choice of a distance metric must be made. The general distance is given by

$$d_{ij} = \left[\sum_{k=1}^{NV} (x_{ik} - x_{jk})^N \right]^{1/N}. \tag{6.55}$$

For $N = 2$, this is the familiar n-space Euclidean distance. Higher values of N will give more weight to smaller distances. For certain cases such as integer data the city block distance may be appropriate.

$$d_{ij} = \sum_{k=1}^{NV} |x_{ik} - x_{jk}|. \tag{6.56}$$

Secondly, a variety of ways to cluster the points have been developed. The *single link* method judges nearness of a point to a cluster on the basis of the distance to the closest point in the cluster. Conversely, the more conservative *complete link* method uses the distance to the farthest point. A more rigorous but computationally slower method is the *centroid* method in which the distance of a point to the center of gravity of the points in a cluster is used. The calculation may be weighted by the number of points in each cluster. These methods are schematically depicted in Figure 6.23.

A schematic illustration of single link clustering and a corresponding dendrogram is shown in Figure 6.24. Note that clusters are *not* the automatic output of this clustering algorithm. The dendrogram is rather an aid to the scientist in learning something about the data. Clusters may be assigned by the scientist or by a criterion imposed on the algorithm. One criterion might be a preset number of clusters. Another alternative is to designate a cluster to be found when the change in similarity coefficient needed to form the cluster exceeds a certain value. Whatever the means of

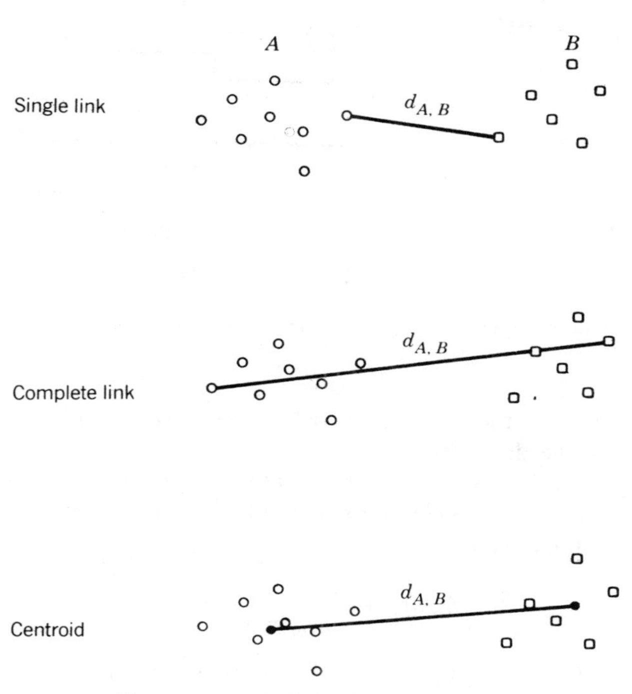

Figure 6.23 Hierarchial clustering methods.

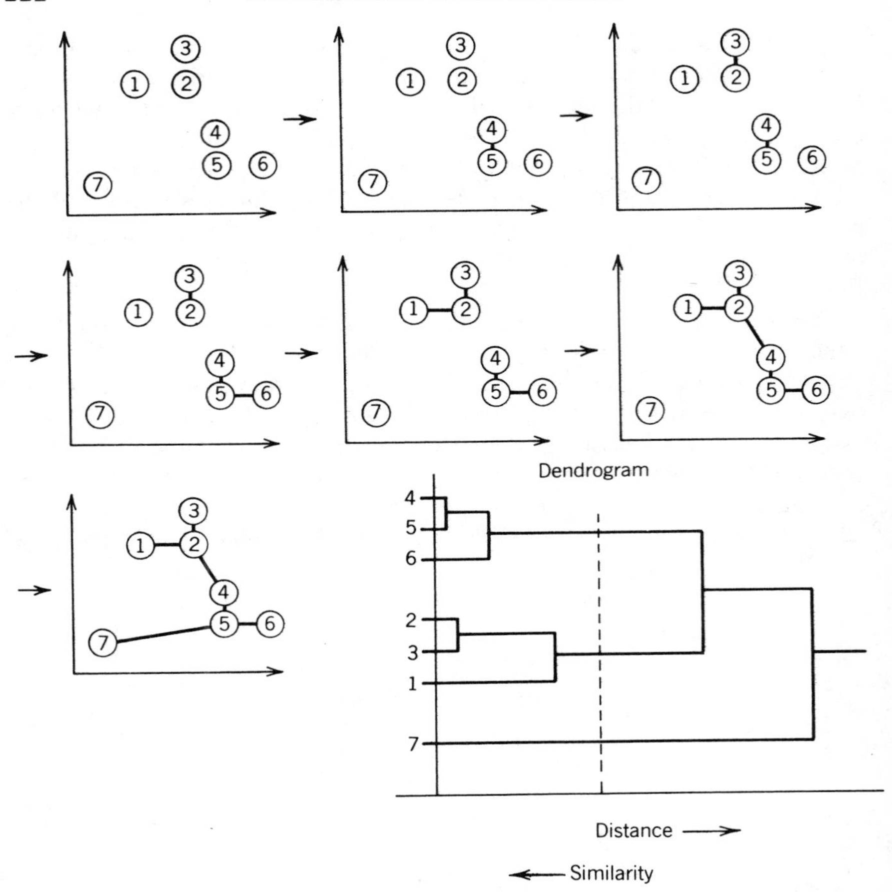

Figure 6.24 Single-link clustering process and corresponding connection dendrogram.

cluster assignment, the remaining task is then to discern the chemical significance of the data structure.

In certain circumstances, it may be appropriate to obtain similarities directly from the samples. While it is possible to calculate a measure of distance from these values as

$$d_{ij} = 1 - \frac{S_{ij}}{S_{ij}(\max)} \tag{6.57}$$

such that $0 \leq d_{ij} \leq 1.0$, and to proceed with some other pattern recognition techniques, it is important to realize that an original "measurement" space and dimensionality are not defined. Nevertheless, some very interesting problems can be tackled in this way.

One application relates to the study of "molecular evolution" via the comparison of amino acid residues in homologous proteins (13). Traditional methods of constructing phylogenetic trees are slow, tedious, arbitrary, and require a great deal of human labor. Clustering algorithms have overcome these difficulties and can be applied automatically and quickly to even large data sets.

Seventy-three cytochrome *c* protein sequences were obtained for a variety of plants and animals. A number was assigned to each of the 20 amino acids found in these proteins and thus, each protein could be represented as a spectrum as shown in Figure 6.25. A cross-correlation between the proteins could then easily be made. A numerical measure of similarity was devised as a function of such quantities as the number of corresponding common amino acids at a specific alignment, the length of the overlapping segments of the proteins, and the total number of base changes required to change one protein to the other. The similarity matrix was analyzed with hierarchical cluster analysis to produce the dendrogram, or phylogenetic tree in this case, shown in Figure 6.26. The similarity calculations and clustering process for these 73 proteins took a total of only about 20 minutes of computer time. It is easier to interpret a dendrogram by turning it on its side and visualizing it as a free-swinging mobile, as shown in the photograph in Figure 6.27. The plants are clearly separated and the bacteria and fungi are also split off as separate clusters.

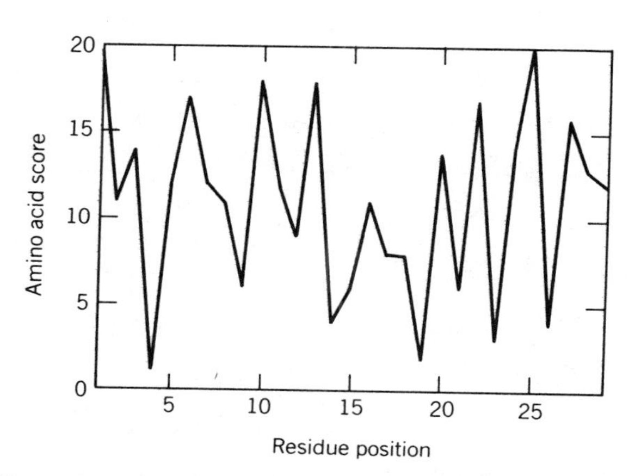

Figure 6.25 Spectral representation of the amino acids in porcine glucagon. [From Sharaf et al., "Construction of Phylogenetic Trees by Pattern Recognition," *Z. Naturforsch*., p. 509 (1980), Fig. 1.]

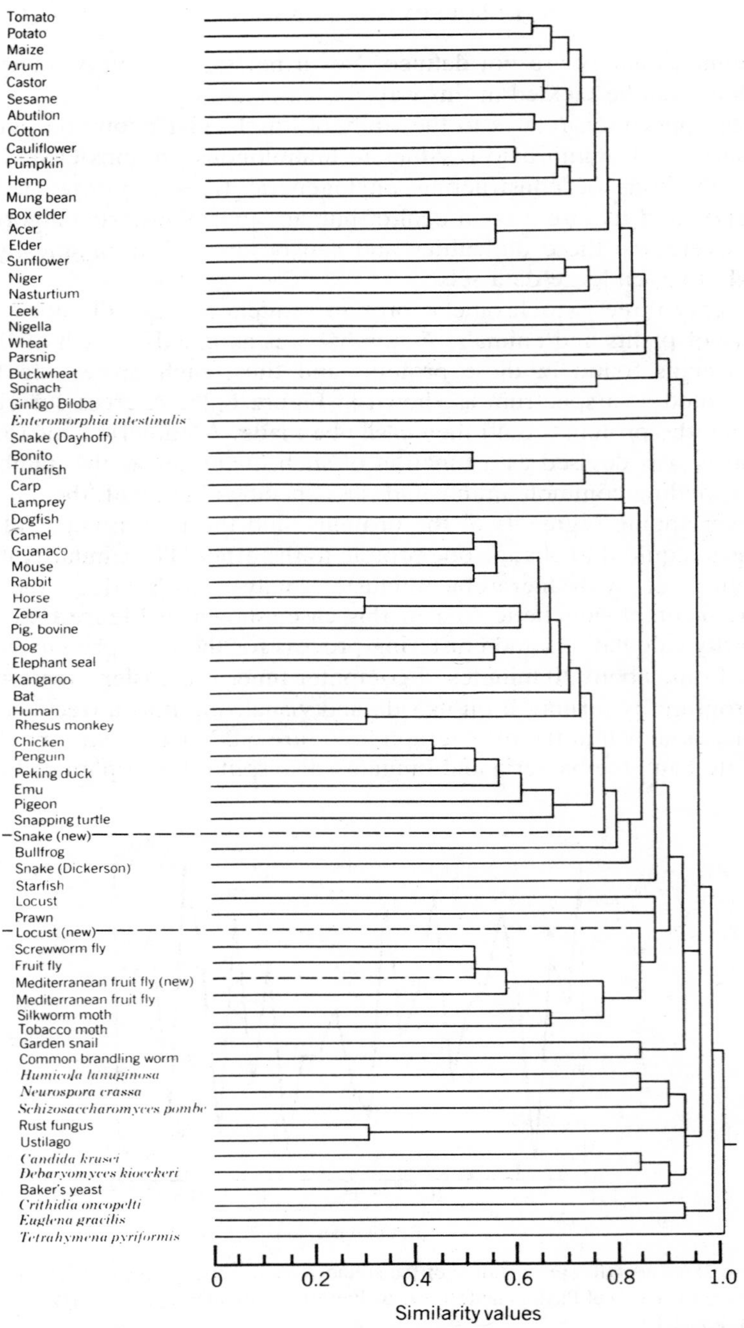

Figure 6.26 Dendrogram, produced by hierarchial cluster analysis, shows the evolution of cytochrome *c*. [From Sharaf et al., "Construction of Phylogenetic Trees by Pattern Recognition," *Z. Naturforsch.*, p. 511 (1980), Fig. 2.]

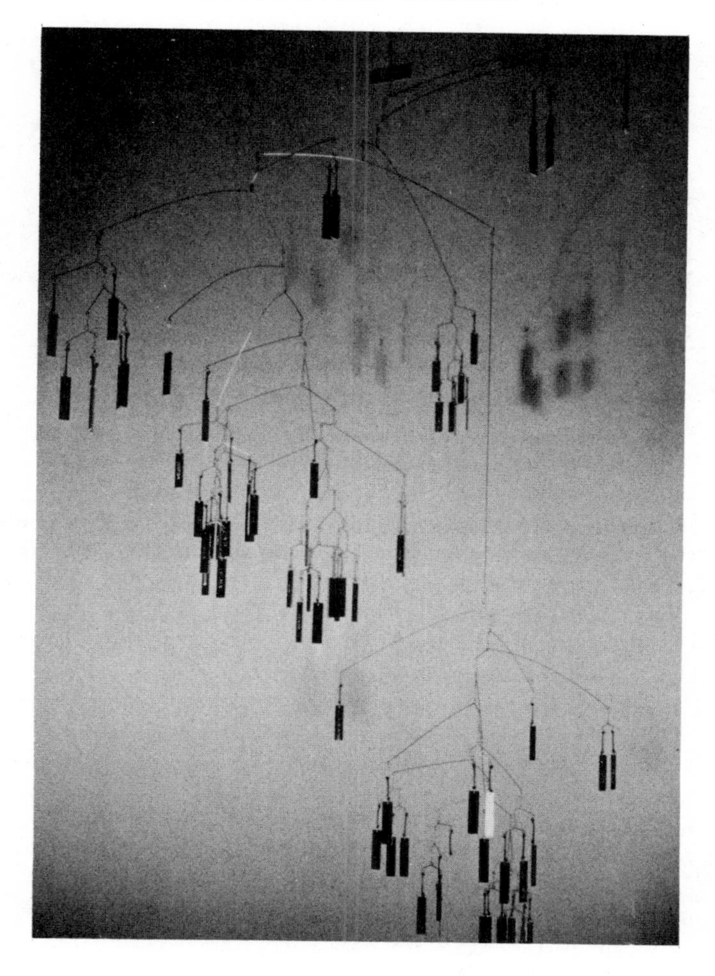

Figure 6.27 A dendrogram can be visualized as a free-swinging mobile, suspended by its single, terminal branch. This mobile represents the cytochrome *c* dendrogram of Figure 6.26; the tags are inscribed with the plant or animal name.

The nonlinear map in Figure 6.28 shows these groups. An interesting result was obtained by mapping the dissimilarity (distance) matrix from its unknown dimensional space successively down to two-, three-, up to n-dimensional space in order to plot the error as a function of dimensionality. In Figure 6.29 it is clear that no significant improvement was obtained beyond six dimensions and thus the intrinsic dimensionality of

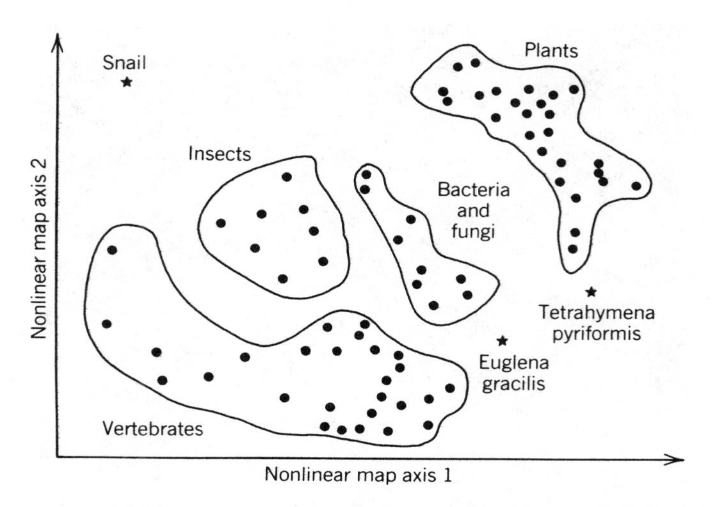

Figure 6.28 Nonlinear map of the cytochrome *c* data. [From Sharaf et al., "Construction of Phylogenetic Trees by Pattern Recognition," *Z. Naturforsch.*, p. 512 (1980), Fig. 4.]

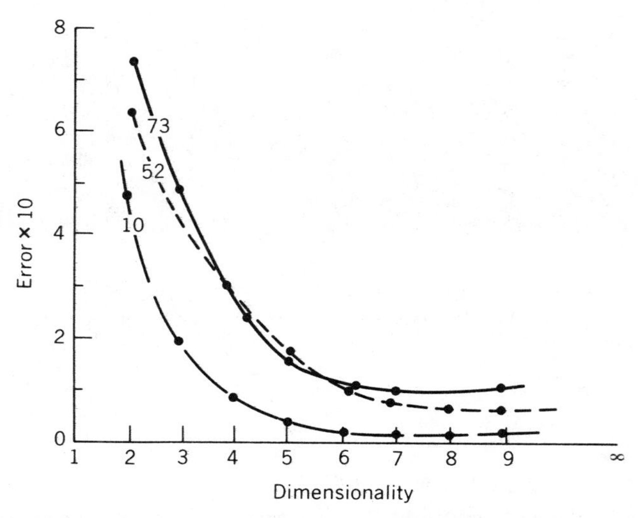

Figure 6.29 "Strain" (error) in the nonlinear mapping as a function of dimensionality. Since no reduction in strain is observed beyond the six dimensions, the latter is suggested to be the intrinsic dimensionality of this protein space. [From Sharaf et al., "Construction of Phylogenetic Trees by Pattern Recognition," *Z. Naturforsch.*, p. 512 (1980), Fig. 3.]

this cytochrome *c* problem is thought to be six. Almost half of the information in the six-dimensional space is represented by the two principal components in the nonlinear map of Figure 6.28.

Anomalies were discovered in the positions of Mediterranean fruit fly, locust, and rattlesnake. It was later discovered that when updated protein sequences were obtained from the more recent literature, the positions changed and made more biological sense.

Another method of cluster analysis generates a *minimal spanning tree* over the data set. A "tree" is formed by connecting the data points together such that each point is a node of the tree, and the sum of all the line segments over the data set is a minimum. An analogous problem would be how to link a set of cities together with roads given a fixed cost per kilometer of road in such a way as to minimize the total cost.

Then, clusters are "pruned" from the tree on the basis of the length of a line segment in comparison to the other line segments nearby. A variety of

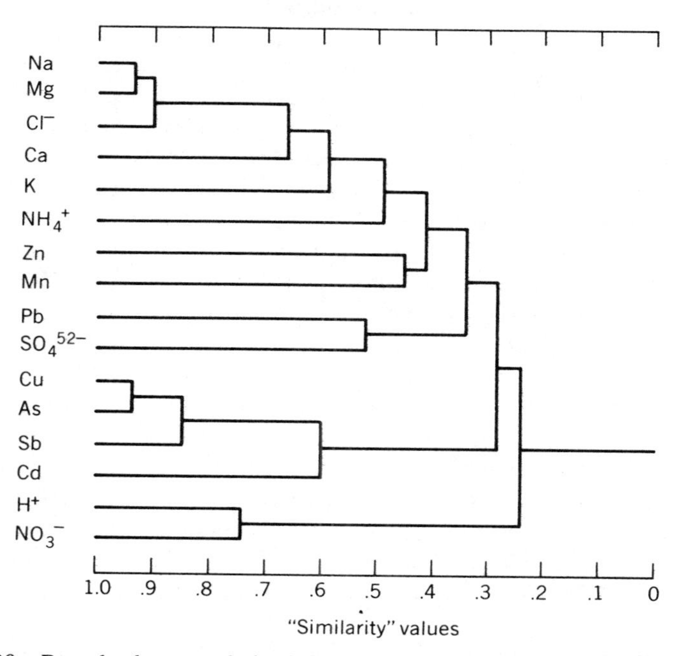

Figure 6.30 *R*-mode cluster analysis of the rainwater study. Groupings of variables here confirm the presence of sea salt, smelter, and urbanization factors. [From Knudson et al., in *Chemometrics; Theory and Application*, B. R. Kowalski, Ed., ACS Symposium Series 52, ACS, Washington, DC., 1977, p. 103, Fig. 12.]

criteria can be used at this point to decide exactly where the branches should be clipped. In this way, clusters are found in the data.

Many more clustering algorithms exist, and the reader is encouraged to consult the Suggested Readings section. However, a different use of hierarchical cluster analysis can yield information about the similarity of the features as opposed to the samples. In *R-mode cluster analysis*, the data matrix is turned on its side, that is, the transpose of the data matrix is obtained, and hierarchical cluster analysis is performed. The resulting dendrogram shows clusters of features and can be used in concert with eigenanalysis to discover underlying factors controlling the system. For example, in the rain study discussed in Varimax rotation, *R*-mode cluster analysis confirmed the existence of the smelter, sea salt, and urban factors as shown in Figure 6.30 (6).

SUPERVISED LEARNING

Introduction

Supervised learning refers to a suite of techniques in which *a priori* knowledge (or assumptions) about the category membership of a set of samples is used to develop a classification rule. The purpose of the rule is usually to predict the category membership for new samples. Sometimes the objective is simply to test the classification hypothesis by evaluating the performance of the rule on a data set.

The classification rule is developed on a *training set* of samples with known classifications. An *evaluation set* is used to rate the performance of the rule by comparing the category predictions with the true classifications. A *test set* is comprised of data vectors whose true categories are unknown.

Usually, one begins a study with just a training set and must decide how to reserve some of the samples for an evaluation set. An effective evaluation method called the *Leave-One-Out* procedure has been developed to address this problem. Given *NP* initial data vectors, each is removed one at a time to be used as an evaluation set while the $NP - 1$ vectors remaining are used as the training set. The rule is trained and evaluated a total of *NP* times. The average performance is a good estimate of the real performance of the classification rule if the training set has been selected carefully.

If the rule is successful for the training set, then the information needed for classification is indeed contained in the data matrix. If not, other measurements should probably be made. If the rule is predictive for an evaluation set, then the training set is a representative sampling of the system. Conversely, a poor predictive ability indicates that more samples are needed.

Another consideration in a classification study is to determine the minimum number of variables that are necessary and sufficient to correctly classify the training set samples. This procedure is called *feature selection* and is useful not only to increase efficiency and reduce costs in a real application but to also provide feedback to the experimental design. Knowing the features that best discriminate between categories also will lead to a better chemical understanding of the system. Feature selection will be discussed later.

Linear Learning Machine

Pattern recognition in chemistry has its origins in the use of the linear learning machine (LLM) (14) to interpret spectroscopic data. In 1969, Jurs et al. (14) applied this technique to mass spectrometry with the objective of determining the molecular formula of molecules containing carbon, hydrogen, nitrogen, and oxygen from their low resolution mass spectra. As the only pattern recognition method used in chemistry before 1972, LLM was also applied to infrared, NMR, and X-ray spectra and polarography during this time (1).

The linear learning machine uses a *linear discriminant function* as the basis for classification. As the name implies, a linear discriminant function is a linear function pertaining to a boundary that divides the n-dimensional space into category regions, and which can be used to make a prediction of category membership for test samples.

The method works in the following way. Training of the LLM is effected by an error-correcting feedback mechanism. The training set data vectors are first *augmented* by adding an $(NV + 1)$th component equal to 1.0 to each vector. Thus, the samples are simply directed distances now in the $(NV + 1)$-dimensional space. This is done because, assuming a two-category problem, we are seeking a boundary, passing through the origin, that correctly separates the samples into their two categories. If more than two categories exist, several linear discriminant functions would be used.

Let us confine our attention to the two-category case and assume that we have only one measurement made on each sample; our original data matrix is one-dimensional as depicted in Figure 6.31*a*. The augmented vectors would appear as in Figure 6.31*b*. In this two-space sample then, the boundary we will find is a $2 - 1 = 1$ dimensional figure, that is, a line in this case. In three-space we would find a plane, and in higher spaces we find hyperplanes of one less dimension than the augmented space itself.

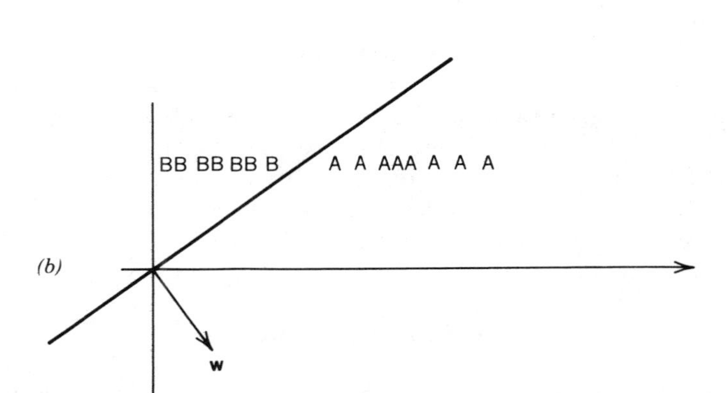

Figure 6.31 Linear learning machine (LLM): (*a*) One-dimensional data set; (*b*) augmented data set, now two-dimensional, showing a final boundary position and associated weight vector **w**.

To find the boundary between the points of categories A and B in our example of Figure 6.31, we will iteratively adjust the elements of a weight vector, **w**, which is the normal to the boundary, such that the dot product of **w** and any vector of **A** is positive, while the dot product of **w** and any vector of **B** is negative.

In other words, the linear discriminant function is expressed by

$$s = \mathbf{w} \cdot \mathbf{y} \tag{6.58}$$

where **y** is any augmented data vector and s is a scalar variable. Remembering that $\mathbf{w} \cdot \mathbf{y} = \|\mathbf{w}\| \|\mathbf{y}\| \cos \theta$, it is easy to see that $\mathbf{w} \cdot \mathbf{A} > 0.0$ means that the angle between those vectors is less than 90°; conversely the angle between **w** and any **B** vector is greater than 90°. Therefore, **w** at the end of the training phase will define a line that separates the two categories.

The goal of the training procedure is therefore to calculate the elements of **w**. The iteration may begin with arbitrary values for the elements of **w**. The first sample is introduced, s is calculated and checked against the correct answer. If the response is correct, no change is made in **w** and the next sample is tested. However, if the response is incorrect, a new weight vector **w** is calculated as

$$\mathbf{w}' = \mathbf{w} + c\mathbf{y}, \tag{6.59}$$

knowing that

$$s' = \mathbf{w}' \cdot \mathbf{y} \tag{6.60}$$

$$c = \frac{s' - \mathbf{w} \cdot \mathbf{y}}{\mathbf{y} \cdot \mathbf{y}} \tag{6.61}$$

which, remembering Equation (6.55) gives

$$c = \frac{s' - s}{\mathbf{y} \cdot \mathbf{y}}. \tag{6.62}$$

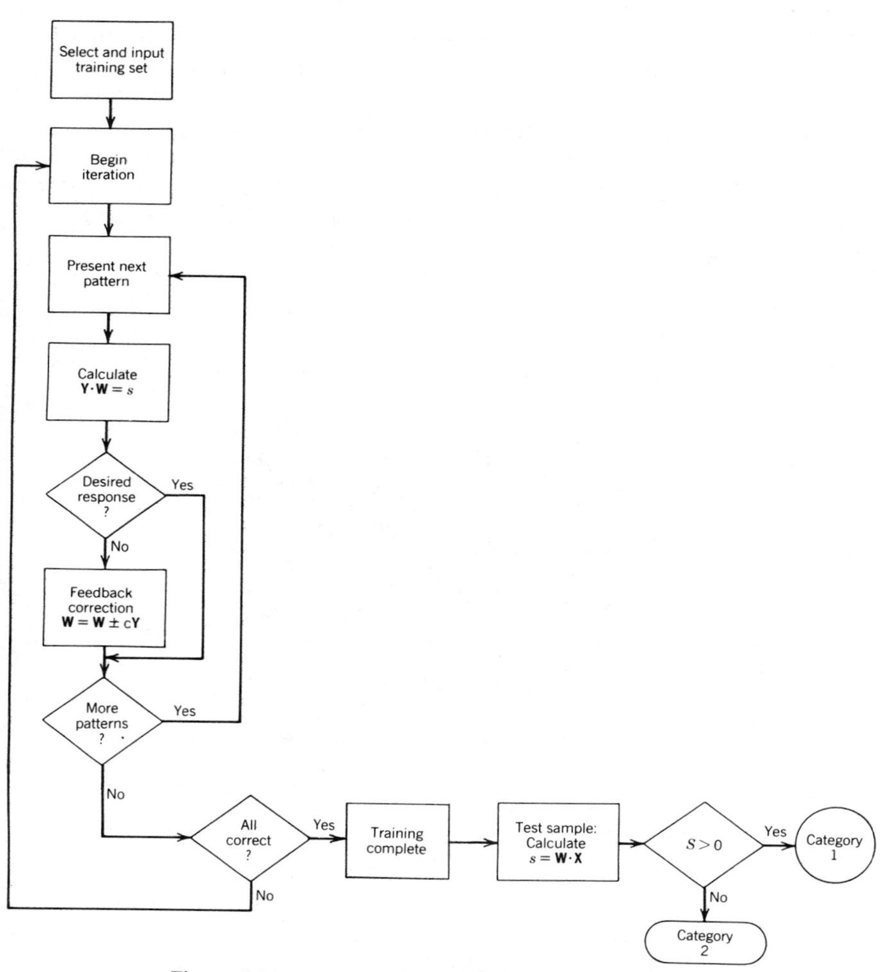

Figure 6.32 Flow chart of the linear learning machine.

An efficient way to calculate c is found in practice to be

$$c = \frac{-2s}{\mathbf{y} \cdot \mathbf{y}} \tag{6.63}$$

which reflects the hyperplane to the correct side of a data point the same distance as it was in error. A flow chart of the entire iterative process is shown in Figure 6.32. In summary, the procedure iteratively changes the position of the hyperplane whose equation is $\mathbf{w} \cdot \mathbf{y} = 0$ until every training set vector is correctly classified.

A major disadvantage of the method is that the solution is nonunique. Unknowns may be classified differently depending on the order in which the training set vectors are presented to the linear learning machine. For example, in Figure 6.33, the final position of the plane separating categories 1 and 2 may be given by any of (a), (b), or (c) depending on the order that these samples were processed in the training phase and on the correction increment used in the feedback process. Obviously, the position is crucial for predicting the category membership of the unknowns (blank circles) in the boundary region.

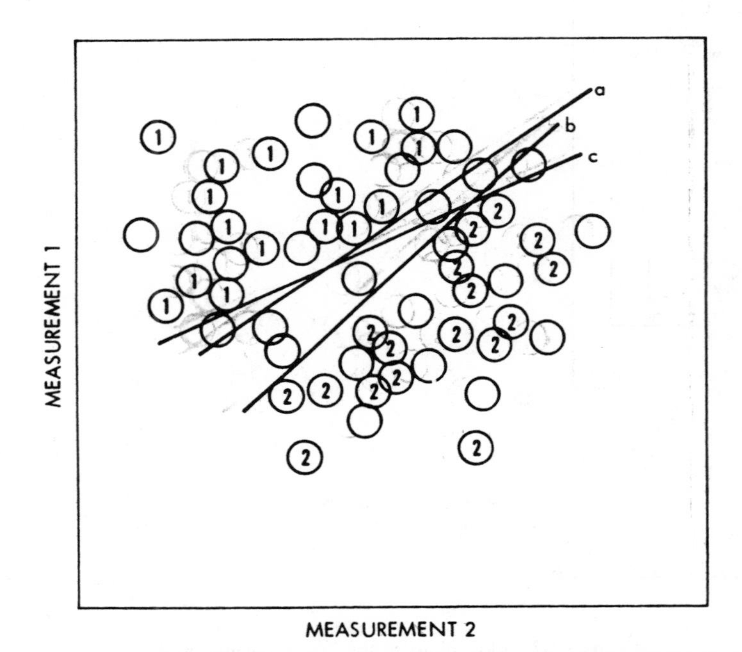

Figure 6.33 Ambiguous solution of the linear learning machine. [From Kowalski and Bender, "The *K*-Nearest Neighbor Classification Rule," *Anal. Chem.* **44**(8), 1406 (1972), Fig. 1.]

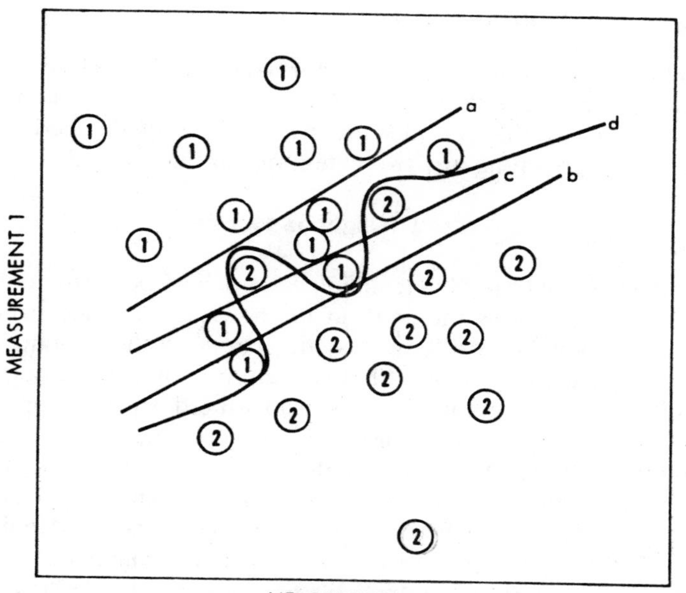

Figure 6.34 Linearly non-separable data set. [From Kowalski and Bender, "The *K*-Nearest Neighbor Classification Rule," *Anal. Chem.* **44**(8), p. 1406 (1972), Fig. 2.]

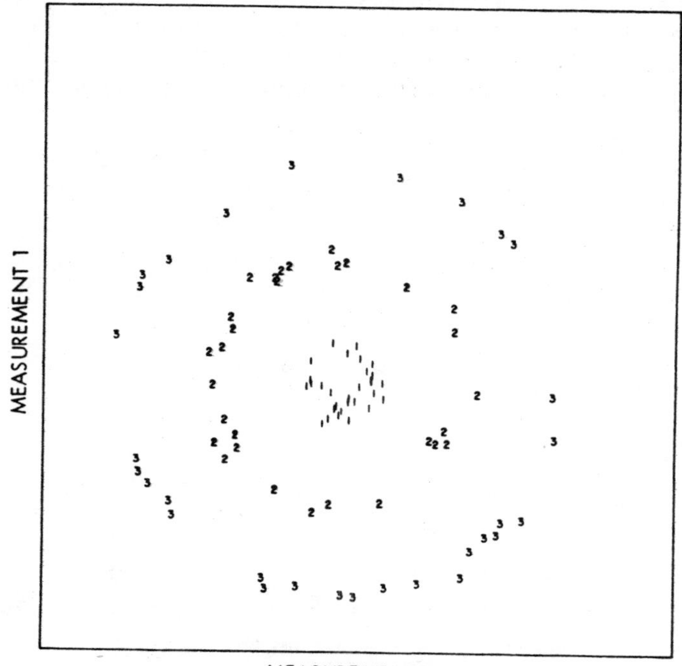

Figure 6.35 Extreme example of linearly non-separable data. [From Kowalski and Bender, "The *K*-Nearest Neighbor Classification Rule," *Anal. Chem.* **44**(8), p. 1407 (1972), Fig. 3.]

Moreover, as a linear method, the performance of LLM is limited by the distribution of the data. Figure 6.34 is an example of a nonlinearly separable data set. In the extreme case, the reader is challenged to find a linear boundary separating any two categories in Figure 6.35.

K-Nearest Neighbor Method

The K-nearest neighbor (KNN) technique (2) is a very powerful tool for its elegant simplicity. In this method, an unknown is classified according to the majority vote of its K-nearest neighbors in the training set in n-dimensional space (Fig. 6.36). Should there be a tie, the closer neighbors are given greater weight. Nearness is measured using an appropriate distance metric as discussed in Unsupervised Learning.

Computationally, KNN is very simple. The matrix of distances of the test and training set points to all other points is computed. Determining the nearest neighbor of any given point amounts to scanning the elements of the upper (or lower) half of the symmetrical distance matrix **D** for the smallest value. In Table 6.5 for example, the first nearest neighbor of test point 4 would be point 2 at a distance of 0.32. Using the 1-NN method point 4 would be assigned the category belonging to point 2. The KNN rule has several advantages over the linear learning machine. First of all, it is inherently a multicategory classifier eliminating the problem of adapting a two-class method to multiclass applications. Secondly, its performance does not depend on the distribution of data points to the degree

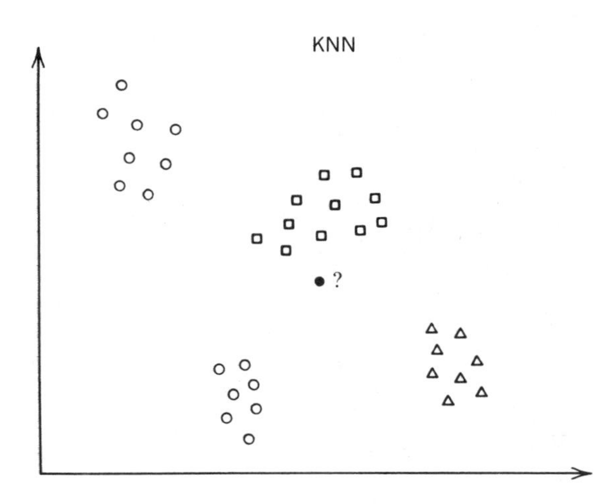

Figure 6.36 The *K*-nearest neighbor method.

Table 6.5 Example of a Distance Matrix

X =	Training set	Sample 1	0.6	0.1	0.2
		2	0.5	0.6	0.7
		3	0.1	0.2	0.3
	Test set	Sample 4	0.5	0.6	0.6

		Sample 1	2	3	4
D =	Sample 1	0	0.71	0.52	0.65
	2		0	0.69	0.32
	3			0	0.64
	4				0

that LLM does. For example, a bimodal distribution within a category (Fig. 6.35) or even more unusual data structure as depicted in Figure 6.35 does not hamper the use of KNN. Furthermore, KNN generates unique solutions; the classification results do not depend on the order in which data are fed to the computer.

Finally, the KNN method has a bounded classification risk. According to parametric pattern recognition techniques, the Bayes decision rule (22) is used to minimize the error associated with classifying unknowns. Assuming that all of the statistics of a data set are known, suffice it to say that the Bayes risk can be calculated exactly and represents the lowest possible classification risk. Given M classes, the risk for KNN can be shown to be bounded; at least half of the classification information is contained in the first nearest neighbor and any other classification method, no matter how sophisticated, can *at most* double the performance of KNN (15).

An example of structure elucidation from NMR spectra using pattern recognition serves to compare the performance of KNN and LLM as well as to illustrate the importance of preprocessing (15). A data base of 198 calculated NMR spectra consists of three categories of compounds: those containing $CH_3CH_2CH_2$—, CH_3CH_2CH—, and CH_3CHCH—. Coupling constants in these spectra were $J_{12} = J_{23} = 7.5$ and $J_{13} = 0.5$. The chemical shifts spanned a typically encountered range of values for such compounds analyzed with a 60 MHz instrument scanned from 0–500 Hz (TMS) and with a 0.5 Hz linewidth and digitized every 0.1 Hz.

Without a crucial preprocessing step the performance of these classification techniques is rather poor. An analysis of the nature of the information representation in an NMR spectrum will reveal why this

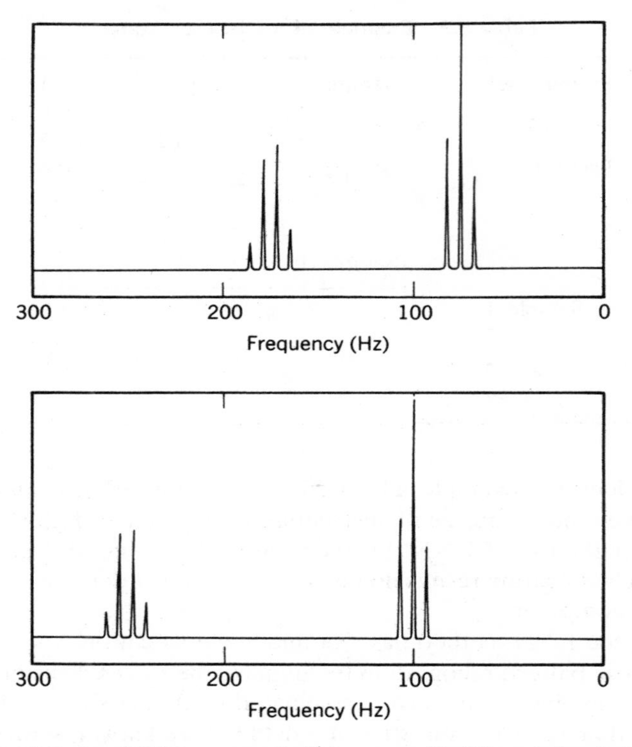

Figure 6.37 Calculated NMR spectra of ethyl groups with different chemical shifts. [From Kowalski and Reilly, "Nuclear Magnetic Resonance Spectral Interpretation by Pattern Recognition," *J. Phys. Chem.* **75**, 140 (1971), Fig. 2.]

should be so. Figure 6.37 contains calculated spectra of ethyl groups with different chemical shifts. Information in these spectra is contained in the positions of the multiplets (i.e., the chemical shifts) and in the fine structure within the multiplet, which tells us about the location and number of other nuclei in the molecule. To the eye it is obvious that these spectra are of ethyl groups despite the large translation differences in the patterns, yet a pattern recognition algorithm would fail to classify these correctly.

Both chemical shifts and splitting information are important to classification and thus they must be present in the data vector in another representation. Preprocessing is necessary in order to map the chemical shift information into a translationally invariant form while preserving the spin splitting information. This can be done using the autocorrelation function of the digitized NMR spectrum. The autocorrelation function

$A(\tau)$ of a digitized NMR spectrum $F(\tau)$ is given by,

$$A(\tau) = \int F(\nu)F(\nu + \tau)\, d\nu. \tag{6.64}$$

For the simple case of an *AB* stick spectrum in Figure 6.38*a*, the autocorrelation spectrum has the form as shown in (b). When applied to the NMR spectra of ethyl groups (Fig. 6.37) the autocorrelation functions (here truncated) in Figure 6.39 are obtained. The translational differences have been eliminated, but not lost since they appear now as differences in the shapes of the autocorrelation functions, a more useful information representation for pattern recognition.

The spectra to be used in the classification study were also normalized to unity to remove the variance in the data due to the difference in the total number of protons.

Linear discriminant analysis was performed on half of the samples and tested with the other half; the training and test sets were then switched and the procedure repeated. The KNN method was evaluated using the Leave-One-Out procedure. Table 6.6 contains the classification results using these two methods. These results suggest that the spectra are not

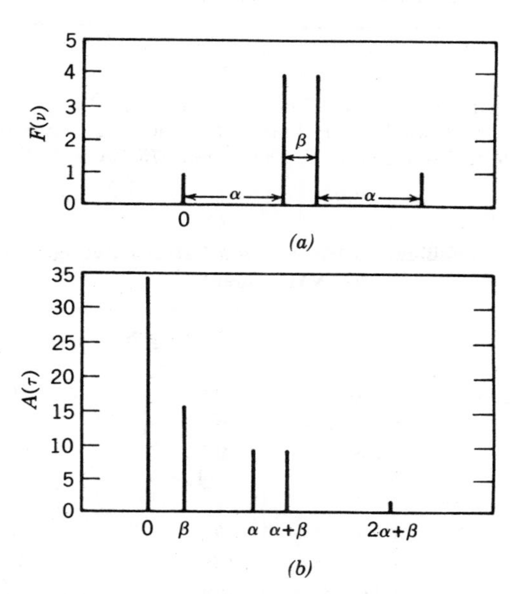

Figure 6.38 (*a*) AB stick spectrum and (*b*) its autocorrelation function. [From Kowalski and Reilly, "Nuclear Magnetic Resonance Spectral Interpretation by Pattern Recognition," *J. Phys. Chem.* **75**(10), p. 1406 (1971), Fig. 3.]

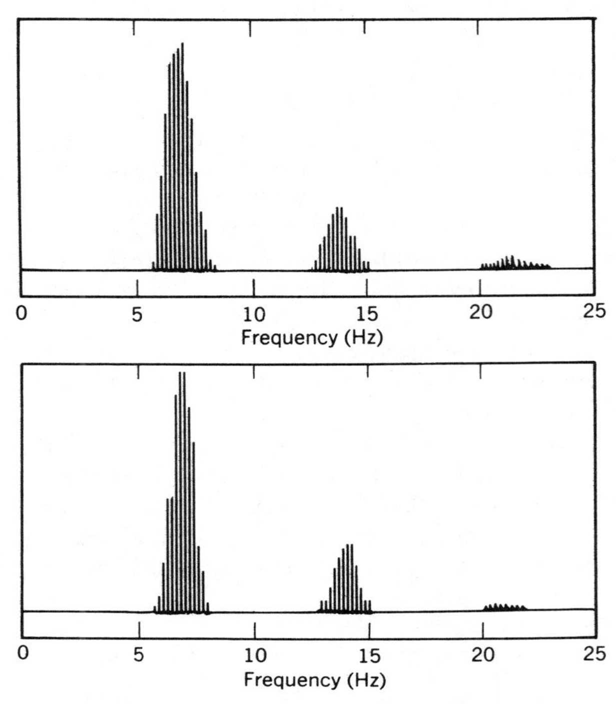

Figure 6.39 Truncated autocorrelation functions for calculated ethyl group NMR spectra of Figure 6.37. [From Kowalski and Reilly, "Nuclear Magnetic Resonance Spectral Interpretation by Pattern Recognition," *J. Phys. Chem.* **75**(10), p. 1408 (1971), Fig. 5.]

Table 6.6 Classification Results in a Pattern Recognition Study of NMR Spectra

		Training Set	Test Set
One nearest neighbor	Class 1	60/66	60/66
	Class 2	60/66	60/66
	Class 3	64/66	64/66
	Total	184/198 = 93%	184/198 = 93%
Learning machine	Class 1	56/66	54/66
	Class 2	12/66	12/66
	Class 3	24/66	23/66
	Total	92/198 = 46%	89/198 = 45%

Source: Reference 15.

linearly separable, because the 46% correct classification obtained by LLM is not too much better than the 33% rate that one could achieve simply by guessing. The KNN method on the other hand does very well on these data at 93% correct classification.

Feature Selection in Classification

A priori, we do not know which variables will be important in a classification problem. The best strategy is always to measure many things at first and then calculate the feature weights to determine the most discriminating variables. As discussed in Feature Weight, one goal of feature weighting is to obtain a quantitative measure of the discriminating ability of a variable in the context of the classification hypothesis posed about the data. Another purpose is then to scale—actually "stretch"—the axes in proportion to their weights so as to emphasize the "good" variables and dilute the "bad." Finally, these weights together with information about the intervariable correlations lead to a means of *feature selection*: to select an optimal set of features that are necessary and sufficient to solve the classification problem.

It is often advisable to reduce the number of features used in a classification problem for a variety of practical reasons. Feature selection saves time and reduces costs, both for the chemical analysis as well as for information acquisition, storage, and processing. It also increases the ease and efficiency of the classification procedure. Feature selection can also lead to an understanding of the essential features that play important roles in the chemistry of the system. And lastly, it is important feedback to the experimental design. It is capable of revealing which measurements are discriminating and which are not, or perhaps that they are all correlated, or that even *none* of them is useful. However, even such a negative result is extremely important, because then the experimental design can at least be altered. The use of such a technique prevents the scientist from beating his or her head against a wall of data which contains no information about the problem.

In designing a method to select an optimal set of features, the reader may imagine that the variables with the highest loadings in the eigenvectors would contain the most information about the data. However, Figure 6.40 shows at a glance that eigenvectors are *modeling* but not necessarily *discriminating*. Obviously, the variance or Fisher weights must be taken into consideration. The problem with using the weight criterion alone is that variables may also be very highly correlated, that is, dependent, and thus redundant. Moreover, it is rather cumbersome and arbitrary to try to eyeball the table of weights and the correlation matrix

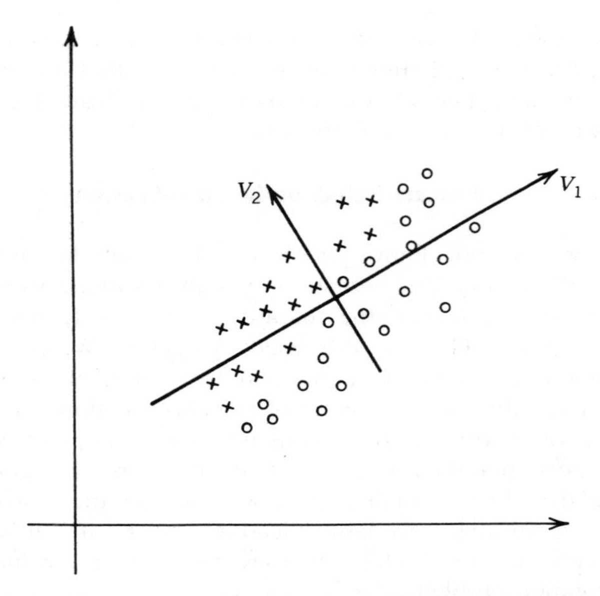

Figure 6.40 Eigenvectors are modeling but not necessarily discriminating.

simultaneously as a feature selection method. A more rigorous approach would be very handy.

One efficient way to perform feature selection, called SELECT (16), begins by choosing the feature, call it x_s, with the highest variance or Fisher weight. The remaining $NV-1$ variables are transformed in such a way as to remove their correlation to x_s. Of the transformed features, the one with the highest weight is chosen as the second selected feature, and the process repeats until either the desired number of features have been selected, or until the highest weight of the transformed features falls below a specified tolerance.

To make vectors uncorrelated is to make them orthogonal, that is, their dot product is zero. Thus, for each of the $NV-1$ vectors remaining after the first feature selection, we impose the following constraints. We wish to find a new vector $\mathbf{x}'_k$ which is constrained to the plane of $\mathbf{x}_s$ and $\mathbf{x}_k$ as in Figure 6.41, that is,

$$\mathbf{x}'_k = a\mathbf{x}_s + b\mathbf{x}_k, \tag{6.65}$$

and so we wish to determine the values a and b. Knowing that $\mathbf{x}_s$ and $\mathbf{x}'_k$ are perpendicular, we can write

$$\mathbf{x}_s \cdot \mathbf{x}'_k = 0, \tag{6.66}$$

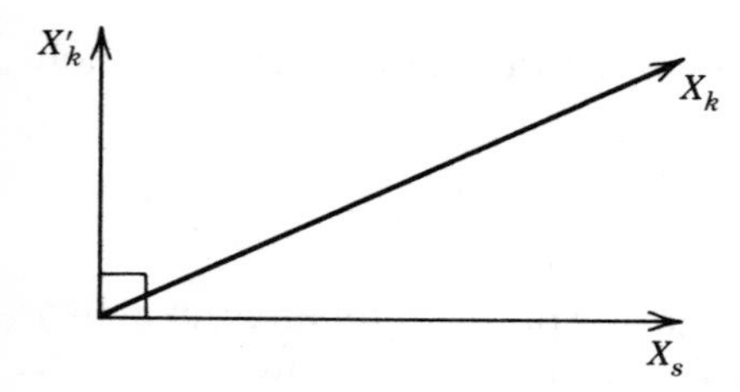

Figure 6.41 Relationship of vectors in feature selection.

or, substituting for $\mathbf{x}_k$ from Equation (6.64),

$$\mathbf{x}_s \cdot (a\mathbf{x}_s + b\mathbf{x}_k) = 0 \tag{6.67}$$

and

$$a(\mathbf{x}_s \cdot \mathbf{x}_s) + b(\mathbf{x}_s \cdot \mathbf{x}_k) = 0. \tag{6.68}$$

However, we impose the constraint that $\mathbf{x}_s$ is normalized to unity, so

$$\mathbf{x}_s \cdot \mathbf{x}_s = 1 \tag{6.69}$$

and therefore,

$$a + b(\mathbf{x}_s \cdot \mathbf{x}_k) = 0. \tag{6.70}$$

We have determined a value for a, since $\mathbf{x}_s \cdot \mathbf{x}_k$ is a calculable quantity:

$$a = -b(\mathbf{x}_s \cdot \mathbf{x}_k). \tag{6.71}$$

We also constrain the vectors $\mathbf{x}_k$ and $\mathbf{x}'_k$ to be normalized to unity:

$$\mathbf{x}_k \cdot \mathbf{x}_k = 1 \tag{6.72}$$

and

$$\mathbf{x}'_k \cdot \mathbf{x}'_k = 1. \tag{6.73}$$

Therefore, by Equation (6.64)

$$(a\mathbf{x}_s + b\mathbf{x}_k)(a\mathbf{x}_s + b\mathbf{x}_k) = 1 \tag{6.74}$$

and

$$a^2 + b^2 + 2ab(\mathbf{x}_s \cdot \mathbf{x}_k) = 1. \tag{6.75}$$

Substituting the value for a from Equation (6.71), we obtain

$$b = \left[\frac{1}{1-(\mathbf{x}_s \cdot \mathbf{x}_k)^2}\right]^{1/2} \quad \text{and} \quad a = -\left[\frac{(\mathbf{x}_s \cdot \mathbf{x}_k)^2}{1-(\mathbf{x}_s \cdot \mathbf{x}_k)^2}\right]^{1/2}. \tag{6.76}$$

It is instructive to generate a very small data matrix of four samples and three correlated variables and manually perform the feature selection described in this section.

SIMCA

In chemistry, often the scope of a data analysis problem requires an ability to go beyond simple classification of samples into known categories. For example, it is very often desirable to have a means to detect outliers—members of a new, unforeseen category or perhaps erroneous data. Similarly, it is often useful to furnish an estimate of the level of confidence in a classification result—how likely is an object to be a member of class A or B? This is something that strictly nonparametric pattern recognition procedures cannot do. Also of interest is the ability to empirically model each category so that it is possible to make quantitative correlations and predictions with external continuous properties. As a result, a modeling and classification method called SIMCA has evolved to provide these capabilities (17–19). The name SIMCA may be thought of as an acronym for simple modeling of class analogy, among other things.

In the SIMCA method, a principal component model is fit to each category and confidence envelopes (volumes or hypervolumes) are constructed around the model to contain the data points. The closed class envelope for each category is derived by first doing a principal component analysis separately on each class. The number of principal components used in the model may be pre-selected or determined by cross-validation (discussed next). If one principal component were used, the data would be modeled by the class mean and a line (the first principal component). If two principal components were used, the class mean and a plane would form the model. These models are schematically depicted in Figure 6.42. Mathematically, a datum $\mathbf{x}_{ik}$ in category q is expressed as

$$(x_{ik})_q = (\bar{x}_k)_q + \sum_{j=1}^{NC} (v_{kj})_q (y_{ij})_q + (e_{ik})_q \tag{6.78}$$

where $\bar{x}_k$ is the category mean for variable k; v_{kj} are loadings of the NC principal components used in the model; y_{ij} are the scores of coordinates of the points within the reference frame of the model; and e_{ik} is the residual describing how the model represents the datum.

In this way, the SIMCA method provides a set of parameters that characterize each category and are the basis for other descriptive quantities that shed light on the data structure, as we shall presently see.

The significance of the coefficients v_{kj} have been discussed earlier in this chapter; they reveal the importance of each variable to the principal component model.

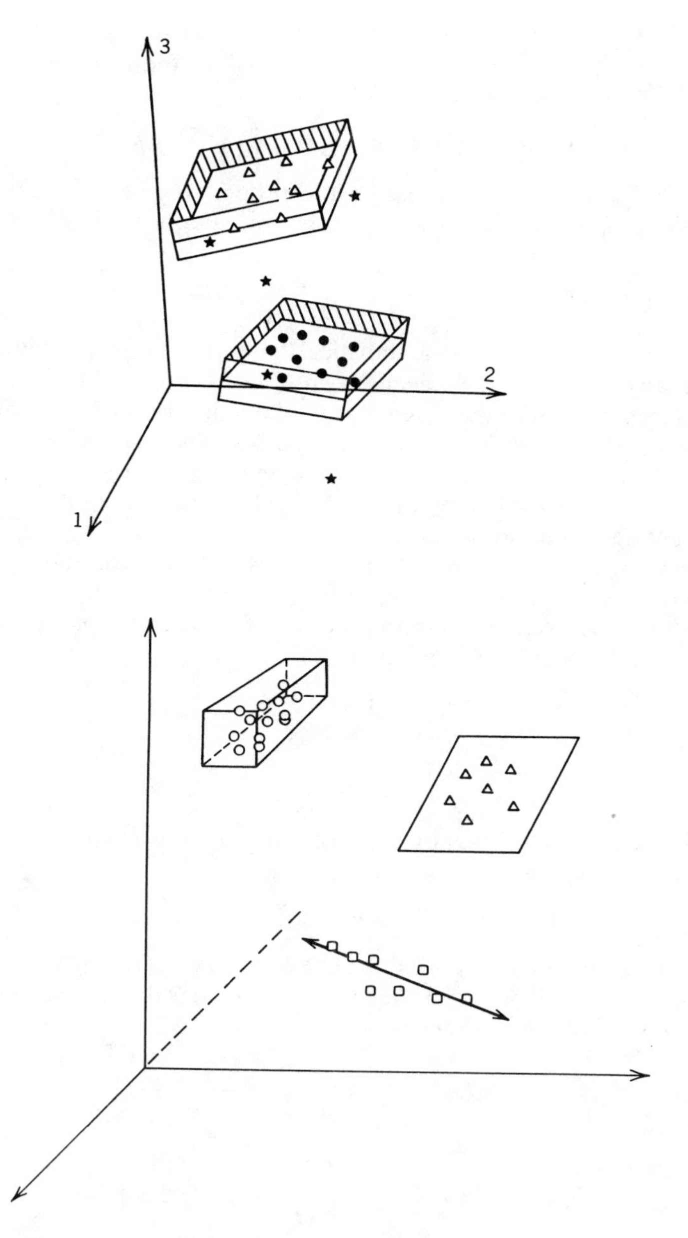

Figure 6.42 SIMCA class models. (upper figure from ref. 19)

The SIMCA method leads to an expression for the *residual variance in class q*, which is a measure of the "tightness" of the category cluster:

$$s_0^2 = \sum_{i=1}^{NP} \sum_{k=1}^{NV} \frac{(e_{ik})^2}{(NP - NC - 1)(NV - NC)}, \tag{6.79}$$

where the denominator represents the proper degrees of freedom. The *residual variance of object i to the model of category q* is given by

$$s_i^2 = \sum_{k=1}^{NV} \frac{(e_{ik})^2}{(NV - NC)}. \tag{6.80}$$

This immediately provides a statistical basis for judging whether a data point is an outlier or not by means of an F test. If s_i^2 is of the same size as s_0^2 the object is typical of the category members; if it is much greater than s_0^2, it is very distant from the model and can be considered an outlier to that model. If the same is true for all categories, then the point is truly an outlier, perhaps belonging to an unforeseen category or perhaps representing some faulty data.

Quantitatively, a "normal range" is established for the scores with respect to each principal component, using the standard deviations s_j where j ranges from 1 to the total number of principal components in the model. The bounds would be given by

$$y_j(\max) + s_j c \tag{6.81}$$

and

$$y_j(\min) - s_j c, \tag{6.82}$$

where c is a coefficient set by the user, typically either 1.0 or varying with the number of objects in the group as the t-statistic;

$$c = c_0 t, \tag{6.83}$$

where c_0 is set typically to 0.5. The SIMCA method also provides a measure of the modeling power of variable k. The *residual variance of variable k* in class q is defined as

$$s_k^2(\text{error}) = \sum_{i=1}^{NP} \frac{e_{ik}^2}{(NP - NC - 1)}. \tag{6.84}$$

The *meaningful variance in variable k* is given by

$$s_k^2(x) = \sum_{i=1}^{NP} \frac{(x_{ik} - \bar{x}_k)^2}{(NP - 1)}, \tag{6.85}$$

which is the familiar expression for the variance. The ratio of $s_k^2(\text{error})$ to

$s_k^2(x)$ gives the "noise to signal" ratio for this variable; as the ratio increases, the information content decreases until in the limit, the ratio approaches 1.0. The *modeling power of variable k* is defined to be

$$R_k = 1 - \frac{s_k(\text{error})}{s_k(x)}. \tag{6.86}$$

As R_k approaches 1.0, the variable is highly relevant; conversely, as it approaches 0.0, the variable approaches zero utility in the model.

A measure of the separation of the categories in n-space can be derived by again making use of the residual variances—tightness—of each category. The *distance between categories p and q* is defined as

$$D_{pq}^2 = \frac{s_{pq}^2 + s_{qp}^2}{(s_0^2)_p + (s_0^2)_q}, \tag{6.87}$$

where s_{pq}^2 is the residual variance of the points in category p fit to the model of q, and vice versa for s_{qp}^2, computed as in Equation (6.84), computed over all of the variables. As this quantity increases, the intercategory separation improves.

The *discriminating power of a variable*, $\phi(k)$, is obtained analogously to the distance between categories except that the residuals are computed for that variable alone. For category pair p and q, the discriminating power of variable k is

$$\phi_{pq}(k) = \left[\frac{[s_{pq}(k)]^2 + [s_{qp}(k)]^2}{[s_{pp}(k)]^2 + (s_{qq}(k)]^2}\right]^{1/2} - 1, \tag{6.88}$$

where the denominator contains the residual variance of the points in p fit to the model of p for variable k, and a similar term for category q. A value of $\phi_{pq}(k)$ close to zero indicates a low discriminating power, and much above one, a good power, much like the variance and Fisher weights. An aggregate measure is obtained over all category pairs by taking the average of their respective values of $\phi(k)$.

Thus far, we have seen just some of the ways in which the SIMCA method may be used to characterize the data structure: to *model* the category data. The SIMCA method may furthermore be used to classify test points in the following way. Given a data vector, x_t, we would like to determine its scores on each of the category models, using linear regression to minimize the sum of the squared errors in each case. Then, by examining the residual variance of x_t with respect to each of the models, a category assignment may (or may not) be made.

Specifically, estimates $\hat{y}$ for category q are obtained in the following expression by linear regression so as to minimize $\sum e_{ij}^2$:

$$x_{tk} - (\bar{x}_k)_q = \sum_{j=1}^{NC} (v_{kj})_q (\hat{y}_{tj})_q + (e_{tk})_q. \tag{6.89}$$

Then, a measure of the fit of this test point is obtained directly from Equation (6.80):

$$s_t^2 = \sum_{k=1}^{NV} \frac{(e_{tk})_q^2}{(NV - NC)}. \tag{6.90}$$

If s_t^2 is of the same order as s_0^2,, then the sample belongs to the class; however, if s_t^2 is much larger, it is not typical of that class and thus would not be assigned to it. Again, an F test may be used as a quantitative measure in this classification criterion. When a test point has score estimates falling outside the normal range for the category, the distance $(d_t)_q$ to the model is calculated from

$$(d_t)_q = (s_t^2)_q + [\hat{y}_{tj} - (y_j)_{\text{bound}}]^2 \phi, \tag{6.91}$$

where j refers to the principal component involved, t to the test point, $(y_j)_{\text{bound}}$ is one of the normal range extrema, and ϕ is a coefficient introduced to make $(s_t^2)_q$ and $[\hat{y}_{tj} - (y_j)_{\text{bound}}]^2$ comparable:

$$\phi = \left(\frac{s_0}{s_y}\right)^2. \tag{6.92}$$

This measure d_t can be seen graphically in Figure 6.43.

Pattern recognition problems in chemistry can be divided into four basic types according to the level and scope of the study (19):

Level 1	Simple classification into predefined categories
Level 2	2-O: Level 1 plus outlier detection
	2-A: Asymmetric case
Level 3	Level 2 plus prediction of one external property
Level 4	Level 2 plus prediction of more than one external property

Although SIMCA can be used at Level 1, it automatically operates at Level 2-O to provide outlier detection. It may also be used at this level to solve asymmetric classification problems. Instead of assuming that all categories can be contained in closed class envelopes, this form of SIMCA allows for the possibility that only one category is modelable with principal components while the other data points may spread randomly through n-space (see Fig. 6.44). This binary classification problem where one alternative is considered against all others is very

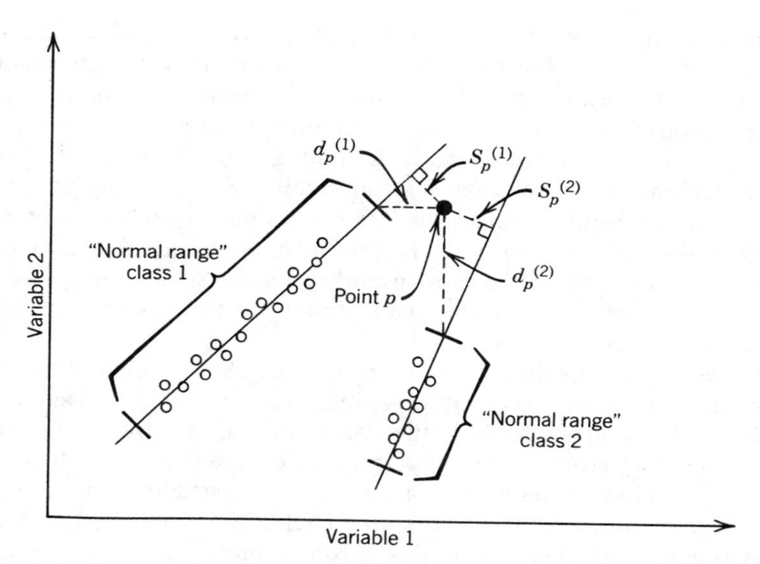

Figure 6.43 Distance of a point to the SIMCA model. [From Wold and Sjostrom, in *Chemometrics: Theory and Application*, B. R. Kowalski, Ed., ACS Symposium Series 52, ACS, Washington, D.C., 1977, p. 263, Fig. 5.]

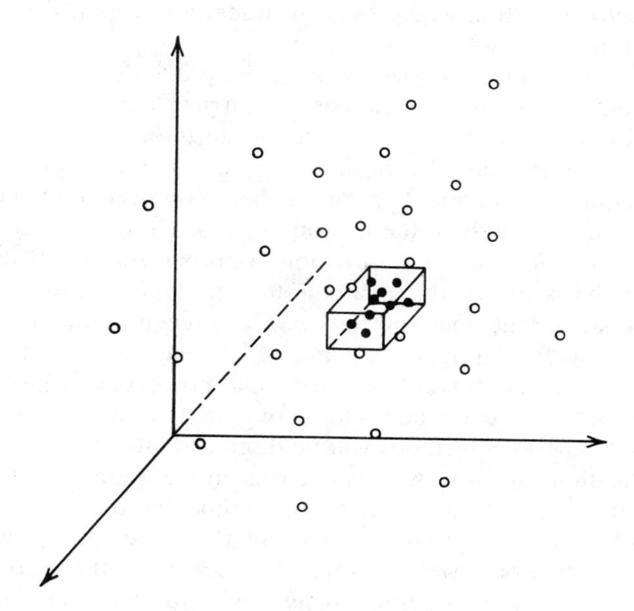

Figure 6.44 The asymmetric case.

commonly encountered in materials science, product quality, and structure-activity studies. For example, there may be many ways to manufacture steel that result in a "poor" product. Consequently, the data points representing these "poor" samples would be spread out through the n-dimensional measurement space, assuming that the chemical or physical measurements made are at least indirectly related to product quality. On the other hand, "good" steel samples must meet stringent specifications and thus may be contained in a small closed class envelope in n-space. Often, one of the most useful results of such a study is to find out if the measurements traditionally employed to evaluate samples are related to product quality at all.

Proceedings to Level 3, it is desirable to be able to relate the value of an external property, such as breaking strength, to the measurements made on the samples, such as the concentration of various elements in the steel. Most structure-activity studies, for example, seek to relate a numerical value of biological activity to the structural features and molecular parameters used to describe each sample (molecule). A Level 3 analysis would accomplish this through multiple regression of the external variable on the scores for the data points of a given category in its SIMCA model. A level B analysis thus permits both the classification of a new sample and the prediction of the magnitude of its associated external property. This concept is reminiscent of the clinical example in Eigenvector Rotation where liver disorders are classified as to *type* and *extent* of disease.

Finally, a Level 4 analysis involves the prediction of several external properties. A two-phase analysis is carried out where closed class envelopes are constructed for a category in the measurement space *and* in the property space. The property space may be considered as simply an additional measurement space. In the second phase of the analysis, a direction in each class of the measurement space is obtained such that it is maximally correlated to a direction in property-space. If the property-space model is linear, this amounts to a multiple regression: if higher-dimensional, a canonical correlation. More recently, the method of Path Modeling by Partial Least Squares (PLS) has been shown to be the method of choice for relating external properties, whether a single variable, or two or even more blocks of data, to a set of measurements on a data set. The PLS methods will be discussed later.

Applications of SIMCA are numerous in the literature. It is important to note that the success of any given method can only be as good as the formulation of the approach. A particularly challenging problem and an early application of SIMCA was the identification of the source (tanker) of oil spills by pattern recognition analysis of their elemental concentrations

(20). Besides serving as an example of SIMCA classification, this study is particularly instructive in *how* to formulate a pattern recognition approach to a complex problem.

The problem to the chemist in this study may be stated as follows: Given a spill sample as received from the field and a group of possible source samples, can the true source be identified (or recognized as unknown) from analytical data, and with what confidence can the identification be reported?

Perhaps the simplest way to solve such a problem, at least from the chemist's point of view, is to have tagged the oils in the tankers from the outset with a known chemical or physical label. Lacking that easy way out, however, the only way to identify a source is to make use of the natural composition of the oils. Complicating this task is the fact that the analysis of the spill sample as received from the field will differ from the analysis of the true source sample for at least three reasons: weathering, contamination, and analytical error.

Weathering of oil may be caused by photolysis, polymerization, emulsification, extraction, evaporation, and biological attack, among others, but the exact conditions to which a sample was exposed remain uncertain. Contamination may be due to salts, sand, water, and organic material. In comparison to these sources of uncertainty, the analytical error is probably insignificant.

The central question is how to design the study so that oil samples may be classified by their source regardless of the history (weathering, contamination, length of time, location, etc.) of the sample. The reader is encouraged to devise one or more possible approaches to this problem before reading on.

In this study, 20 crude and 20 residual fuel oils were chosen to be representative of oils shipped on U.S. waterways (Table 6.7). Ten different weathering treatments (Table 6.8) were given to aliquots of each oil type. The data set was therefore comprised of 40 categories (oil type) with ten treatments performed on each oil type, yielding 400 total data vectors. The samples were analyzed after their weathering treatment with Neutron Activation Analysis for the elements in Table 6.9; unfortunately, there was a considerable amount of missing data where the concentrations were below the detection limit. The starred elements in Table 6.9, occurring for essentially all of the samples, were used in a separate data analysis to provide a sort of "lower bound" for the data interpretation. These variables are the most reliable, but perhaps not the most useful or discriminating. An "upper bound" estimate was obtained using all 22 variables.

The variance weights computed from the autoscaled data are shown in

Table 6.7 Source of Samples

Crude	Residual	Origin	Supplier
1		East Texas Field, Texas	Bureau of Mines
2		El Morgan, Egypt	Bureau of Mines
3		Wilmington, California	Bureau of Mines
4		Minas Field, Sumatra	Bureau of Mines
5		Duri Field, Sumatra	Bureau of Mines
6		Export, Neutral Zone, Saudi Arabia–Kuwait	Bureau of Mines
7		Kuwait Export, Kuwait	Bureau of Mines
8	21	Ceuta, East Shore Lake Maracaibo, Venezuela	Gulf
9	22	Mesa, Orinoco Basin, Venezuela	Gulf
10	23	Timbalier Bay, Offshore Louisiana	Gulf
11	24	Sarir, Sirte Basin, Libya	Gulf
12	26	Agha Jari, Iran	Gulf
13	27	Orito, Putumayo Basin, Columbia	Gulf
14	28	Kuwait, Export Blend	Gulf
15	29	Arabian Light, Export Blend	Gulf
16		Ward-Estes, North Field, Texas	Bureau of Mines
17		Goldsmith Field, Texas	Bureau of Mines
18		Kelly-Snyder Field, Texas	Bureau of Mines
19		Sprayberry (Trend Area), Texas	Bureau of Mines
20		Headlee, Texas	Bureau of Mines
	25	Minas Field, Sumatra	Gulf
	30	Philadelphia No. 6	Gulf
	31	Santa Fe No. 6	Gulf
	32	Vengref Menez Bunker C	Gulf
	33	Port Arthur No. 6	Gulf
	34	Caribbean 200 vis.	Gulf
	35	Vengref Menez No. 6	Gulf
	36	Zakum	Gulf
	37	Nigerian	Gulf
	38	Cabinda	Gulf
	39	Oficina	Gulf
	40	Sumatran	Chevron Research

[a]*Source:* Reference 20.

Table 6.8 Weathering Treatments Given to Aliquots of Each Oil Type

Aliquot	Treatment
0	Unweathered
1	12-hour gentle contact with fresh water
2	24-hour gentle contact with fresh water
3	12-hour vigorous contact with fresh water
4	24-hour vigorous contact with fresh water
5	12-hour gentle contact with sea water
6	24-hour gentle contact with sea water
7	48-hour gentle contact with sea water
8	12-hour vigorous contact with sea water
9	24-hour vigorous contact with sea water

Source: Reference 20.

Table 6.9 Elements Assayed by Neutron Activation Analysis in Oil Samples

Element[a]	Measurement Means, $\bar{x}$	Percent of Categories Having Data	Variance Weights
*Na	54.89	100.0	2.97
*Al	2.088	100.0	1.77
*S	11870.	100.0	8.61
*Cl	94.92	100.0	2.23
*V	48.25	100.0	51.37
*Mn	0.09758	97.5	7.46
Co	0.2621	15.0	2.18
Ni	20.77	65.0	7.81
Zn	4.359	17.5	2.42
Ga	0.07844	40.0	3.72
As	0.07084	42.5	4.32
*Br	0.5611	100.0	2.14
Sr	0.03493	2.5	1.02
Mo	0.1450	7.5	1.26
In	0.000105	12.5	1.81
Sn	0.1921	17.5	1.93
Sb	0.006205	12.5	1.07
I	0.2524	50.0	3.47
Ba	0.3086	40.0	4.65
La	0.01839	7.5	1.50
Sm	0.000470	5.0	1.33
Au	0.000024	2.5	1.05

Source: Reference 20.

[a]Starred elements are detectable for essentially all oils.

Table 6.9. Vanadium has the highest weight at 51.37; the descriminating ability of this variable may indeed be due to its ability to survive intact during weathering because it is tightly bound in the porphyrin rings in the oil. Sulfur has the next highest weight followed by Nickel and Barium.

An eigenvector projection of the variance weighted, preprocessed data using all 22 elements is shown in Figure 6.45. Nine of the data points in each of the 40 categories are not shown, but enclosed by a circle, with a solid dot indicating the position of the unweathered oil. The effect of weathering has a directional character, seen usually as a trailing away from the unweathered sample: such anisotropy is to be expected since elements should concentrate or be diluted with washing.

The KNN and SIMCA results are compared in Table 6.10 for both

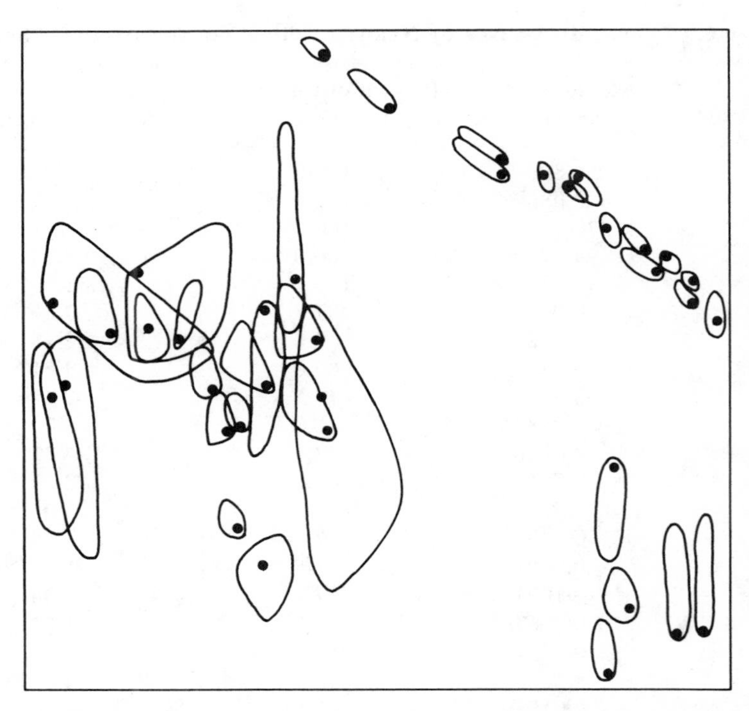

Figure 6.45 Eigenvector projection of the variance weighted, 22-variable oil data containing 99% of the variance. Category regions are enclosed by circles with the unweathered oil data point represented by (•). Numbers indicate category numbers. [From Duewer et al., "Source Identification of Oil Spills by Pattern Recognition Analysis," *Anal. Chem.* **47**(9), p. 1577 (1975), Fig. 2.]

Table 6.10 KNN and SIMCA Misclassification Results in the Oil Study

		KNN[a]		SIMCA[a]	
egory	# Variables	7	22	7	22
1		1(24)	0	1(24)	0
2		3(3)	0	0	0
3		5(2,22,27,30)	0	5(2,6,28)	0
4		2(20)	0	0	0
5		1(25)	1(7)	1(37)	0
6		0	1(7)	0	0
7		3(14)	1(6)	2(14)	0
8		0	0	0	0
9		4(12,27,35)	0	3(27,33)	0
10		2(18)	0	0	0
11		3(19,25)	0	2(19)	0
12		2(9,35)	1(35)	0	0
13		2(20,27)	1(5)	2(14)	0
14		4(7,16,28)	1(28)	0	0
15		2(14,16)	2(17,36)	2(16,36)	1(36)
16		3(17)	1(33)	4(17)	0
17		3(16)	0	4(15,16)	0
18		0	0	0	0
19		5(4,11,24,25)	0	6(11,24)	0
20		4(3,4,13)	2(4,21)	0	0
21		8(8,12,32,34)	5(8,12)	3(8,32)	2(8)
22		4(3,34,39)	2(9,21)	3(10,32,39)	1(9)
23		0	0	0	0
24		1(1)	0	1(19)	1(11)
25		2(19,40)	0	1(5)	0
26		1(9)	1(21)	2(8,9)	0
27		2(3,35)	0	0	0
28		2(6,14)	0	2(14)	0
29		1(31)	0	1(8)	0
30		0	0	1(11)	0
31		0	0	6(8,27,29,33,34,35)	0
32		6(21,34)	2(21,24)	6(8,21,34)	2(34)
33		4(29,31,36)	0	2(10)	0
34		6(21,32)	1(32)	5(8,21)	1(32)
35		1(27)	0	4(10,21,34,38)	0
36		0	0	0	0
37		1(1)	0	1(11)	0
38		0	0	0	0
39		2(22)	1(22)	2(22)	0
40		2(5,25)	0	2(5)	0
Missed		92	23	74	8
Correct		77.0%	94.3%	81.5%	98.0%

rce: Reference 20.

[a] mbers in parentheses give the category numbers that were confused with the true category. mbers preceding parentheses give the number of samples misclassified.

the seven-element and 22-element studies. The SIMCA results are superior and moreover provide a measure of goodness of fit. Because SIMCA considers each class individually, it allows for variables to have different modeling powers for each category. It therefore performs better than KNN for the case where the categories are not isotropic with respect to the n-space variables; this may account for the superior results in this study.

The lower bound estimate in this analysis indicates that the "worst-case scenario" is a classification accuracy of about 80%; on the other hand, classification may be as good as 99%.

In conclusion, the use of "fuzzy" as opposed to "point" sources forms the basis for the strategy to solve an oil-spill identification problem. SIMCA is seen to be the superior method in this case for its classification accuracy and for a measure of goodness of fit; the latter aspect is very important in practical and legal considerations.

Cross-Validation

A number of situations have arisen in which it is desirable to determine the number of statistically significant eigenvectors: in curve resolution (to determine the number of components giving rise to a chromatographic peak), in principal component analysis of an entire data set (as in display methods), and in SIMCA. The method of *cross-validation* (21) is used to calculate these based on a criterion of best predictive ability, as opposed to best fit. In this method the minimum number of eigenvectors needed to span the meaningful variance is determined; it is not attempted to fit the noise.

Cross-validation is performed by an iterative procedure in which elements are first removed from the data matrix; some percentage γ, usually 10%, of the data is deleted in a prescribed manner. Some versions of cross-validation simply delete 10% of the data vectors in the data matrix, although this approach is not preferred by the authors. It is preferable to delete diagonal lines of elements as in Figure 6.46. Beginning the iteration with the number of principal components, NC, equal to one, a principal component analysis is performed on the reduced data set to extract NC components. Then, the deleted data values are predicted using these principal components (in this first case only 1 since $NC = 1$) as

$$\hat{x}_{ij} = \bar{x}_j + \sum_{k=1}^{NC} y_{ik} v_{kj} . \qquad (6.93)$$

Of course, using less than the truly significant number of principal

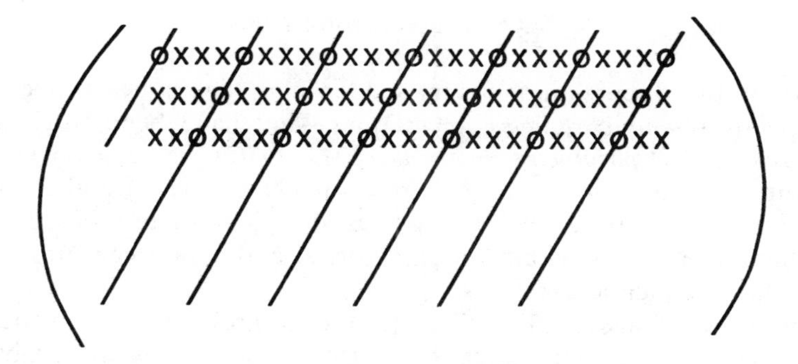

Figure 6.46 Data deletion pattern for cross-validation.

components will mean that the prediction x_{ij} will be in error by an amount e_{ij}:

$$e_{ij} = x_{ij} - \hat{x}_{ij}. \tag{6.94}$$

These errors are calculated and stored as the same procedure is repeated $100/\gamma$ times, changing the deletion pattern position so that each point is left out once and only once. A quantity called the PRESS (predictive residual error sum of squares) is calculated and stored for later use:

$$\text{PRESS}(NC) = \sum_i \sum_j e_{ij}^2. \tag{6.95}$$

The number of components, NC, is incremented by one and the entire procedure is repeated to arrive at a PRESS($NC + 1$). In other words, for $NC + 1 = 2$, two principal components would be extracted and used to predict the deleted data values.

As NC approaches the true number of significant components, the prediction should improve and thus, the PRESS should decrease. However, as the truly significant number is passed, we begin to include noise within the model, which certainly has poor predictive ability. Thus, the PRESS should start to increase again at this point. This leads to the criterion used in cross-validation to select NC; when

$$\frac{\text{PRESS}(NC + 1)}{\text{PRESS}(NC)} > 1, \tag{6.96}$$

the number of significant principal components is selected as NC.

Bayes Classification Rule

We began this discussion of supervised learning techniques at one end of the spectrum—*nonparametric* techniques—and have been working our way toward the *parametric methods* (22). SIMCA is seen an in intermediate method between the extremes of KNN or LLM and the Bayes Classification Rule, because it involves some parameters based on the distribution of points within the categories, and some measures of *how likely* the classification result is.

The Bayes Classification Rule (22) is a probabilistic approach to supervised learning; it requires that certain probabilities and probability density functions be known. Such information is only available when very large populations of objects have been studied. For this reason it is not very often in chemistry that the Bayes method is an appropriate classification technique. It is, however, the best classification method *if* (and only if) the probability information is available.

Let $P(C_n)$ be the a priori probability of the occurrence of objects belonging to class C_n, and let $p(x_i \mid C_n)$ be the probability density function of that class. Our goal is to derive the probability that, given a data vector x_i, the object associated with this data vector belongs to class C_n (as opposed to some other class C_j). The latter is expressed as $p(C_n \mid x_i)$ and is obtained as

$$p(C_n \mid x_i) = \frac{p(x_i \mid C_n)P(C_n)}{\sum_{n=1}^{N} p(x_i \mid C_n)P(C_n)}, \tag{6.97}$$

where the summation in the denominator is over all N classes, and is included to normalize the sum of all the $p(C_j \mid x_i)$ to unity:

$$\sum_{n=1}^{N} p(C_n \mid x_i) = 1.0. \tag{6.98}$$

An object is therefore classified as belonging to C_n if $p(C_n \mid x_i)$ is greater than $p(C_j \mid x_i)$; moreover, the probability indicates how likely it is that a sample belongs to that class.

Very often a risk or weighting factor is incorporated into the Bayes calculation when there is a penalty or cost attached to incorrectly predicting an outcome. As an extreme example, let us imagine that patients are being screened for cancer based on a battery of sophisticated blood tests, x_i. Suppose that the consequence of a positive result (having cancer) is immediate surgery. On the other hand, the consequence of a negative result is simply having another test in a month. It is highly undesirable to subject a normal patient to surgery, so the cost of being wrong in predicting a positive result is very great. We would like to amplify the possibility that someone is normal, by weighting that probability with a factor $L(C_n)$ greater than 1:

$$p(\text{normal} \mid x_i) = \frac{p(x_i \mid \text{normal}) P(\text{normal}) L(\text{normal})}{p(x_i \mid \text{normal}) P(\text{normal}) L(\text{normal}) + p(x_i \mid \text{cancer}) P(\text{cancer})} . \quad (6.99)$$

This makes our prediction of cancer more conservative. Obviously, the weights are assigned by the analyst depending on the context of the problem and may even be quite subjective.

Our discussion so far has assumed prior knowledge about infinite populations. In the real world, finite populations must first be smoothed by one of a variety of methods before proceeding.

PATTERN RECOGNITION ANALYSIS IN PRACTICE: CLASSIFICATION OF ARCHAEOLOGICAL ARTIFACTS ON THE BASIS OF TRACE ELEMENT DATA

It can be said that there are as many approaches to pattern recognition analysis as there are data sets. Each study requires its own unique formulation, preprocessing and selection of methods. However, some general guidelines have emerged that are useful in approaching any data set. In this section, a summary of these guidelines for the general classification problem will be presented in the context of a specific example from the literature concerning archaeological artifacts (23). Results of several pattern recognition methods, as contained in the computer program ARTHUR (24) will be discussed for this example. There are currently a number of software packages commercially available for various types of pattern recognition analyses (see ref. 25). It is intended that this discussion assist the reader in developing a framework for approaching any data set using any of these software tools.

The example we will work with here comes from archaeology. In recent years, quantitative chemical methods such as elemental analysis have been increasingly used to help relate the composition of artifacts to some property of interest to the archaeologist such as origin, authenticity, cultural distribution, and so on. The patterns of trace elements in artifacts such as coins, glass, and pottery have been the subject of several studies of this nature.

Obsidian is a volcanic glass which has been used by ancient peoples to construct weapons, tools, and jewelry among other things. Because the elemental composition of a given volcanic flow of this material tends to be quite homogeneous, there is a good chance that the composition of an obsidian artifact will be very similar to the source material from whence it came.

Suppose that Indian artifacts made of obsidian glass were collected from five archaeological sites in northern California. Let us further suppose that archaeologists would like to know where a given artifact was actually made in order to learn something about migration patterns, trading routes, and so on.

The central questions in this study can be posed as follows: (1) Can different sources of obsidian be distinguished on the basis of some set of chemical measurements? (2) Can the artifacts be classified as to the source of the obsidian from which they were made?

The archaeologists went to four natural geological sources, or quarries, of obsidian in the San Francisco Bay area and collected many samples of each. The first task facing the scientist is to decide what measurements to make. One does not know, a priori, the optimal kinds of measurements to make. Thus, it is generally very useful to conduct an exploratory phase in which different analytical methods are scrutinized for their utility in the problem.

In this case, it was decided to analyze the glass samples by X-ray fluorescence spectroscopy for ten elements—Fe, Ti, Ba, Ca, K, Mn, Rb, Sr, Y, and Zr—whose concentrations ranged from 40 to 1000 ppm. A total of 63 samples from the four geological sources of obsidian were analyzed in this way; these data form the training set in this study. Twelve unknown samples (the artifacts) suspected to come from these locations were similarly treated and form the test set.

Input and Validate the Data. The first step in a data analysis, regardless of the type of software tools employed, is to input and validate the data. In other words, the data matrix should be checked for missing data, constant/redundant measurements, and any invalid or suspect data (outliers, etc.). The latter may be performed by autoscaling and then examining the first few eigenvector plots, as well as the results of cluster analysis of the data. This will very often reveal typing errors, experimental problems, and a host of other unanticipated, annoying, and sticky problems that will need to be cleared up before proceeding.

Exploratory Examination of the Data. In our archaeological example, let us assume these preliminary matters have already been cleaned up. We may continue with what amounts to an initial exploratory examination of the data.

First, the univariate measurement statistics and distributions of the training set should be analyzed. Table 6.11 contains some of the summary statistics for each variable in this data set. The data should be scaled in some meaningful way—usually autoscaling, as in this case.

Second, useful information may be obtained by seeing which features covary. The interfeature correlations should be studied by a variety of methods.

1. Examine the correlation matrix. The interfeature correlation of features i and j is given by

$$C_{ij} = \frac{\sum_{k=1}^{NP} (x_{ik} - \bar{x}_i)(x_{jk} - \bar{x}_j)}{\left[\sum_{k=1}^{NP} (x_{ik} - \bar{x}_i)^2 \sum_{k=1}^{NP} (x_{jk} - \bar{x}_j)^2 \right]^{1/2}}.$$

A confidence interval, $(CI)_{ij}$, about the correlation can be obtained using the Fisher Z-transform (26):

$$\mp (CI)_{ij} = \tanh(Z \pm t\sigma_z)$$

where $Z = \tanh^{-1}(C_{ij})$ is approximately normally distributed with standard deviation $\sigma_z = (NP - 3)^{-1/2}$, and t is the Student's t value. The lower and upper limits of this interval appear to the left and right, respectively, of the calculated correlations in Table 6.12. One can see that the most highly correlated features here are Mn-Fe (0.933), Ca-Ti (0.905), Sr-Ti (0.859), Zr-Ba (0.825), and so on.

2. Perform principal component analysis. The significant eigenvectors should be analyzed to see which constituent variables have the highest loadings. Table 6.13 shows a breakdown, from largest to smallest, of the loadings of each eigenvector. With 53.4% of the toal variance in the data set, eigenvector number 1 is comprised predominately of Fe, Mn, Ba, and Ti, whereas number 2 (21.1%) is largely K, Ca, Rb, and Sr. Note that number 3 is mostly Y.

The goal in this analysis is to interpret the significance of these covariations if possible. The geochemists no doubt may have a much easier time interpreting this obsidian data than the analytical chemist. Varimax rotation may assist the scientist in some cases—results for this data set are shown in Table 6.14. Varimax rotation of the first five eigenvectors shows that a reordering has occurred to yield mainly Ca, Ti, and Sr comprising varivector number 1; Rb and K comprising number 2, and Zr and Ba comprising number 3. Lonely Yttrium now makes up

Table 6.11 Summary Statistics for Obsidian Data Set (ARCH)

Feature	Mean	Scatter	Standard Deviation	Normal Standard Deviation	Minimum	Maximum	Range	Skewness	Kurtosis
1 Fe	1.209E+03	9.851E+07	3.233E+02	2.675E−01	6.890E+02	1.720E+03	1.031E+03	4.927E−01	1.616E+00
2 Ti	2.602E+02	5.041E+06	1.118E+02	4.295E−01	1.140E+02	4.410E+02	3.270E+02	−4.391E−02	1.371E+00
3 Ba	4.248E+01	1.310E+05	1.670E+01	3.931E−01	7.000E+00	6.600E+01	5.900E+01	−8.600E−01	2.765E+00
4 Ca	6.808E+02	3.402E+07	2.786E+02	4.093E−01	2.720E+02	1.010E+03	7.380E+02	−4.246E−01	1.319E+00
5 K	3.928E+02	9.906E+06	5.451E+01	1.388E−01	2.900E+02	5.150E+02	2.250E+02	5.262E−02	2.349E+00
6 Mn	4.652E+01	1.492E+05	1.441E+01	3.097E−01	2.600E+01	8.300E+01	5.700E+01	5.339E−01	2.026E+00
7 Rb	1.075E+02	7.451E+05	1.684E+01	1.568E−01	7.100E+01	1.450E+02	7.400E+01	2.532E−01	2.417E+00
8 Sr	3.159E+01	8.155E+04	1.736E+01	5.497E−01	1.000E+01	7.400E+01	6.400E+01	3.240E−01	1.815E+00
9 Y	5.602E+01	2.007E+05	6.987E+00	1.247E−01	3.800E+01	7.100E+01	3.300E+01	−2.590E−01	2.809E+00
10 Zr	1.568E+02	1.667E+06	4.373E+01	2.790E−01	5.300E+01	2.240E+02	1.710E+02	−8.205E−01	3.110E+00
Average	2.983E+02	1.504E+07	8.842E+01	3.058E−01	1.570E+02	4.349E+02	2.779E+02	−7.516E−02	2.160E+00

Source: Data described in ref. 23; analyzed with ARTHUR (24).

Table 6.12 Interfeature Correlations for ARCH

Feature[a]	Interfeature Correlations														
	J	LO	COR	HI	Prob[b]	J	LO	COR	HI	Prob[b]	J	LO	COR	HI	Prob[b]
1 Fe															
2 Ti	1	.567	.717	.821	(.000)										
3 Ba	1	.428	.614	.751	(.000)	2	.382	.579	.725	(.000)					
4 Ca	1	.542	.699	.809	(.000)	2	.846	.905	.943	(.000)	3	.109	.352	.555	(.005)
5 K	1	−.801	−.687	−.526	(.000)	2	−.420	−.187	.068	(.141)	3	−.640	−.462	−.237	(.000)
	4	−.340	−.095	.161	(.458)										
6 Mn	1	.891	.933	.960	(.000)	2	.540	.697	.808	(.000)	3	.419	.608	.746	(.000)
	4	.554	.707	.814	(.000)	5	−.744	−.606	−.417	(.000)					
7 Rb	1	−.761	−.629	−.448	(.000)	2	−.434	−.204	.051	(.109)	3	−.636	−.456	−.230	(.000)
	4	−.400	−.164	.093	(.199)	5	.767	.854	.910	(.000)	6	−.729	−.584	−.388	(.000)
8 Sr	1	.360	.561	.713	(.000)	2	.774	.859	.913	(.000)	3	.453	.633	.764	(.000)
	4	.683	.798	.874	(.000)	5	−.287	−.037	.217	(.772)	6	.421	.609	.747	(.000)
	7	−.320	−.073	.183	(.571)										
9 Y	1	−.233	.021	.272	(.872)	2	−.325	−.079	.177	(.538)	3	−.262	−.010	.243	(.935)
	4	−.290	−.040	.215	(.755)	5	−.208	.047	.296	(.716)	6	−.175	.081	.327	(.526)
	7	−.166	.091	.336	(.480)	8	−.179	.077	.324	(.547)					
10 Zr	1	.404	.596	.737	(.000)	2	−.022	.232	.458	(.067)	3	.723	.825	.892	(.000)
	4	−.192	.064	.312	(.618)	5	−.775	−.650	−.475	(.000)	6	.389	.585	.729	(.000)
	7	−.710	−.558	−.355	(.000)	8	.041	.291	.506	(.021)	9	−.135	.122	.363	(.341)

[a]The correlation of the feature name in the left most column with feature J ($J = 1, \ldots, 9$) is given in the middle under the heading "COR" between the lower and upper bounds on the confidence interval.
[b]The probability in parentheses is the chance that there is a correlation of 0.0 between the features (23, 24).

Table 6.13 Eigenvalues and Loadings of the First Five Eigenvectors of ARCH[a]

Index	Eigenvalue	Variance Preserved Each	Variance Preserved Sum	Feature	Component	Variance Preserved Each	Variance Preserved Total
1	5.341E+00	53.41	53.41	1 Fe	−.407	16.58	8.86
				6 Mn	−.402	16.12	8.61
				3 Ba	−.345	11.93	6.37
				2 Ti	−.345	11.87	6.34
				8 Sr	−.311	9.69	5.18
				4 Ca	−.309	9.55	5.10
				10 Zr	−.297	8.81	4.71
				5 K	.280	7.85	4.19
				7 Rb	.276	7.59	4.06
				9 Y	−.005	.00	.00
Index	Eigenvalue	Variance Preserved Each	Variance Preserved Sum	Feature	Component	Variance Preserved Each	Variance Preserved Total
2	2.112E+00	21.12	74.53	5 K	−.462	21.31	4.50
				4 Ca	−.428	18.34	3.87
				7 Rb	−.419	17.56	3.71
				8 Sr	−.403	16.24	3.43
				2 Ti	−.373	13.92	2.94
				10 Zr	.345	11.92	2.52
				3 Ba	.077	.60	.13
				1 Fe	.034	.12	.02
				6 Mn	−.004	.00	.00
				9 Y	−.002	.00	.00
Index	Eigenvalue	Variance Preserved Each	Variance Preserved Sum	Feature	Component	Variance Preserved Each	Variance Preserved Total
3	1.096E+00	10.96	85.50	9 Y	−.871	75.92	8.32
				10 Zr	−.291	8.48	.93
				7 Rb	−.211	4.45	.49
				3 Ba	−.187	3.48	.38
				8 Sr	−.146	2.14	.24
				4 Ca	.146	2.12	.23
				5 K	−.138	1.92	.21
				2 Ti	.102	1.04	.11
				1 Fe	.066	.44	.05
				6 Mn	−.009	.01	.00
Index	Eigenvalue	Variance Preserved Each	Variance Preserved Sum	Feature	Component	Variance Preserved Each	Variance Preserved Total
4	8.330E−01	8.33	93.83	3 Ba	−.579	33.56	2.80
				9 Y	.435	18.91	1.58
				10 Zr	−.400	15.99	1.33
				4 Ca	.248	6.13	.51
				6 Mn	.241	5.82	.48

Table 6.13 (*Continued*)

Index	Eigenvalue	Variance Preserved Each	Variance Preserved Sum	Feature	Component	Variance Preserved Each	Variance Preserved Total
				1 Fe	.241	5.81	.48
				7 Rb	−.222	4.93	.41
				8 Sr	−.213	4.55	.38
				5 K	−.207	4.27	.36
				2 Ti	−.021	.05	.00

Index	Eigenvalue	Variance Preserved Each	Variance Preserved Sum	Feature	Component	Variance Preserved Each	Variance Preserved Total
5	2.553E−01	2.55	96.38	7 Rb	.639	40.89	1.04
				6 Mn	.397	15.80	.40
				1 Fe	.353	12.43	.32
				10 Zr	.342	11.70	.30
				8 Sr	−.310	9.62	.25
				3 Ba	−.195	3.80	.10
				9 Y	−.177	3.12	.08
				2 Ti	−.154	2.37	.06
				5 K	.051	.26	.01
				4 Ca	.012	.01	.00

[a]See refs. 23 and 24.

essentially all of varivector number 5; it does not covary with any of the other measurements in this data set.

3. Cluster analysis of variables (*R*-mode cluster analysis). By transposing the data matrix and performing hierarchical cluster analysis, a measure of the "similarity" of the variables over all of the samples is obtained, which can provide a complementary insight into the interfeature correlations. In our example, this analysis produced the dendrogram of Figure 6.47, confirming our conclusions that Ti, Ca, and Sr are similar, that Ba and Zr group together, that K and Rb are similar and that Y is definitely in a class by itself.

Display the data for visual inspection and "human pattern recognition." First, a plot of the scores of these data for the first two eigenvectors reveals an excellent initial separation of these data points by source. A data vector is plotted in eigenvector space (v_2 versus v_1) in Figure 6.48 by its category number in this case instead of by its sample number. It is essential to note that the computer had no idea of the category membership of these points when they were plotted, yet one can see in Figure 6.48 that categories 3 and 4 are specially well separated and that 1 and 2 are reasonably distinct.

Table 6.14 Varimax Rotated Eigenvectors in ARCH

Index	Eigenvalue	Variance Preserved Each	Sum	Feature	Component	Variance Preserved Each	Total
1	3.640E+00	37.76	37.76	4 Ca	−.508	25.81	9.75
				2 Ti	−.496	24.62	9.30
				8 Sr	−.466	21.75	8.21
				6 Mn	−.344	11.82	4.46
				1 Fe	−.333	11.09	4.19
				3 Ba	−.215	4.63	1.75
				7 Rb	.047	.22	.08
				10 Zr	−.024	.06	.02
				5 K	.010	.01	.00
				9 Y	.006	.00	.00

Index	Eigenvalue	Variance Preserved Each	Sum	Feature	Component	Variance Preserved Each	Total
2	2.607E+00	27.05	64.81	7 Rb	−.592	34.99	9.47
				5 K	−.557	31.03	8.39
				1 Fe	.364	13.25	3.58
				6 Mn	.322	10.35	2.80
				10 Zr	.261	6.79	1.84
				3 Ba	.168	2.51	.68
				2 Ti	.064	.40	.11
				4 Ca	.059	.35	.10
				8 Sr	−.051	.26	.07
				9 Y	−.025	.06	.02

Index	Eigenvalue	Variance Preserved Each	Sum	Feature	Component	Variance Preserved Each	Total
3	1.921E+00	19.94	84.75	10 Zr	−.622	38.75	7.72
				3 Ba	−.617	38.08	7.59
				8 Sr	−.254	6.47	1.29
				6 Mn	−.209	4.35	.87
				5 K	.208	4.33	.86
				1 Fe	−.201	4.04	.81
				7 Rb	.139	1.93	.39
				2 Ti	−.131	1.71	.34
				4 Ca	.050	.25	.05
				9 Y	−.029	.08	.02

Index	Eigenvalue	Variance Preserved Each	Sum	Feature	Component	Variance Preserved Each	Total
4	1.035E+00	10.73	95.48	9 Y	−.980	96.11	10.32
				10 Zr	−.102	1.05	.11
				6 Mn	−.097	.95	.10
				2 Ti	.080	.64	.07
				8 Sr	−.079	.62	.07

Table 6.14 (*Continued*)

Index	Eigenvalue	Variance Preserved Each	Variance Preserved Sum	Feature	Component	Variance Preserved Each	Variance Preserved Total
				7 Rb	−.053	.29	.08
				3 Ba	.032	.10	.01
				1 Fe	−.032	.10	.01
				5 K	−.027	.07	.01
				4 Ca	.025	.06	.01

Figure 6.49 and 6.50 give different views, that is, planes of projection, of this data set.

In all of these figures the test point positions are indicated by the numbers 5, 6, and 7. The archaeologists assigned these numbers in this way because they *suspected* that the artifacts came from sources 1, 2, and 3, respectively, and wished to test this hypothesis.

In addition, a nonlinear map of the data provides the best 2-dimensional view possible (at the expense of more computer time). Such a view of our archaeological data is shown in Figure 6.51.

The hierarchical cluster analysis (Q-mode) is useful in a classification problem in addition to its obvious importance to unsupervised learning. In a classification problem one may combine the training and test sets, perform cluster analysis and thereby test the validity of the intitial category assumptions. Sometimes a more suitable categorization of the data may be discovered.

The natural clustering seen in Figure 6.52 of the ARCH data set shows a high degree of consistency with the assumed categories. The test points are also classified for the most part according to the archaeologists' suspicions, with the exception of sample number 64, which is most similar to category 3 (as opposed to category 1). Notice that point 75 is comparatively dissimilar to category 3; it is possible that this point could represent some undiscovered (i.e., unrepresented in this study) source of obsidian.

We have discussed so far an exploratory examination of the data that should be performed regardless of whether the problem involves category data or not. In unsupervised learning, the true categories of the data are not known; a cluster analysis on the training set is followed by breaking the set up into the determined groupings and evaluating the internal consistency of these groupings on the basis of the accuracy of one or more classification algorithms. This provides a quantitative comparison of alternative groupings and ultimately some evaluation of the final cluster assignment.

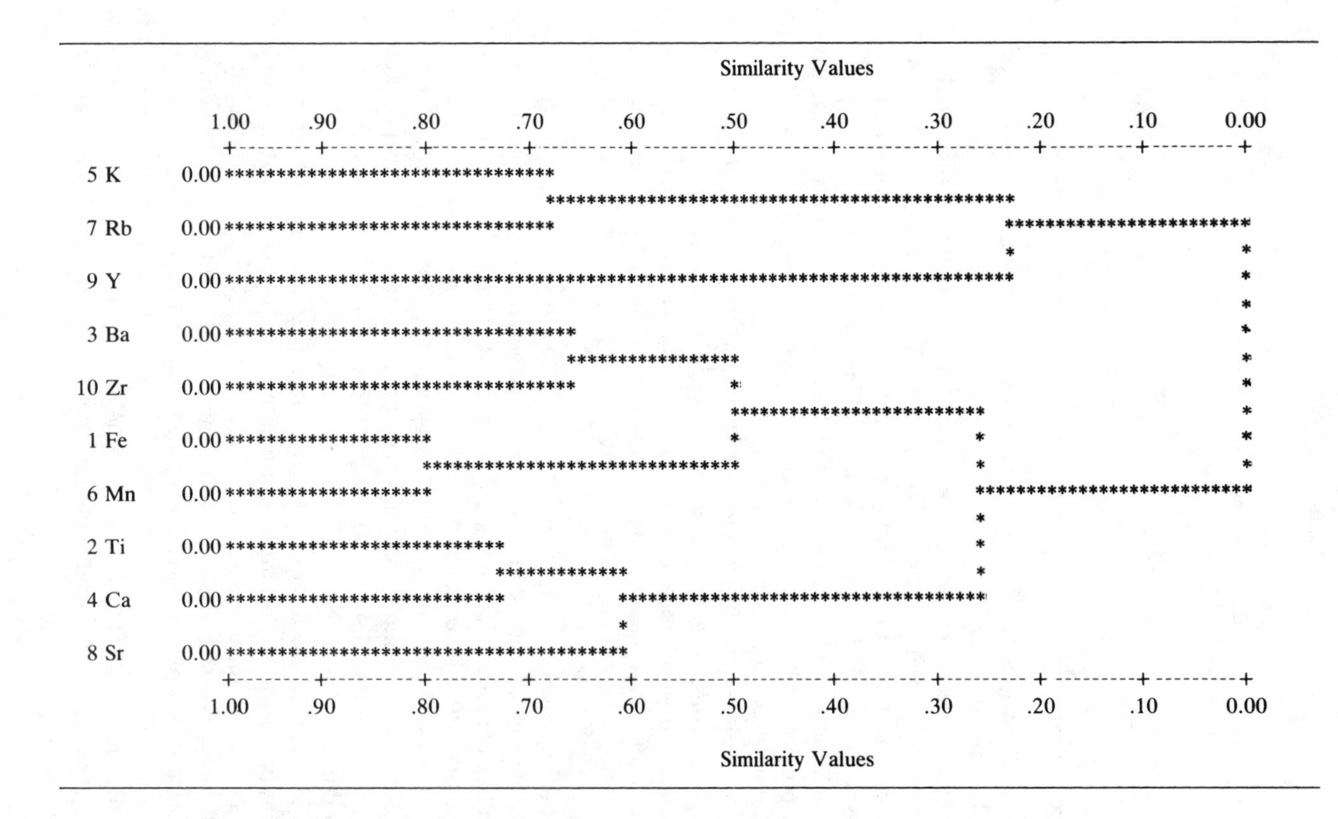

Figure 6.47 Cluster analysis of variables in the ARCH data set.

Classification Analysis. If category data are present, a number of techniques may be applied to perform classification analysis. First of all, the *variance and Fisher weights* should be calculated to see which of the features are most (and least) useful to classify the training set. These values for our obsidian data are given in Table 6.15. Note that the variance weight of Ba of 37.01 is extremely good, and that with the exception of Y, the weights are all rather good for this data set. Yttrium has nearly no discriminating power at all—it is beginning to appear that this measurement is essentially irrelevant to the problem and really need not be even measured at all. Also, it is interesting to examine the pair-wise weights for the individual features. For example, note that the variation in magnitude of the pair-wise values for Ba, shown in Table 6.16, are quite dramatic. For the lowest weight (category pair 1–4), it can be seen in Table 6.17 that the best feature to separate just these two categories is Fe with a variance weight of 48.73.

The data may be weighted by these variance or Fisher weights in order to enhance the performance of any *classification algorithm* that one may now apply (discriminant analysis, KNN, etc.).

Results of the Leave-one-out procedure performed on the training set for KNN are tabulated for 1, 3, and up to 10 nearest neighbors in Table 6.18. A misclassification matrix is a useful way to summarize these results since it allows one to see at a glance which categories if any are being confused. Obviously, the ARCH data set is a scientist's dream come true, having perfect classification accuracy. Most real data sets fall somewhat short of this picture.

Again, for the test set (Table 6.19) we see all of the points but number 64 classified in agreement with the archaeologists' initial suspicions. KNN classifies point number 64 as a 3, just as we saw in cluster analysis.

Using a one-component (i.e., the mean plus the first principal component) model in SIMCA, the data for each category are modeled independently with a residual variance shown in Table 6.20. The results for the training set are again shown in a misclassification matrix in Table 6.21. It is also useful to examine the category distance matrix (Table 6.22) in order to gain some insight into the data structure. Notice that the diagonal elements of this matrix are simply the residual variances of Table 6.19. The discriminating power of each variable (Table 6.23) roughly parallels the variance weights seen earlier. The classification results for the test samples are shown in Table 6.24. An estimate of the probability of class membership is provided by the stars next to each classification; the greater the number of stars, the more likely the test point belongs to that class. Notice that the anomalous point 64 does not fit well to any category by the SIMCA method.

```
YMAX = 2.524E-01  +*********I*********I*********I*********I*********I*********I*********I*********I*********I*********I****+
                  *                                                                               7                        *    1       0
                  *                                                                                                        *    1       0
                  *                                                                              3                         *    0       0
                  -                                                                               3     3                  -    2       0
                  *                  4                                                     3        3                      *    2       0
                  *                                                                              33                        *    2       1
                  *                4                                                              3 7  3                   *    3       0
                  *                                                                                  7                     *    3       1
                  -                                                                              33   3                    -    1       1
                  *                                                                                                        *    3       0
                  *                                                                                3      3                *    0       0
                  *                                                                                7                       *    2       0
                  *          4    444                                             5              33      3                 *    5       0
                  -       4    4                                                                  37                       -    6       0
                  *           4                                                                       3                    *    3       0
                  *    4       4                                                                                           *    3       0
                  *     4 4                                                                                                *    2       0
                  *                       4 4                                                                              *    2       0
                  -       4       444                                                                                      -    4       0
                  *                                                                                                        *    0       0
                  *                                                                                                        *    0       0
                  *                                                                                                      6 *    1       0
                  *              4                                                                                         *    1       0
Y = (2(  2)       -                                                                                                        -    0       0
                  *                    4                                                                             2     *    2       0
                  *                                                                                                        *    0       0
                  *                                                                                                        *    0       0
                  *                                                                                                        *    0       0
                  -                                                                                              6         -    1       0
                  *                                                                                                        *    0       0
                  *                                                                                                        *    0       0
```

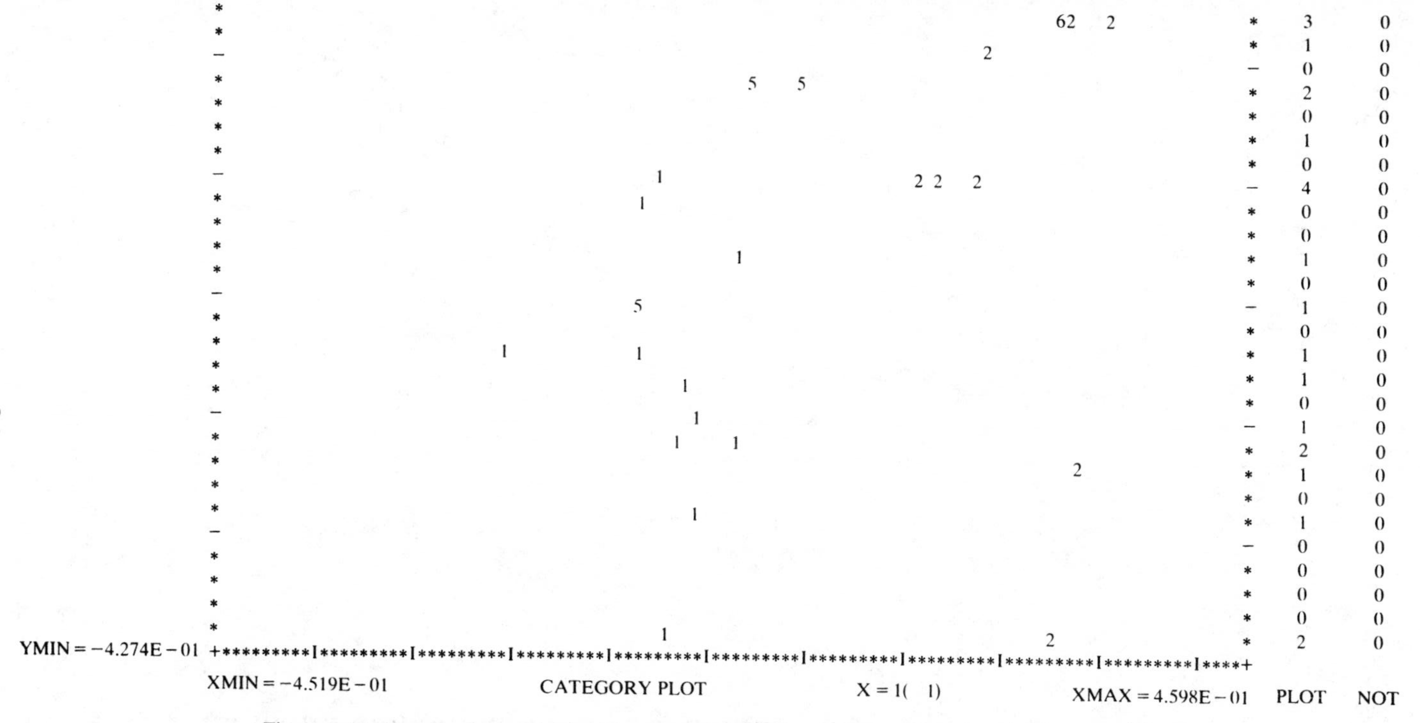

Figure 6.48 Scores of the ARCH data plotted in the plane determined by eigenvectors $X = V_1$ and $Y = V_2$. Points are plotted by their category number. Overlapping points indicated in the "Not" (not-plotted) column.

```
YMAX = 3.135E-01  +*********I*********I*********I*********I*********I*********I*********I*********I*********I*********I****+
                  *                                                                          2                             *   1     0
                  *                                                                                                        *   0     0
                  *              4                                                                                         *   1     0
                  *                                                                                                        *   0     0
                  -                                                                                                        -   0     0
                  *    4                                                                                                   *   1     0
                  *         4                                                                                              *   1     0
                  *                  4                                                                                     *   1     0
                  *                                                                                                        *   0     0
                  -                                                                                                     2  -   1     0
                  *                                            1                                2                          *   0     0
                  *              4                                                                                    6    *   2     0
                  *                                                                                            2           *   2     0
                  -                                          1                 5                  2                        -   1     0
                  *        4                                                                    3       3                  *   3     0
                  *                                              1                       2                                 *   3     0
                  *                                                                                                        *   2     0
                  *                                                                                3                       *   0     0
                  -                      4                              1                             3                    -   1     0
                  *                                                                                3    33         2       *   3     0
                  *                                            1                                     3                   6 *   4     0
                  *                                                                      3        3                        *   3     0
                  *      4       4                                                                     3                   *   2     0
Y = 3(   3)       -                4                                                                                       -   3     0
                  *                 44                                                                                     *   1     0
                  *                                                                            3      2                    *   2     0
                  *          4                                5                                                6           *   2     0
                  *     4      4    4                                                           3   7                      *   3     0
                  -       4                                      1                                  3                      -   5     1
```

270

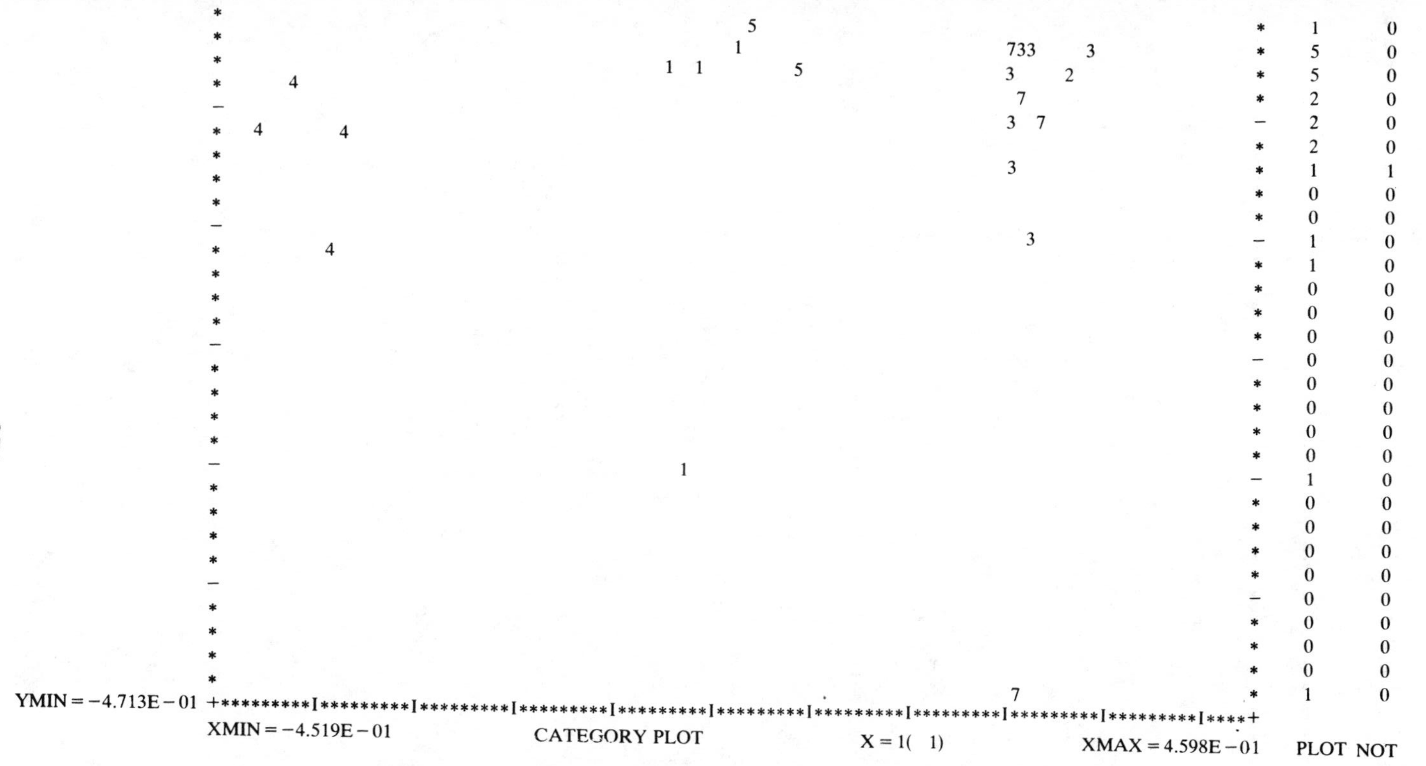

Figure 6.49 Eigenvector plot of ARCH data using $X = V_1$ and $Y = V_3$.

```
YMAX = 3.135E-01  +*********I*********I*********I*********I*********I*********I*********I*********I*********I*********I****+
                  *                                    2                                                                   *   1   0
                  *                                                                                                        *   0   0
                  *                                                                               4                        *   1   0
                  *                                                                                                        *   0   0
                  -                                                                                                        -   0   0
                  *                                                                        4                               *   1   0
                  *                                                                           4                            *   1   0
                  *                                                                                              4         *   1   0
                  *                                                                                                        *   0   0
                  -                                                           2                                            -   1   0
                  *                                                                                                        *   0   0
                  *                                    1                                                                   *   1   1
                  *                                                      6                        4                        *   2   0
                  *                                                2                                                       *   1   0
                  -               1                             2                               5                          -   3   0
                  *                                                              4               3         3               *   3   0
                  *             1                    2                                                                     *   2   0
                  *                                                                                                        *   0   0
                  *                                                                                 3                      *   1   0
                  -                             1                                      4                     3             -   3   0
                  *                                              2                                            3  3 3       *   4   0
                  *                                    1                             6                3                    *   3   0
                  *                                                                                    3       3           *   2   0
                  *                                                                    4    3     4                        *   3   0
Y = 3(   3)       -                                                                                           4            -   1   0
                  *                                                              4       4                                 *   2   0
                  *  2                                                                                                3    *   2   0
                  *                         5                        6                           4                         *   3   0
                  *                                                                  4          434         7     3        *   6   0
                  -                      1                                             4                          3        -   3   0
                  *                                          5                                                             *   1   0
                  *                 1                                                            3   3   3               7 *   5   0
                  *           1  2 1                        5                                         3                   *   5   0
```

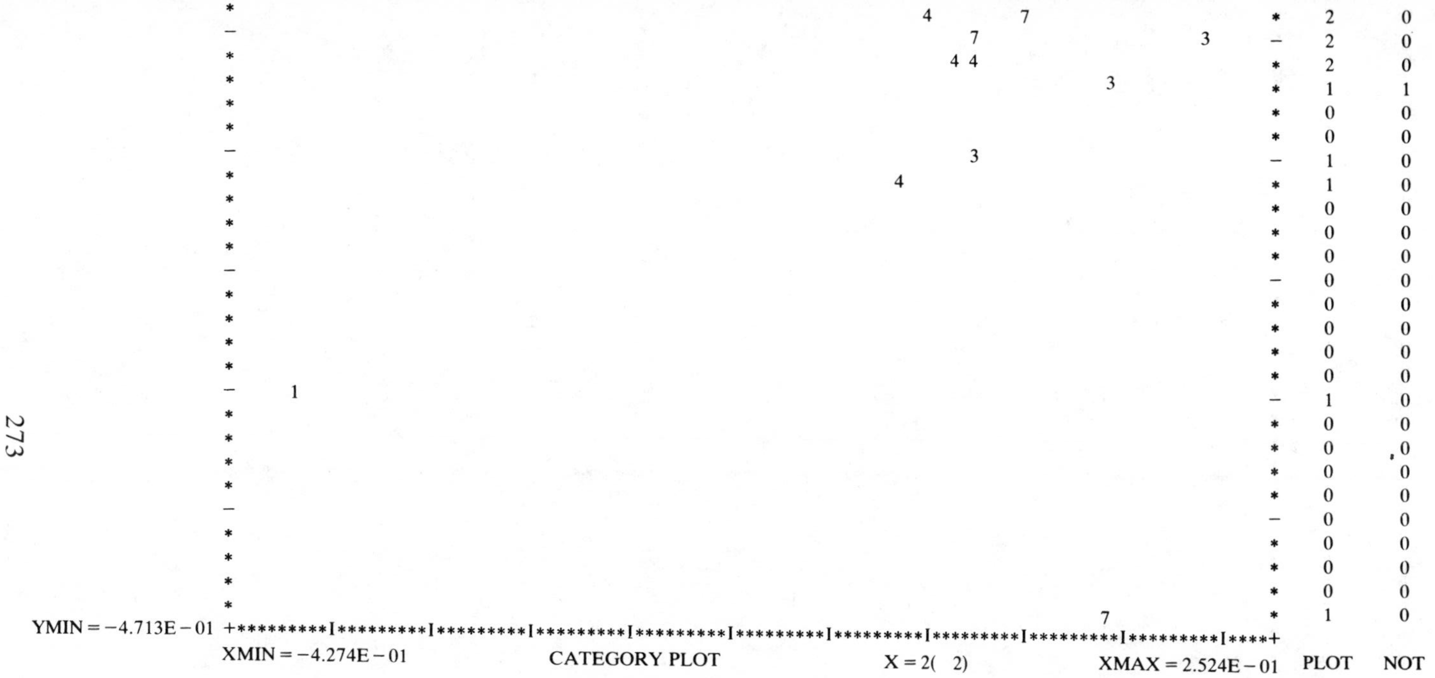

Figure 6.50 Eigenvector plot of ARCH data using $X = V_2$ and $Y = V_3$.

```
YMAX = 5.524E-01 +*********I*********I*********I*********I*********I*********I*********I*********I*********I*********I****+
                 *                                                                 4                                      *  1    0
                 *                                                                                                        *  0    0
                 *                                                                                                        *  0    0
                 *                                                                                                        *  0    0
                 -                                                        4           4                                   -  2    0
                 *                                                                            4               4           *  2    0
                 *                                                                                                        *  0    0
                 *                                                                                                        *  0    0
                 *                                                                               4                        *  1    0
                 -                                                                                         4              -  1    0
                 *                                                             4           4                              *  2    0
                 *                                                                                                        *  0    0
                 *                                                                                 4                     4*  2    0
                 *                                                                      4                 4               *  2    0
                 -                                                                                                        -  0    0
                 *                                                                                             4          *  1    0
                 *                                                                                  4                     *  1    0
                 *                        3                                                              4                *  2    0
                 *                                                                              4              4          *  2    0
                 -                  3           3                                                                         -  2    1
                 *                                                                      4             4                   *  2    0
                 *                        3                                                               4               *  2    0
                 *                                       3                                                                *  1    0
                 *                          3  3                                                                          *  2    0
Y = 2(NLM)       -         3                      3                                                                       -  2    0
                 *              3                                                                                         *  1    0
                 *                    3                                                                                   *  1    0
                 *                             3    3                                                                     *  2    0
                 *                  3                                                                                     *  1    0
                 -                                     3                                                                  -  1    0
```

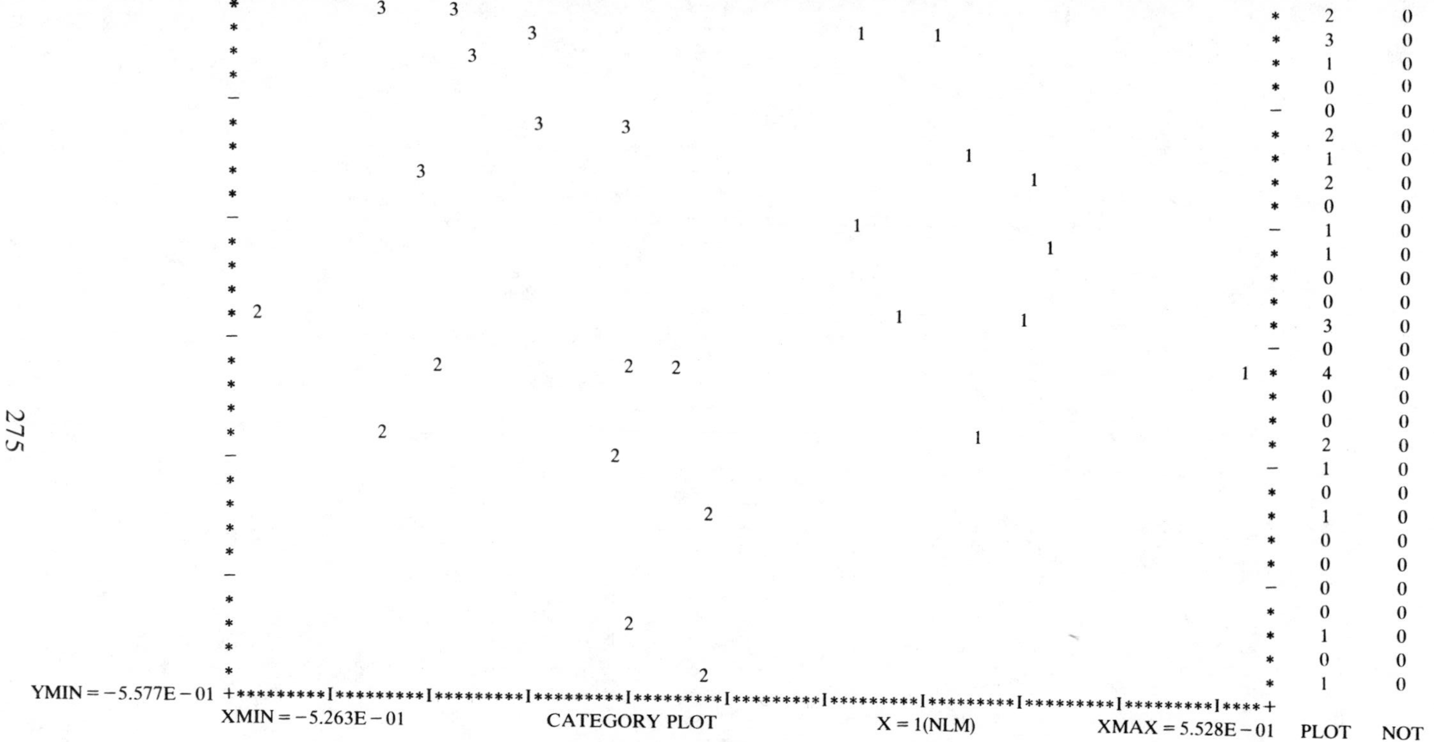

Figure 6.51 Nonlinear map of ARCH data.

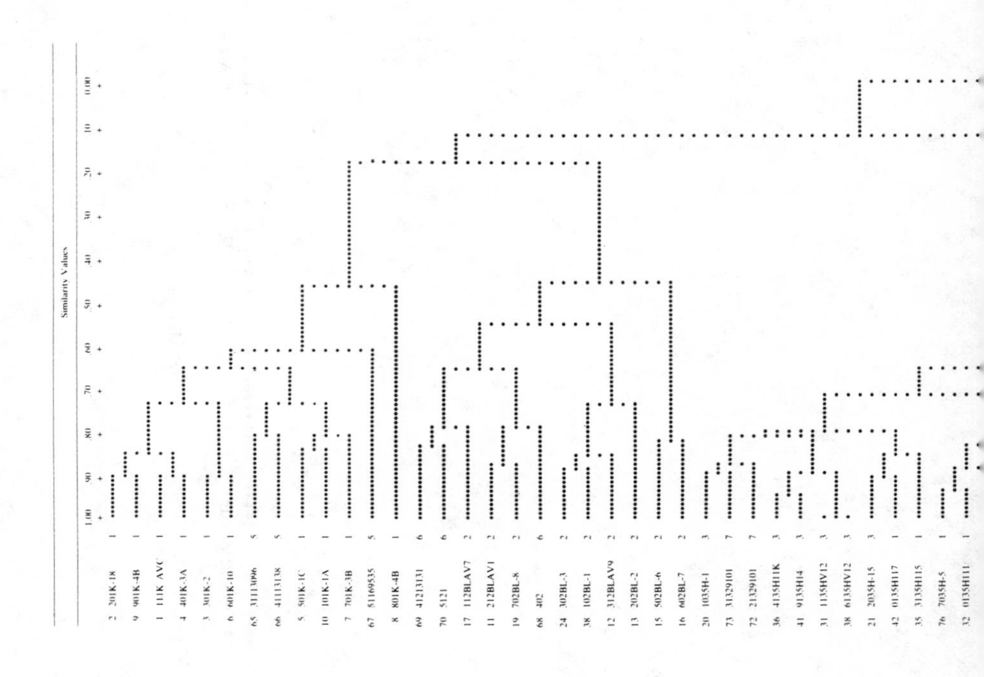
Similarity Values
1.00
.90
.80
.70
.60
.50
.40
.30
.20
.10
0.00
2 201K-18 1
9 901K-4B 1
1 111K AVC 1
4 401K-3A 1
3 301K-2 1
6 601K-10 1
65 31113096 5
66 41113138 5
5 501K-1C 1
10 101K-1A 1
7 701K-3B 1
67 51169535 5
8 801K-4B 1
69 41213131 6
70 5121 6
17 112BLAV7 2
11 212BLAV1 2
19 702BL-8 2
68 402 6
24 302BL-3 2
38 102BL-1 2
12 312BLAV9 2
13 202BL-2 2
15 502BL-6 2
16 602BL-7 2
20 1035H-1 3
73 31329101 7
72 21329101 7
36 4135H11K 3
41 9135H14 3
31 1135HV12 3
38 6135HV12 3
21 2035H-15 3
42 0135H117 1
35 3135H115 1
76 7035H-5 1
32 0135H111 1

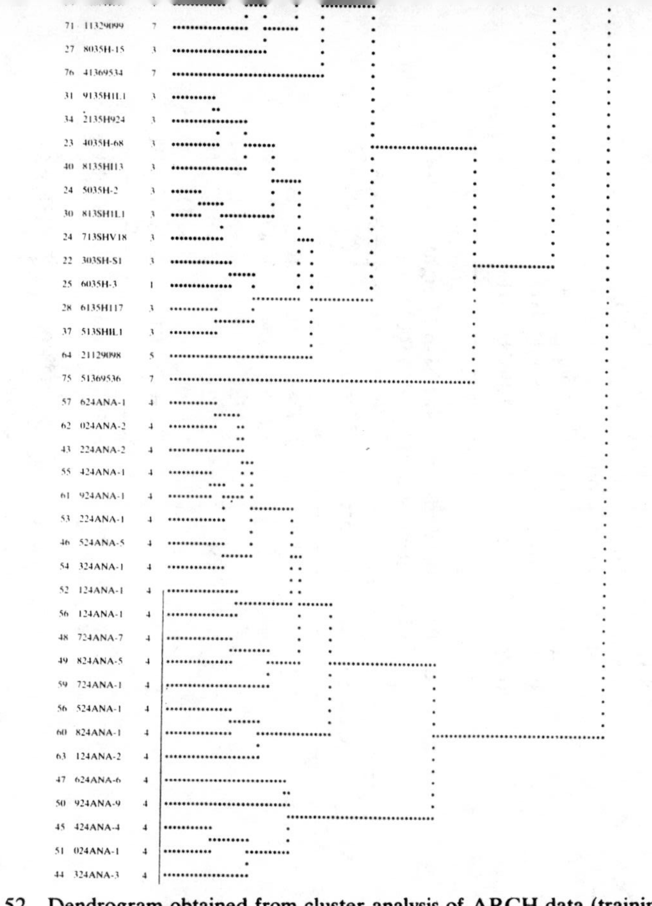

Figure 6.52 Dendrogram obtained from cluster analysis of ARCH data (training and test sets).

Table 6.15 Variance and Fisher Weights for the ARCH Example

Feature	Variance	Fisher
1 Fe	8.414E+00	2.195E+01
2 Ti	9.005E+00	3.055E+01
3 Ba	3.701E+01	8.768E+01
4 Ca	8.830E+00	4.749E+01
5 K	4.957E+00	5.876E+00
6 Mn	4.792E+00	6.128E+00
7 Rb	2.680E+00	2.254E+00
8 Sr	6.305E+00	7.096E+00
9 Y	1.016E+00	1.506E−02
10 Zr	1.111E+01	1.601E+01

Table 6.16 Pairwise Variance and Fisher Weights for the First Three Variables of ARCH

	Category vs.	Category	Variance	Fisher
Fe	1	2	1.078E+00	6.913E−02
	1	3	5.818E+00	4.470E+00
	1	4	4.873E+01	4.452E+01
	2	3	1.302E+00	2.701E−01
	2	4	1.150E+01	9.411E+00
	3	4	7.750E+01	7.298E+01
			8.414E+00	2.196E+01
Ti	1	2	6.847E+00	5.203E+00
	1	3	1.684E+02	1.526E+02
	1	4	1.609E+00	5.761E−01
	2	3	4.026E+00	2.694E+00
	2	4	3.201E+00	2.007E+00
	3	4	2.230E+01	2.029E+01
			9.005E+00	3.055E+01
Ba	1	2	2.864E+02	2.552E+02
	1	3	2.730E+01	2.444E+01
	1	4	1.162E+00	1.520E−01
	2	3	9.551E+01	8.912E+01
	2	4	1.468E+02	1.379E+02
	3	4	2.015E+01	1.826E+01
			3.701E+01	8.768E+01

Table 6.17 Category-Pair Variance Weights for ARCH

Category-Pair (1,2) Feature	Weight	Category-Pair (2,3) Feature	Weight
1 Fe	1.078E+00	1 Fe	1.302E+00
2 Ti	6.847E+00	2 Ti	4.025E+00
3 Ba	2.864E+02	3 Ba	9.551E+01
4 Ca	1.489E+00	4 Ca	1.101E+01
5 K	1.139E+00	5 K	2.117E+00
6 Mn	1.365E+00	6 Mn	1.358E+00
7 Rb	1.002E+00	7 Rb	1.450E+00
8 Sr	1.099E+01	8 Sr	2.046E+00
9 Y	1.010E+00	9 Y	1.017E+00
10 Zr	2.789E+01	10 Zr	2.294E+01
	3.392E+01		1.428E+01

Category-Pair (1,3) Feature	Weight	Category-Pair (2,4) Feature	Weight
1 Fe	5.818E+00	1 Fe	1.150E+01
2 Ti	1.684E+02	2 Ti	3.201E+00
3 Ba	2.730E+01	3 Ba	1.468E+02
4 Ca	2.005E+02	4 Ca	1.263E+00
5 K	5.685E+00	5 K	8.097E+00
6 Mn	3.331E+00	6 Mn	1.013E+01
7 Rb	1.467E+00	7 Rb	6.146E+00
8 Sr	2.067E+01	8 Sr	4.954E+00
9 Y	1.044E+00	9 Y	1.005E+00
10 Zr	1.671E+00	10 Zr	4.381E+01
	4.359E+01		2.369E+01

Category (1,4) Feature	Weight	Category-Pair (3,4) Feature	Weight
1 Fe	4.873E+01	1 Fe	7.750E+01
2 Ti	1.609E+00	2 Ti	2.230E+01
3 Ba	1.162E+00	3 Ba	2.015E+01
4 Ca	1.224E+00	4 Ca	9.325E+01
5 K	2.041E+01	5 K	6.550E+00
6 Mn	1.042E+01	6 Mn	1.863E+01
7 Rb	5.680E+00	7 Rb	4.980E+00
8 Sr	2.425E+00	8 Sr	1.125E+01
9 Y	1.022E+00	9 Y	1.001E+00
10 Zr	9.678E+00	10 Zr	4.156E+00
	1.024E+01		2.598E+01

Table 6.18 Summary Results for the Leave-One-Out Procedure Performed with KNN on the Training Set of ARCH

	1-NN	3-NN			· · ·				10-NN
Total missed	0	0	0	0	0	0	0	0	0
Percent correct	100.0	100.0	100.0	100.0	100.0	100.0	100.0	100.0	100.0

Misclassification Matrix

Computed class		1	2	3	4
True class	1	10 100.0	0 0.0	0 0.0	0 0.0
	2	0 0.0	9 100.0	0 0.0	0 0.0
	3	0 0.0	0 0.0	23 100.0	0 0.0
	4	0 0.0	0 0.0	0 0.0	21 100.0

Percent correct: Total 100.0
Average 100.0

For problems with enough samples to make Bayesian classification feasible, the classification results are accompanied by a probability of belonging to a class as opposed to a goodness of fit measure as in SIMCA.

It is often desirable to reduce the number of measurements retained in a classification problem using a *feature selection* algorithm. The results of SELECT are shown in Table 6.25. In iteration number 1, the feature number 3, Ba, with weight 37.01, was selected and the other features were decorrelated to it. The coefficients A and B refer to the Equations (6.65–6.76). Then, from among these decorrelated weights (listed under the second iteration) the value 9.893 corresponding to Ca (variable number 4) was selected next. Iron (number 1) was the third selection, and so on. In this way, an optimal reduced set of features may be chosen—either a preset number of them, or those with a weight above a certain user-specified tolerance value (such as 1.10, for instance). This reduced data set should be analyzed with the classification methods used previously to compare classification accuracy.

Finally, an estimate of the degree of confidence that can be placed on the results should be attempted. For example, when the experimental uncertainties are known or can be estimated, a Monte Carlo technique of uncertainty perturbation may be useful in this regard (see Suggested Readings, Duewer et al., 1976).

PARTIAL LEAST-SQUARES PATH MODELING

Thus far we have treated the case of a single data matrix arising from samples which may be characterized by discrete categories. When the samples are instead associated with one or more continuous properties (dependent variables), a variety of techniques such as multiple regression, stepwise regression, and canonical correlation are available. These topics will not be discussed further here; the reader is referred to the extensive number of books on these subjects (see, e.g., refs. 27–28).

A new, general method of handling regression problems, called partial least-squares path modeling with latent variables (PLS), will be presented in this section. Developed by Herman Wold (29–33), this method allows the relationships between many blocks of data (i.e., data matrices) to be characterized. Multiple and other regression techniques can be considered special cases of the PLS method.

The driving forces behind the development of this methodology are at least threefold. First of all, in many studies it is desirable to understand the interrelationships between several parts of a large, complex system. One example might be the modeling of a river system in which the influence of various sources of water on the downstream flow is evaluated (see ref. 34 and 35). Another example might be to determine how the quality of a chemical product is related to feedstock characteristics and physical process conditions such as temperature, pressure, flow rate, stirring rate, and so on (36). In such problems, blocks of data are connected together by some predetermined scheme or causal pathway. The PLS method allows the influence of each block to be evaluated with respect to the path model.

Secondly, it is not uncommon to encounter the unfortunate situation where the number of variables in a regression problem is greater than the number of samples. For the purpose of prediction, this problem is often approached by using stepwise regression, despite the fact that the variables eliminated in this way may still contain some useful information. Also, variables are often selected that have only a spurious correlation. The PLS method allows all of the variables to be retained in the problem by extracting "latent variables." These are found by an iterative procedure and are mutually orthogonal (within each block), linear com-

Table 6.19 Test Set Results of KNN for ARCH

Index	Sample Name	True Category	Committee Votes 1-NN	3-NN	4-NN	5-NN	6-NN	7-NN	8-NN	9-NN	10-NN	Index	Category	Distance	Index	Category	Distance
64	21129098	5	3	3	3	3	3	3	3	3	3	37	3	2.525E+00	36	3	2.891E+00
												27	3	2.907E+00	33	3	2.981E+00
												38	3	2.931E+00	41	3	3.088E+00
												34	3	3.047E+00	10	1	3.216E+00
												23	3	3.222E+00	22	3	3.248E+00
65	31113096	5	1	1	1	1	1	1	1	1	1	4	1	1.876E+00	7	1	2.095E+00
												3	1	2.664E+00	8	1	2.737E+00
												1	1	2.585E+00	5	1	2.922E+00
												2	1	3.289E+00	9	1	3.475E+00
												6	1	3.476E+00	46	4	3.546E+00
66	41113138	5	1	1	1	1	1	1	1	1	1	7	1	9.465E−01	4	1	1.151E+00
												5	1	1.701E+00	3	1	1.884E+00
												1	1	1.936E+00	8	1	2.344E+00
												2	1	2.394E+00	6	1	2.487E+00
												9	1	2.543E+00	10	1	2.805E+00
67	51169535	5	1	1	1	1	1	1	1	1	1	7	1	1.868E+00	4	1	2.167E+00
												5	1	2.352E+00	1	1	2.548E+00
												3	1	2.657E+00	9	1	2.799E+00
												2	1	2.875E+00	6	1	2.971E+00
												10	1	3.007E+00	8	1	3.029E+00
68	402	6	2	2	2	2	2	2	2	2	2	11	2	7.528E−01	19	2	9.976E−01
												12	2	1.234E+00	17	2	1.981E+00
												18	2	1.927E+00	15	2	2.108R+00
												13	2	2.122E+00	14	2	2.146E+00
												16	2	2.329E+00	30	3	7.810E+00

69	41213131	6	2	2	2	2	2	2	2	2	2	17	2	1.314E+00	11	2	1.335E+00
												19	2	1.709E+00	18	2	2.355E+00
												12	2	2.506E+00	15	2	2.583E+00
												13	2	2.589E+00	14	2	2.711E+00
												16	2	2.757E+00	24	3	6.201E+00
70	5121	6	2	2	2	2	2	2	2	2	2	17	2	1.280E+00	11	2	1.568E+00
												19	2	2.041E+00	12	2	2.961E+00
												18	2	3.082E+00	15	2	3.141E+00
												13	2	3.325E+00	14	2	3.423E+00
												16	2	3.463E+00	24	3	6.485E+00
71	11329099	7	3	3	3	3	3	3	3	3	3	23	3	8.349E−01	34	3	8.451E−01
												41	3	9.059E−01	22	3	9.699E−01
												36	3	1.001E+00	25	3	1.041E+00
												33	3	1.062E+00	38	3	1.062E+00
												39	3	1.217E+00	21	3	1.248E+00
72	21329101	7	3	3	3	3	3	3	3	3	3	20	3	5.273E−01	35	3	7.185E−01
												22	3	1.001E+00	25	3	1.109E+00
												28	3	1.110E+00	42	3	1.125E+00
												21	3	1.149E+00	39	3	1.161E+00
												41	3	1.268E+00	32	3	1.288E+00
73	31329101	7	3	3	3	3	3	3	3	3	3	36	3	7.897E−01	33	3	8.365E−01
												38	3	8.355E−01	37	3	9.414E−00
												22	3	9.726E−01	23	3	1.118E+00
												41	3	1.133E+00	25	3	1.218E+00
												20	3	1.225E+00	34	3	1.286E+00
74	41369534	7	3	3	3	3	3	3	3	3	3	31	3	1.019E+00	29	3	1.068E+00
												24	3	1.126E+00	26	3	1.164E+00
												30	3	1.253E+00	40	3	1.367E+00
												34	3	1.358E+00	28	3	1.396E+00
												32	3	1.911E+00	39	3	1.570E+00
75	51369536	7	3	3	3	3	3	3	3	3	3	23	3	6.385E−01	41	3	7.364E+00
												34	3	8.571E−01	22	3	9.008E−01
												33	3	9.097E−01	38	3	9.097E−01
												36	3	9.291E−01	25	3	9.458E−01
												21	3	9.956E−01	39	3	1.005E+00

Table 6.20 Residual Variances of Each Category Using SIMCA

Category	NC	Residual Variance
1	2	2.400E−01
2	2	2.721E−01
3	2	2.213E−01
4	2	3.254E−01

Table 6.21 SIMCA Results for ARCH Training Set

Misclassification Matrix

Computed class		1	2	3	4
True class 1		10	0	0	0
		100.0	0.0	0.0	0.0
2		0	9	0	0
		0.0	100.0	0.0	0.0
3		0	0	23	0
		0.0	0.0	100.0	0.0
4		0	0	0	21
		0.0	0.0	0.0	100.0

Percent correct: Total 100.0
Average 100.0

Table 6.22 Category Distance Matrix for ARCH

Category	Distance Matrix: $D(I, J)$ = Standard Deviation for All Objects in Category (I) Fitted to Model (J)			
1	2.400E−01	2.445E+00	8.927E−01	6.360E−01
2	1.354E+00	2.721E−01	6.759E−01	1.139E+00
3	1.006E+00	1.682E+00	2.213E−01	7.909E−01
4	6.561E−01	2.598E+00	8.794E−01	3.254E−01

Table 6.23 Discriminating Power of Each Variable Used in the SIMCA Analysis of ARCH

Feature	Discriminating Power
1 Fe	7.692E+01
2 Ti	4.294E+01
3 Ba	1.214E+03
4 Ca	1.672E+02
5 K	2.635E+01
6 Mn	2.173E+01
7 Rb	1.717E+01
8 Sr	2.715E+01
9 Y	2.290E+00
10 Zr	3.456E+01

binations of all the original variables. The power of PLS is due to the fact that the latent variables simultaneously 1) describe the maximum predictive variance of a block, and 2) provide maximal fit to the path model.

One of the assumptions made in multiple regression is that the independent variables are truly independent. To the degree that this assumption is invalid, the resulting model parameters will be more affected by noise, eventually leading to loss of full rank (28). Attempts to eliminate this *colinearity problem* have lead to such developments as principal components regression and ridge regression. PLS provides a solution to this problem in the case of PLS regression since the latent variables derived from the independent variable block are constrained to be orthogonal.

Another aspect of the PLS method is the signal-to-noise advantage gained by making use of all of the measurements. Furthermore, by using only the significant number of latent variables in the procedure, a noise filtering effect is obtained which results in the improved predictive ability of PLS.

The simplest type of application is PLS regression, depicted in Figure 6.53. Here a relationship is sought between a single response (or dependent) variable, which is the vector **Y**, and a data matrix **X**. Latent variables T_x are extracted both to model **X** and to correlate with **Y**. The reader may note the contrast to principal component analysis, in which the latent variables (eigenvectors) only model **X**.

The iterative procedure is begun on mean-centered data, scaled to unit

Table 6.24 Test Set Results Using SIMCA on the ARCH Data

		True	Calculated							
Index	Name	Category	Category . . . Fit . . .	Distance	Category . . . Fit . . .	Distance	Category . . . Fit . . .	Distance	Category . . . Fit . . .	Distance
64	21129098	5	3 *	8.333E−01	4 *	1.066E+00	1	1.172E+00	2	3.538E+00
65	31113096	5	1****	4.081E−01	4 *	1.145E+00	3	1.746E+00	2	5.159E+00
66	41113188	5	1****	3.435E−01	4 *	1.190E+00	3	1.425E+00	2	4.725E+00
67	51169535	5	1 **	7.158E−01	4 *	1.250E+00	3	1.283E+00	2	4.546E+00
68	402	6	2****	3.053E−01	3	9.665E−01	4	1.932E+00	1	2.540E+00
69	41213131	6	2****	4.274E−01	3 *	7.017E−01	4	1.972E+00	1	2.531E+00
70	5121	6	2****	4.648E−01	3 *	6.792E−01	4	1.903E+00	1	2.667E+00
71	11329099	7	3****	3.080E−01	4	1.367E+00	1	1.693E+00	2	3.038E+00
72	21329101	7	3****	3.306E−01	4	1.326E+00	1	1.699E+00	2	2.978E+00
73	31329101	7	3****	2.624E−01	4	1.400E+00	1	1.578E+00	2	3.225E+00
74	41369534	7	3****	3.580E−01	4	1.555E+00	1	1.868E+00	2	2.699E+00
75	51369536	7	3****	2.503E−01	4	1.405E+00	1	1.734E+00	2	2.959E+00

Summary results for the 12 test set data vectors:

Category	Number	Percent
1	3	25.00
2	3	25.00
3	6	50.00
4	0	0.00

Table 6.25 Results of Feature Selection[a]

Iteration	(*I*)	Weight (*I*)	(*J*)	Weight (*J*)	Corr (*I*, *J*)	*A*	*B*
1	3	3.701E+01	1	8.414E+00	.514402	−7.787E−01	1.267E+00
			2	9.005E+00	.578740	−7.097E−01	1.226E+00
			4	8.830E+00	.351729	−3.757E−01	1.068E+00
			5	4.957E+00	−.461896	5.208E−01	1.127E+00
			6	4.792E+00	.507660	−7.651E−01	1.259E+00
			7	2.680E+00	−.456437	5.130E−01	1.124E+00
			8	6.305E+00	.533484	−8.187E−01	1.292E+00
			9	1.016E+00	−.010434	1.043E−02	1.000E+00
			10	1.111E+01	.824667	−1.458E+00	1.768E+00
2	4	9.893E+00	1	7.153E+00	.653600	−8.636E−01	1.321E+00
			2	5.813E+00	.919470	−2.339E+00	2.543E+00
			5	3.960E+00	.080962	−8.123E−02	1.003E+00
			6	3.910E+00	.564033	−8.881E−01	1.337E+00
			7	2.128E+00	−.004006	4.006E−03	1.000E+00
			8	3.817E+00	.794257	−1.307E+00	1.646E+00
			9	1.016E+00	−.039003	3.903E−02	1.001E+00
			10	3.676E+00	−.425878	4.720E−01	1.106E+00
3	1	9.133E+00	2	1.342E+00	−.134157	1.354E−01	1.009E+00
			5	4.020E+00	−.834559	1.515E+00	1.815E+00
			6	3.580E+00	.812727	−1.395E+00	1.716E+00
			7	2.126E+00	−.653523	8.634E−01	1.321E+00
			8	1.480E+00	−.515260	6.012E−01	1.167E+00
			9	1.010E+00	.079137	−7.939E−02	1.003E+00
			10	2.453E+00	.699012	−9.775E−01	1.398E+00
4	2	1.138E+00	5	1.035E+00	−.036988	3.701E−02	1.001E+00
			6	1.015E+00	−.262822	2.724E−01	1.036E+00
			7	1.029E+00	.177697	−1.806E−01	1.016E+00
			8	1.009E+00	.161984	−1.642E−01	1.013E+00
			9	1.001E+00	−.127804	1.289E−01	1.006E+00
			10	1.025E+00	−.421288	4.645E−01	1.103E+00
5	5	1.039E+00	6	1.004E+00	−.052475	5.255E−02	1.001E+00
			7	1.021E+00	.672415	−9.085E−01	1.351E+00
			8	1.004E+00	.174062	−1.768E−01	1.016E+00
			9	1.001E+00	.209524	−2.143E−01	1.023E+00
			10	1.015E+00	.060101	−6.021E−02	1.002E+00
6	10	1.015E+00	6	1.003E+00	.222244	−2.279E−01	1.026E+00
			7	1.003E+00	.218013	−2.234E−01	1.025E+00
			8	1.005E+00	.093288	−9.370E−02	1.004E+00
			9	1.000E+00	.216249	−2.215E−01	1.024E+00
7	8	1.005E+00	6	1.004E+00	.168956	−1.714E−01	1.015E+00
			7	1.004E+00	.188274	−1.917E−01	1.018E+00
			9	1.000E+00	.308423	−3.242E−01	1.051E+00

[a]See text for explanation.

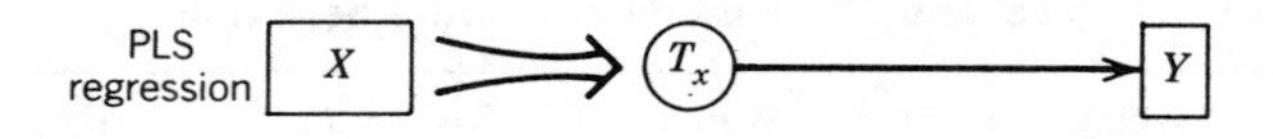

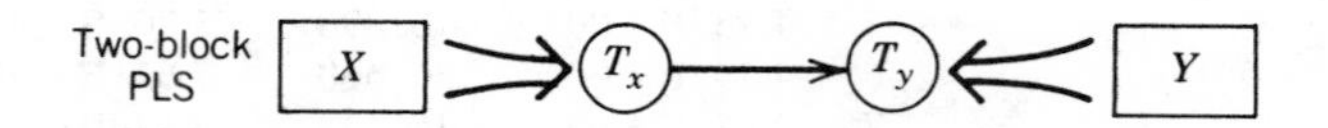

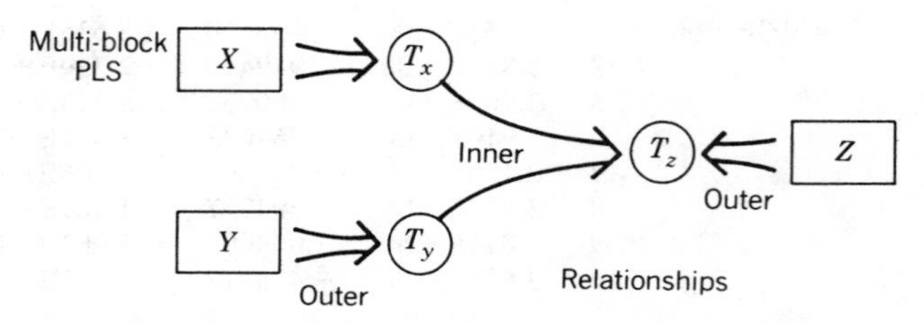

Figure 6.53 Partial least-squares path modeling.

variance and normalized to unity, by initializing the latent variables of the response matrix, $\mathbf{u}_{ih}$, to be equal to the response matrix:

$$\mathbf{u}_h = \mathbf{Y},$$

where h refers to the number of dimensions ($h = 1, a$).

The weights for the $\mathbf{X}$ variables are calculated as

$$\mathbf{w}_h = \mathbf{u}'_h \mathbf{X},$$

and $\mathbf{w}$ is normalized to unity. Then, the latent variable $\mathbf{t}_h$ is calculated as

$$\mathbf{t}_h = \mathbf{X}\mathbf{w}'_h,$$

and the coefficient v_h is determined which relates the latent variable of $\mathbf{X}$ to the latent variable of $\mathbf{Y}$:

$$\mathbf{u}_h = v_h \mathbf{t}_h + d$$

$$v_h = \frac{\mathbf{t}'_h \mathbf{u}_h}{\|\mathbf{t}_h\|^2}.$$

The loading vector for $\mathbf{X}$, given by $\mathbf{b}_{hx}$, is computed as

$$\mathbf{b}_{hx} = \mathbf{t}'_h \mathbf{X},$$

and is then normalized to unity. Finally, the residuals $\mathbf{E}_x$ and $\mathbf{E}_y$ are computed:

$$\mathbf{E}_x = \mathbf{X} - \mathbf{t}_h \mathbf{b}_{hx}$$

$$\mathbf{E}_y = \mathbf{Y} - v_h \mathbf{t}_h$$

For the next dimension, $\mathbf{E}_x$ and $\mathbf{E}_y$ are used in place of $\mathbf{X}$ and $\mathbf{Y}$ and the procedure begins again back at the beginning. Thus the meaningful variance in $\mathbf{X}$ that describes $\mathbf{Y}$ is successively "stripped off," stopping at dimension (a) when the test of cross validation indicates that dimension $(a+1)$ is not significant. This result would converge to the least squares solution after a number of iterations equal to the number of variables.

To use the PLS model for prediction of the y values associated with a test set, $\mathbf{X}_t$, the data are first scaled and centered as above. The first iteration begins at $h = 0$ with an initial guess for $\mathbf{y} = \bar{\mathbf{y}}$, that is, the average value of the training set data. In the next pass $h = h + 1$, $\mathbf{t}_h$ is computed as

$$\mathbf{t}_h = \mathbf{X}\mathbf{w}'_h$$

and thus

$$\mathbf{Y} = \mathbf{Y} + v_h \mathbf{t}_h \mathbf{b}_{hy}$$

$$\mathbf{X} = \mathbf{X} - \mathbf{t}_h \mathbf{b}_{hx}.$$

These steps are repeated until $h > a$. The predicted values of y are thus obtained in scaled and centered form and may then be back-transformed.

The fit of a test data vector i to the model may be evaluated by the standard deviation of the residual $\mathbf{e}_i$:

$$\mathbf{e}_i = \mathbf{x}_i - \mathbf{x}_i \mathbf{b}'\mathbf{b}$$

$$s^2 = \frac{\|e\|^2}{p-a},$$

where $p - a$ indicates the proper degrees of freedom (number of variables − number of dimensions in model). A criterion of acceptable fit to the model may be defined as

$$s^2 < s_a^2(F)$$

where s_a^2 is the residual standard deviation of the training set at (a) dimensions and F is the appropriate F statistic.

A very simple example of PLS regression can be worked out using the data set of Figure 6.12. Let $\mathbf{X}$, the data matrix of "independent variables," be given by

$$\mathbf{X} = \begin{pmatrix} 2 & 1 \\ 3 & 2 \\ 4 & 3 \end{pmatrix},$$

and $\mathbf{Y}$, the "response," or "dependent variable," be given by

$$\mathbf{Y} = \begin{pmatrix} 10 \\ 20 \\ 30 \end{pmatrix}.$$

The $\mathbf{X}$ and $\mathbf{Y}$ matrices are mean-centered and scaled to unit variance as well as normalized to unity:

$$\mathbf{X} = \begin{bmatrix} -1/\sqrt{2} & -1/\sqrt{2} \\ 0 & 0 \\ 1/\sqrt{2} & 1/\sqrt{2} \end{bmatrix} \quad \text{and} \quad \mathbf{Y} = \begin{bmatrix} -1/\sqrt{2} \\ 0 \\ 1/\sqrt{2} \end{bmatrix}.$$

Then the weights $\mathbf{w}_1$ for the first dimension ($h = 1$) are given by

$$\mathbf{w}_1 = (-1/\sqrt{2} \quad 0 \quad 1/\sqrt{2}) \begin{pmatrix} -1/\sqrt{2} & -1/\sqrt{2} \\ 0 & 0 \\ 1/\sqrt{2} & 1/\sqrt{2} \end{pmatrix} = (1 \quad 1).$$

The vector $\mathbf{w}_1$ is normalized to unity, yielding

$$\mathbf{w}_1 = (1/\sqrt{2} \quad 1/\sqrt{2}).$$

The latent variable $\mathbf{t}_1$ is computed by

$$\mathbf{t}_1 = \mathbf{x}\mathbf{w}'_1 = \begin{pmatrix} -1/\sqrt{2} & -1/\sqrt{2} \\ 0 & 0 \\ 1/\sqrt{2} & 1/\sqrt{2} \end{pmatrix} \begin{pmatrix} 1/\sqrt{2} \\ 1/\sqrt{2} \end{pmatrix} = \begin{pmatrix} -1 \\ 0 \\ 1 \end{pmatrix},$$

and so the coefficient $\mathbf{v}_1$ relating the latent variable of $\mathbf{x}$ to the latent variable of $\mathbf{Y}$ (which is just $\mathbf{Y}$) is found to be

$$\mathbf{v}_1 = \frac{\mathbf{t}'_1\mathbf{y}}{\|t\|^2}$$

$$= \frac{(-1 \quad 0 \quad 1)\begin{pmatrix} -1 \\ 0 \\ 1 \end{pmatrix}}{2}.$$

$$= \frac{1}{\sqrt{2}}.$$

In other words, there is a perfect correlation between the latent variable of $\mathbf{X}$ and the response variable $\mathbf{Y}$, which is what we expected to find in this case.

The loading for $\mathbf{X}$, given by $\mathbf{b}_{1x}$, is computed as

$$\mathbf{b}_{1x} = \mathbf{t}_1'\mathbf{X} = (-1 \quad 0 \quad 1)\begin{pmatrix} -1/\sqrt{2} & -1/\sqrt{2} \\ 0 & 0 \\ 1/\sqrt{2} & 1/\sqrt{2} \end{pmatrix} = (2/\sqrt{2} \quad 2/\sqrt{2})$$

which, after normalization yields $\mathbf{b}_{1x} = (1/\sqrt{2} \quad 1/\sqrt{2})$.

We can now compute the residuals $\mathbf{E}_x$ and $\mathbf{E}_y$:

$$\begin{aligned} \mathbf{E}_x &= \mathbf{X} - \mathbf{t}_1\mathbf{b}_{1x} \\ &= \begin{pmatrix} -1/\sqrt{2} & -1/\sqrt{2} \\ 0 & 0 \\ 1/\sqrt{2} & 1/\sqrt{2} \end{pmatrix} - \begin{pmatrix} -1 \\ 0 \\ 1 \end{pmatrix}(1/\sqrt{2} \quad 1/\sqrt{2}) \\ &= \begin{pmatrix} 0 & 0 \\ 0 & 0 \\ 0 & 0 \end{pmatrix} \\ \mathbf{E}_y &= \mathbf{Y} - \mathbf{v}_1\mathbf{t}_{1x} \\ &= \begin{pmatrix} -1/\sqrt{2} \\ 0 \\ 1/\sqrt{2} \end{pmatrix} - \frac{1}{\sqrt{2}}\begin{pmatrix} -1 \\ 0 \\ 1 \end{pmatrix} \\ &= \begin{pmatrix} 0 \\ 0 \\ 0 \end{pmatrix}. \end{aligned}$$

Since the null matrices were obtained for $\mathbf{E}_x$ and $\mathbf{E}_y$, all the information in this case is described by the latent variable $\mathbf{t}_1$, which we have seen to be totally correlated to $\mathbf{Y}$. Thus, there is no noise in this data set.

The two-block PLS algorithm is depicted schematically in Figure 6.53. The procedure begins in the same way as PLS regression:

1. Autoscale to unit variance and normalize $\mathbf{X}$ and $\mathbf{Y}$ to unity, $h = 1, a$.
2. Starting values: $\mathbf{u}_{ih} = \mathbf{Y}_{i1}$.
3. Weights for x variables: $\mathbf{w}_h = \mathbf{u}_h'\mathbf{X}$.
4. Normalize $\mathbf{w}$ to $\|\mathbf{w}\| = 1$.

5. Latent variables $\mathbf{t}_h = \mathbf{X}\mathbf{w}'_h$. At this point, the loadings for the $\mathbf{Y}$ variables, $\mathbf{b}_y$ are computed as:
6. $\mathbf{b}_y = \mathbf{t}'_h\mathbf{Y}$ and normalized to unit length. Then, new latent variable values are found for $\mathbf{Y}$:
7. $\mathbf{u}_h = \mathbf{Y}\mathbf{b}'_h$.

A convergence check is performed here to see if $\|\mathbf{u} - \mathbf{u}_{\text{old}}\| < \delta$ where δ is some preset tolerance such as $10^{-6}\|\mathbf{u}\|$; if it is less, then we continue to the next step, number 8. If convergence has not been reached, we return to step number 3 using the new value of $\mathbf{u}_h$ from step number 7.

Assuming convergence is reached, the coefficient v_h is calculated as before:

8. $v_h = \mathbf{t}'_h\mathbf{u}_h / \|\mathbf{t}_h\|^2$. The loadings $\mathbf{b}_{hx}$ are again found by
9. $\mathbf{b}_{hx} = \mathbf{t}'_h\mathbf{X}$ and normalized to unit length. Finally, the residuals are calculated as
10. $\mathbf{E}_x = \mathbf{X} - \mathbf{t}_h\mathbf{b}_{hx}$ and $\mathbf{E}_y = \mathbf{Y} - v_h\mathbf{t}_h\mathbf{b}_{hy}$ and the procedure starts again at step number 2 for the next dimension until the test of cross-validation indicates we should stop.

Applications of the PLS method in chemistry range from multivariate calibration in analytical chemistry (37–39) to problems as previously mentioned in environmental (35) and industrial chemistry (36) as well as food science (37–39). As a new technique the full potential of PLS continues to be explored.

REFERENCES

1. B. R. Kowalski, "Pattern Recognition in Chemical Research," in *Computers in Chemical and Biochemical Research*, Vol. 2, C. E. Klopfenstein and C. L. Wilkins (Eds.), Academic Press, New York, 1974.
2. B. R. Kowalski and C. F. Bender, "Pattern Recognition. A Powerful Approach to Interpreting Chemical Data," *J. Am. Chem. Soc.* **94**, (1972) 5632.
3. B. R. Kowalski and C. F. Bender, "Pattern Recognition. II. Linear and Nonlinear Methods for Displaying Chemical Data," *J. Am. Chem. Soc.* **95**, 686 (1973).
4. D. L. Duewer and B. R. Kowalski, "Forensic Data Analysis by Pattern Recognition. Categorization of White Bond Papers by Elemental Composition," *Anal. Chem.* **47**, 526 (1975).

5. B. R. Kowalski, "Measurement Analysis by Pattern Recognition," *Anal. Chem.* **47**, 1152A (1975).
6. E. J. Knudson, D. L. Duewer, G. D. Christian, and T. V. Larson, in *Chemometrics: Theory and Application*, B. R. Kowalski (Ed.), ACS Symposium Series 52, Washington, D.C., 1977.
7. Horst, *Factor Analysis of Data Matrices*, Holt, Rinehart and Winston, New York, 1965.
8. R. J. Rummel, *Applied Factor Analysis*, Northwestern University Press, 1970.
9. J. B. Kruskal, *Psychometrika* **29**, 115 (1964).
10. R. N. Shepard and J. D. Carroll in *International Symposium of Multivariate Analysis*, P. R. Krishnaiah (Ed.), Academic Press, New York, 1966.
11. C. Jochum and B. R. Kowalski, *Anal. Chim. Acta* **133**, 583 (1981).
12. J. W. Sammon, Jr., *I.E.E.E. Trans. Comput.* **C-20**, 68 (1971).
13. M. A. Sharaf, B. R. Kowalski, and B. Weinstein, *Z. Naturforsch.* **35C**, 508 (1980).
14. N. J. Nilsson, *Learning Machines*, McGraw-Hill, New York, 1965.
15. B. R. Kowalski and C. F. Bender, *Anal. Chem.* **44**, 1405 (1972).
16. B. R. Kowalski and C. F. Bender, *J. Pattern Recog.* **8**, 1 (1976).
17. S. Wold, *J. Pattern Recog.* **8**, 127 (1976).
18. S. Wold and M. Sjostrom, in *Chemometrics: Theory and Application* (B. R. Kowalski (Ed.), ACS Symposium Series 52, Washington, D.C., 1977.
19. C. Albano, W. Dunn III, U. Edlund, E. Johansson, B. Norden, M. Sjostrom, and S. Wold, *Anal. Chim. Acta* **103**, 429 (1978).
20. D. L. Duewer, B. R. Kowalski, and T. F. Schatzki, "Source Identification of Oil Spills by Pattern Recognition Analysis of Natural Elemental Composition," *Anal. Chem.* **47**, 1573 (1975).
21. W. Stone, *J. R. Stat. Soc.* **B38**, 111 (1974).
22. T. W. Anderson, *An Introduction to Multivariate Statistical Analysis*, Wiley, New York, 1958.
23. B. R. Kowalski, T. F. Schatzki, and F. H. Stross, *Anal. Chem.* **44**, 2176 (1972).
24. A. M. Harper, D. L. Duewer, B. R. Kowalski, and J. L. Fasching, "ARTHUR and Experimental Data Analysis: the Heuristic Use of a Polyalgorithm," in *Chemometrics: Theory and Application*, B. R. Kowalski (Ed.), ACS Symposium Series 52, Washington, D.C., 1977.
25. D. D. Wolff and M. L. Parsons, *Pattern Recognition Approach to Data Interpretation*, Plenum Press, New York, 1983.
26. O. L. Davies and P. L. Goldsmith, *Statistical Methods for Research and Production*, Hafner, New York, 1972.
27. R. Ghanadesikan, *Methods for Statistical Data Analysis of Multivariate Observations*, Wiley, New York, 1977.

28. N. Draper and H. Smith, *Applied Regression Analysis*, 2nd ed., Wiley, New York, 1981.
29. H. Wold, in *Multivariate Analysis*, P. R. Krishnaiah (Ed.), Academic Press, New York, 1966.
30. H. Wold, in *Quantitative Sociology*, H. M. Blalock (Ed.), Academic Press, New York, 1975.
31. H. Wold, in *Perspectives in Probability and Statistics*, J. Gani (Ed.), Academic Press, New York, 1975.
32. H. Wold, in *Mathematical Economics and Game Theory: Essays in Honor of Oskar Morgenstern*, R. Henn and O. Moeschlin, (Eds.), Springer, Berlin, 1977.
33. K. G. Joreskog and H. Wold, *Systems Under Indirect Observation*, North Holland, Amsterdam, 1982.
34. S. D. Brown, R. K. Skogerboe, and B. R. Kowalski, *Chemosphere* **9**, 265 (1980).
35. R. W. Gerlach, B. R. Kowalski, and H. Wold, *Anal. Chim. Acta* **112**, 417 (1979).
36. I. E. Frank, J. Feikema, N. Constantine, and B. R. Kowalski, "Prediction of Product Quality from Spectral Data Using the Partial Least Squares (PLS) Method," *J. Chem. Info. Comp. Sci.*, **24**, 20 (1984).
37. W. Lindberg, J. A. Persson, and S. Wold, *Anal. Chem.* **55**, 643 (1983).
38. M. Sjostrom, S. Wold, W. Lindberg, J. Persson, and H. Martens, *Anal. Chim. Acta* **150**, 61 (1983).
39. I. E. Frank, J. H. Kalivas, and B. R. Kowalski, *Anal. Chem.* **55**, 1500 (1983).

SUGGESTED READINGS

H. C. Andrews, *Introduction to Mathematical Techniques in Pattern Recognition*, Wiley-Interscience, New York, 1972.

J. C. Davis, *Statistics and Data Analysis in Geology*, Wiley, New York, 1973.

R. O. Duda and P. E. Hart, *Pattern Classification and Scene Analysis*, Wiley-Interscience, New York, 1973.

D. L. Duewer, B. R. Kowalski, and J. L. Fasching, "Improving the Reliability of Factor Analysis of Chemical Data by Utilizing the Measured Analytical Uncertainty," *Anal. Chem.* **48**, 2002 (1976).

E. A. Feigenbaum and J. Feldman, Eds., *Computers and Thought*, McGraw-Hill, New York, 1963.

K. S. Fu, *Sequential Methods in Pattern Recognition and Machine Learning*, Academic Press, New York, 1968.

K. S. Fu, *Applications of Pattern Recognition*, CRC Press, Boca Raton, 1982.

K. Fukunga, *Introduction to Statistical Pattern Recognition*, Academic Press, New York, 1972.

A. J. Goldstein, L. D. Harmon, and A. B. Lesk, "Identification of Human Faces," *Proc. IEEE*, May 1971.

J. A. Hartigan, *Clustering Algorithms*, Wiley, New York, 1975.

B. R. Kowalski, "Analytical Chemistry as an Information Science," *Tr.A.C.* **1**, 71 (1981).

B. R. Kowalski (ed.), *Chemometrics: Mathematics and Statistics in Chemistry*, NATO ASI Series, D. Reidel Publishing Company, Dordrecht, 1984.

P. J. Lewi, *Multivariate Data Analysis in Industrial Practice*, Research Studies Press, John Wiley, Chichester, 1982.

D. L. Massart, A. Dijkstra, and L. Kaufman, *Evaluation and Optimization of Laboratory Mehods and Analytical Procedures*, Elsevier, New York, 1978.

J. M. Mendel and K. S. Fu, *Adaptive Learning and Pattern Recognition Systems*, Academic Press, New York, 1970.

E. A. Patrick, *Fundamentals of Pattern Recognition*, Prentice-Hall, Englewood Cliffs, N.J., 1972.

G. S. Sebestyen, *Decision-Making Processes in Pattern Recognition*, MacMillan, New York, 1962.

R. C. Tryon and D. E. Bailey, *Cluster Analysis*, McGraw-Hill, New York, 1970.

K. Varmuza, *Pattern Recognition in Chemistry*, Springer-Verlag, New York, 1980.

D. D. Wolff and M. L. Parsons, *Pattern Recognition Approach to Data Interpretation*, Plenum Press, New York, 1983.

Chemometrics Reviews in *Anal. Chem.* **52**, 112R (1980); **54**, 232R (1982); and **56**, 261R (1984).

CHAPTER

7

AN INTRODUCTION TO CONTROL AND OPTIMIZATION

Previous chapters have dealt with methods for designing experiments, tools for calibration, resolution and, in general, getting more useful information from measurements. Another way of thinking about the methods introduced in those chapters is as chemometric tools that can add to our knowledge base on chemical systems. In this chapter, methods will be introduced that allow one to control or optimize a chemical system using what we have learned about it. These mathematical methods are called control and/or optimization methods. The student of chemometrics should realize that these topics are very important to several engineering fields and make up a significant portion of the engineering college curriculum. A number of journals are dedicated to mathematics that can be used to control systems of all descriptions and their application to such diverse problems as aircraft tracking and the optimization of an industrial production process.

Chemists use control and optimization methods either directly or indirectly but are usually unfamiliar with the breadth and depth of available technology. For example, chromatographers take direct advantage of so-called PID (proportional-integral-derivative) control for oven temperature control, perhaps without really knowing the mathematics involved. The organic chemist is making an attempt at system optimization when the parameters of a new synthetic reaction are changed in an attempt to optimize product yield.

In this chapter, some powerful methods that can be used to control and/or optimize chemical systems will be introduced. The chemical systems can range from analytical instruments to chemical reactions with, as in previous chapters, an emphasis on multivariate systems. Very few of the methods available to the chemist will be covered in this chapter. However, it should be useful as a springboard that can lead a chemist to an appropriate control or optimization strategy.

Figure 7.1 shows the simplest type of system with which the chemist usually deals. Examples range from an oven, where heat is the input and temperature is the controlled output, to a simple reactor where stirring

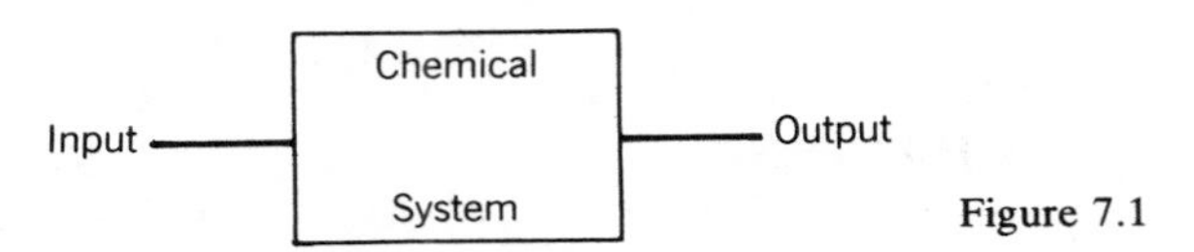

Figure 7.1

rate is the only adjustable parameter (input) and product yield is the only output. The former case obviously involves a control situation where current to a resistance heater is changed in order to achieve a preset or programmable oven temperature: the control *set point*. The challenge is to hold the temperature as close to the set point as possible; the level of success depends on the heat loss time constant, the characteristics of the energy input device, and the selection of an appropriate control strategy. The latter example involves optimization rather than control, the goal being to find the stirring rate that offers the best product yield. Controlling the input stirring rate is not generally considered to be a challenge.

The methods used to control or optimize Figure 7.1-type systems are many and represent the simplest case covered in this chapter. The simplest goal is optimization and will be covered first. Following this, a control strategy will be introduced followed by an introduction to methods that can handle more complex systems.

SINGLE INPUT/SINGLE OUTPUT: OPTIMIZATION

Usually, inputs to chemical systems are bounded. For the stirred reaction example, the lower bound for stirring would be no stirring and the upper bound would either be the maximum stirring rate achievable or the point at which an instability is reached (e.g., the reaction mixture begins to spill out of the vessel). The unknown, so-called *response surface* (see Chapter 3 for more on response surfaces) relating stirring rate to reaction yield can have a variety of shapes between these bounds as shown in Figure 7.2. These shapes correspond to conditions also listed in the figure that can be detected with a simple algorithm.

The goal is to find the optimum (minimum or maximum) value of the output, y, which is a function of the input, x, over some preset range of x, using the minimum number of experiments. The method of choice for this univariate optimization problem is the Fibonacci search (1). While the search method itself was developed in the early 1950s, it is based on the Fibonacci number theory which dates back to the early thirteenth century.

For the purpose of introducing this powerful search method, it will be assumed that one is looking for a maximum for an experimental output

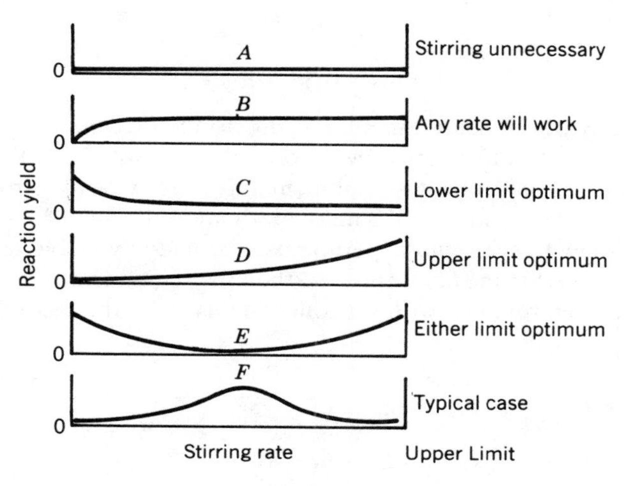

Figure 7.2

variable $y(x)$ over a range of the input variable, $a \leq x \leq b$. The modified Fibonacci search method will be demonstrated since it does not assume that the total number of experiments is fixed at the beginning of the search. For the complete theory behind this search method and its efficiency one should consult the literature (1).

If $L_1 = b_1 - a_1$ is the total input variable range, then the method (for maximization) begins by evaluating the output at two points on the range, x_1 and x_2, given by

$$x_1 = a_1 + m_1 \tag{7.1}$$

and

$$x_2 = b_1 - m_1, \tag{7.2}$$

with

$$m_1 = 0.3820L_1. \tag{7.3}$$

Next, the region containing the largest value for $y(x)$ would be retained in the search. If $y(x_1) > y(x_2)$, then the region from a_1 to $(b_1 - m_1)$ would be retained and the process repeated with the following new definitions:

$$a_2 = a_1, \tag{7.4}$$

$$b_2 = b_1 - m_1, \tag{7.5}$$

$$L_2 = b_2 - a_2, \tag{7.6}$$

and

$$m_2 = 0.3820L_2. \tag{7.7}$$

The efficiency of this search method is due to the fact that one of the two new points to test in this new range is already available from the previous test. Therefore, only one test point, in this case $x_3 = a_2 + m_2$, need be tested to find $y(x_3)$ and so on. The reader can verify that the other point needed to span L_2 with two experiments is already available as x_1.

If $y(x_2) > y(x_1)$ in the first range L_1, then the region from $(a_1 + m_1)$ to b_1 would have been retained with the following used for the second iteration:

$$a_2 = a_1 + m_1, \tag{7.8}$$

$$b_2 = b_1, \tag{7.9}$$

$$L_2 = b_2 - a_2, \tag{7.10}$$

$$m_2 = 0.3820L_2, \tag{7.11}$$

and

$$x_3 = b_2 - m_2. \tag{7.12}$$

Continuing this procedure will decrease the range to some value small enough to use the midpoint of the range as the final result. However, for improved efficiency, the modified search should be abandoned in favor of the normal Fibonacci search when the range is about one order of magnitude larger than the final desired range. It will then take five more tests (iterations) to complete the search.

The maximum number, N, of experiments or iterations must be known in advance in order to use the normal Fibonacci search. If the modified procedure is employed, then $N = 5$ and it is understood that only five more iterations (experimental determinations of $y(x)$) are required. The search uses the Fibonacci number series given by

$$F_0 = F_1 = 1 \tag{7.13}$$

and

$$F_n = F_{n-1} + F_{n-2}, \quad n \geq 2, \tag{7.14}$$

where each number in the series is the sum of the two preceding numbers. If k is the step number starting at $k = 1$ and ending at $k = N$, and the total length of the interval from a_k to b_k is L_k, then

$$m_k = \frac{F_n - (k+1)L_k}{F_n - (k-1)} \tag{7.15}$$

defines the length used to position the test points in the region at $a_k + m_k$ and $b_k - m_k$. The procedure is the same as used in the modified search described above except that m_k does not involve a constant. It can now be seen that the constant, 0.3820, comes from the asymptotic approach of the ratio of Fibonacci numbers in the search sequence.

The search only proceeds until k reaches $(N-1)$ since, when $k = N$, the midpoint of the region is used for the final value of x. Also, the Fibonacci search always yields the advantage that only one new experiment is required for each iteration. At this point it is instructive to find the minimum for $y = x^2$ over the range -1.0 to $+1.0$. The reader will note the similarity of this method to the simplex method.

SINGLE INPUT/SINGLE OUTPUT: CONTROL

Engineers have long known how to use feedback strategies to control single input/single output systems. The control strategies can be either analog or digital and come under the general heading of proportional plus integral plus derivative (PID) control (2). The strategy involves maintaining the system output at some set level (set point) or according to some preprogrammed function by adjusting the input so that the difference between the desired output and the actual measured output at any point in time is small. This requires that the output be controlled by the input but *does not* assume that the relationship is exclusive. Using the oven temperature control example, the goal might be to establish the temperature profile shown in Figure 7.3. This would be accomplished by

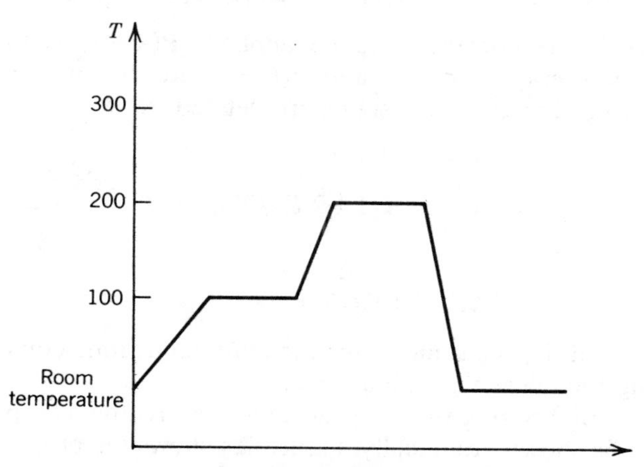

Figure 7.3 A temperature profile for an oven.

measuring the temperature with an acceptable sampling frequency, and then comparing the measured temperature to the desired temperature from a function generator in order to establish an error signal. The error signal $e(k)$ at the kth instant in time generates a corrective action $P(k)$ that will supply that level of power to the oven heaters.

The temperature rises, the increase is measured at time $k+1$, and the iteration continues. Obviously, the larger the error $e(k)$, the larger should be the correction, $P(k)$. However, it is important to have the appropriate sampling frequency and to use the proper algorithm so as not to overshoot the set point and possibly even cause wild temperature fluctuations.

The continuous form of the time dependent PID equation is

$$P(t) = K\left[e(t) + \frac{1}{T_1}\int e(t)\,dt + T_D\frac{de(t)}{dt}\right], \tag{7.16}$$

and shows that the corrective term $P(t)$ is proportional (K is the overall gain) to a function of the error itself, the integral of the error with respect to time (T_I is the integral time constant) and the derivative of the error with respect to time (T_D is the derivative time constant). In the oven example, the function would be used in a discontinuous mode of analog control since no power would be added to the heater if the oven temperature were above the set point at time t. The task of the chemist is to tune the PID controller by adjusting the sampling frequency, K, T_I, and T_D to the desired performance level.

The discrete form of the PID controller commonly used with digital control is

$$P(k) = P(k-1) + A_1e(k) + A_2e(k-1) + A_3e(k-2), \tag{7.17}$$

where $P(k)$ is the correction to be applied, $P(k-1)$ is the previous correction and $e(k)$, $e(k-1)$, and $e(k-2)$ are the present and two previous errors. The three constants are defined by

$$A_1 = K(1 + T_s/T_I + T_D/T_s), \tag{7.18}$$

$$A_2 = -K(1 + 2T_D/T_s), \tag{7.19}$$

and

$$A_3 = K(T_D/T_s) \tag{7.20}$$

with K, T_I, and T_D the same as for the continuous time equation and T_s representing the chosen sampling time.

The effect of ignoring the integral (I) and derivative (D) parts of PID control can be shown graphically. Figure 7.4 shows the proportional-only control response to a linear ramp control function. Using a low to

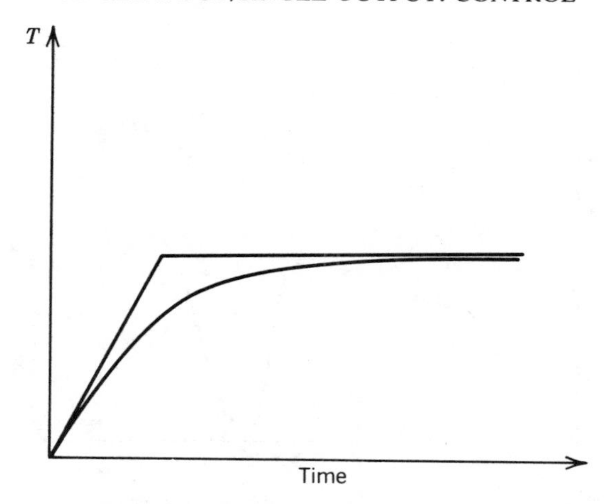

Figure 7.4 Proportional-only control showing the steady state offset.

moderate gain, K, the temperature never reaches the desired level, the difference being known as the *steady state offset*. As the gain is increased, the offset becomes smaller but then it will begin to oscillate. At infinite gain, the oven is switched from full on to completely off and the oscillations are at a maximum.

As the integral part is added, the very commonly used PI controller is now able to reach the control set point and eliminate the steady state offset. However, the PI controller cannot respond rapidly to a major perturbation in the system, such as the door of the oven being opened (Fig. 7.5).

With only the derivative part added, the PD controller would be able to respond rapidly to perturbations but may overshoot the set point even more than proportional-only control and possibly even cause the oscillations to amplify. This would most certainly occur in a "bucked" control system where, for example, a cooling device is added to buck the heating system for more precise oven temperature control as shown in Figure 7.6. The D part of PID control is not often used in practice. It is especially not used in very noisy systems as derivatives amplify noise.

A properly tuned PID control system can respond rapidly to perturbations in the error and dampen instabilities or oscillations. For this reason, simple PID control is used for the vast majority of control applications. It can be referred to as a "dumb" controller since it uses minimum knowledge of process dynamics and is model free. While it is a relatively sluggish controller, it is remarkably robust in that it is resistant to changes in process characteristics. In a later section in this chapter, a brief

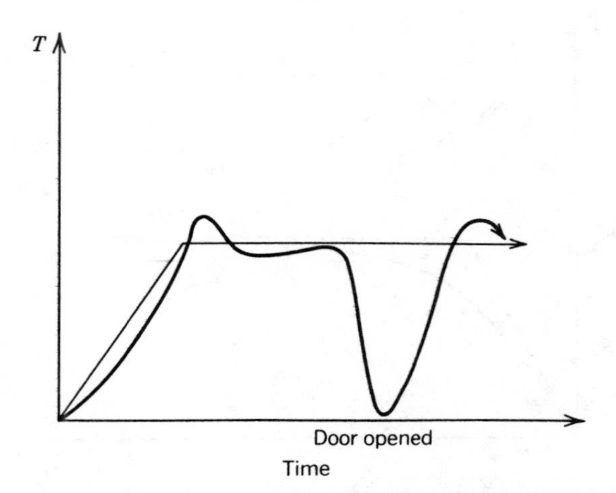

Figure 7.5 The PI controller eliminates the steady state offset but cannot respond rapidly to a major perturbation in the system.

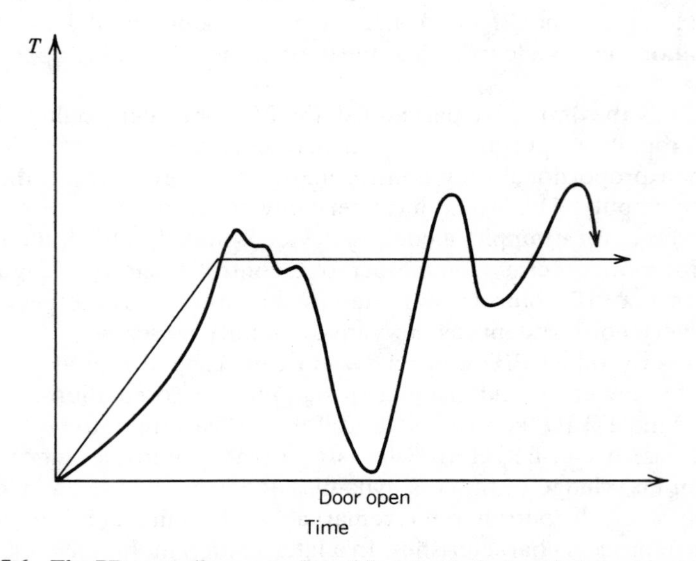

Figure 7.6 The PD controller responds rapidly to perturbations but may be unstable.

introduction to Model Based Control (MBC) is given. MBC is often introduced as intelligent control because state models are used and a fundamental understanding of the process often results.

An excellent example of PID control applied to an instrumentation problem in analytical chemistry can be found in the temperature control of a direct inlet probe for a high resolution mass spectrometer (3). The tuning sequence and additional references in this article should provide the interested reader with the tools required to solve rather complex control problems.

MULTIPLE INPUT/SINGLE OUTPUT: OPTIMIZATION

In this section, methods for optimizing systems of the general type shown in Figure 7.7 are presented. Systems of this type are the most likely to be of interest to the chemist. For the analytical chemist, Figure 7.7 could represent an instrument with the goal to adjust several instrumental controls (inputs) in order to arrive at a maximum sensitivity for an analyte. For the organic chemist, the goal might be to optimize the yield of a synthetic reaction by varying the reaction parameters.

In the rare cases where the inputs are independent from one another, each input can be adjusted individually as with single input/single output systems. However, it is more common that the inputs are dependent, or coupled; simple one-at-a-time optimization will usually fail. Figure 7.8 is a contour plot of a response surface for a hypothetical two input system. The contour lines represent equal output values. The elongated ellipses indicate the correlated inputs. If one were to select a level of I_1 corresponding to I_1^* and adjusted I_2 to obtain a maximum output, I_2^* would be found to yield the best value for the output. A test of I_1 would yield I_1^* as the best value for I_1 and the erroneous (I_1^*, I_2^*) pair would be

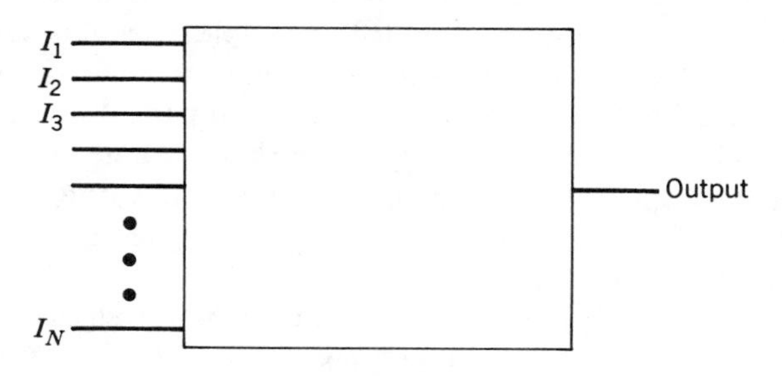

Figure 7.7

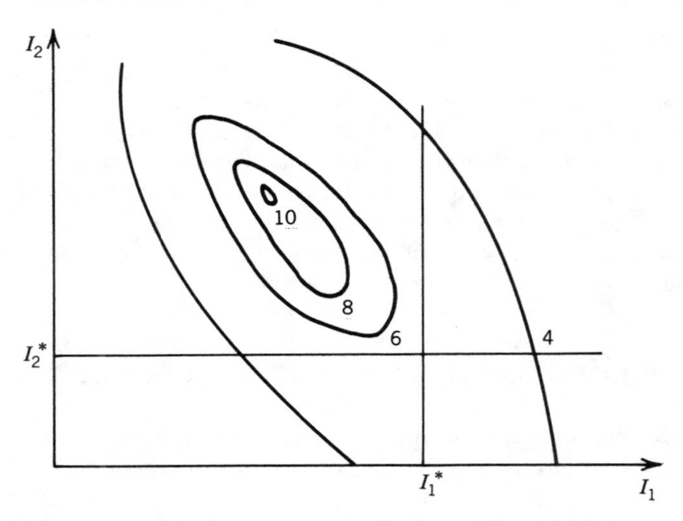

Figure 7.8 Contour plot of a response surface for a hypothetical two input system.

selected for the optimal output. If the system were a synthetic polymer reaction with temperature, pressure, flow rate, and percent composition as inputs and polymer yield as the output, the selected reaction conditions would be far from the best conditions, thereby yielding an inferior product.

In 1957, George Box, a chemist/statistician, offered a way out of the dilemma of being stranded on ridges leading to false optima. The solution to the problem came in the form of Evolutionary Operation or EVOP (4, 5). The reader is already familiar with this method, since it was discussed under experimental design in Chapter 2. A newer method, and the one discussed in this chapter, begins optimizing the process with as few experiments as possible. Then, if desired, the relative effects of the parameters and their importance to control can be investigated. The optimization methods under discussion are those involving one or more variations on Simplex optimization (6). A simplex is a geometric structure defined by $N+1$ points in an N-dimensional space. The initial simplex is formed by selecting $N+1$ sets of input values for the N inputs, evaluating the output value for each set of inputs and then connecting the response points in the N-space. The object is simply to use one of a number of algorithms to move the simplex through the N-space until the best value for the output is achieved.

To expand on our discussion of the simple simplex algorithm of Chapter 2, the modified simplex method (7) will be presented here. Before beginning, however, it is important to realize that the ever important response surface that relates the inputs to the single output is assumed to be unknown. If it is known and a mathematical model is available that can be fit to the response surface to obtain the model parameters, then optimization involves setting the derivatives of the model function to zero and solving for the optimum input parameters (8). If the model function is known or can be approximated but the parameters are unknown, then gradient optimization algorithms (9) can be used to find the optimum. As these cases are rare in chemometrics, *simplex optimization*, a useful method for optimization without derivatives, is presented here.

Consider the case of optimizing a function (output) of two variables (inputs). It might be helpful to use an example system: the optimization of emission intensity for a standard sample in atomic flame emission spectrometry by controlling the fuel and oxidant flow rates. As the flow rates are varied, the emission of light at a selected wavelength with a steady aspiration of a standard sample into the flame will increase or decrease depending on the effect of the energy of the flame on the desolvation, atomization, excitation, and ionization processes inherent in atomic emission spectrometry.

Using x_1 as the fuel flow rate, x_2 as the oxidant flow rate, and $F(x_1, x_2)$ as the light emission, the optimization process begins with $F(x_1, x_2)$ evaluated at three sets $(N+1=3)$ of x_1, x_2 combinations. The three responses are then ranked and identified as best, $\mathbf{B}$, next worst, $\mathbf{N}$, and worst, $\mathbf{W}$ with $\mathbf{B}$, $\mathbf{N}$, and $\mathbf{W}$ being the vectors of (x_1, x_2) coordinates used.

Figure 7.9 gives a representation of the simplex formed by the three

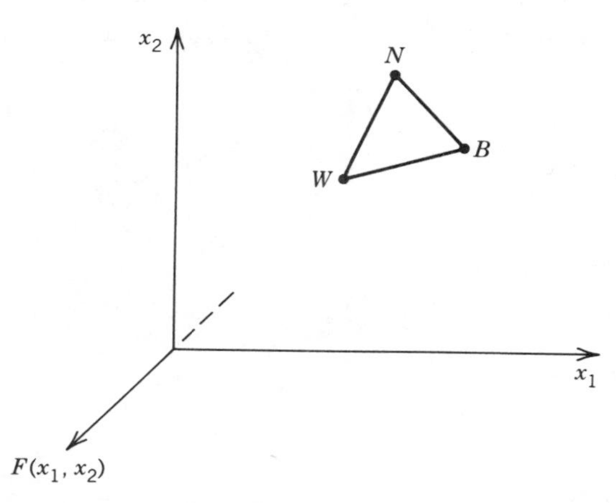

Figure 7.9 A simplex formed by three experimental observations.

experimental observations. The following definitions are then made:

$$F(\mathbf{B}) > F(\mathbf{N}) > F(\mathbf{W}), \tag{7.21}$$

$$\mathbf{P} = \tfrac{1}{2}(\mathbf{B}+\mathbf{N}), \tag{7.22}$$

and

$$\mathbf{R} = \mathbf{P} + (\mathbf{P} - \mathbf{W}). \tag{7.23}$$

The next step in the algorithm is to compute a new set of inputs to be tested that will hopefully yield an even better value for the output. The strategy of the simple simplex is to move away from the worst output at **W** in the direction of the perpendicular bisector of the line between **B** and **N**. A simple reflection of the simplex away from **W** gives **R**, which is the next point (set of input values) to be tested. For the atomic emission example, the values in *R* for fuel and oxidant flow rates would be used and the emission of the standard-measured, **F(R)**. If $\mathbf{F(R)} > \mathbf{F(B)}$, the following definitions would be made and the process repeated:

$$\left.\begin{aligned} \mathbf{F(W)}_{\mathrm{NEW}} &= \mathbf{F(N)}_{\mathrm{OLD}}, \\ \mathbf{F(N)}_{\mathrm{NEW}} &= \mathbf{F(B)}_{\mathrm{OLD}}, \\ \mathbf{F(B)}_{\mathrm{NEW}} &= \mathbf{F(R)}. \end{aligned}\right\} \tag{7.24}$$

Otherwise, the process would end and the values for **B** used as the optimum input parameters. It should be obvious that, in this simple case, the simplex size would be a constant and that a compromise would have to be made between rapid movement (large simplex) and accuracy (small simplex).

The modified simplex method avoids this compromise by employing the following algorithm once **B**, **N**, **W**, **P**, and **R** have been obtained.

1. If $\mathbf{F(R)} > \mathbf{F(B)}$ use:

   $\mathbf{B}_{(\mathrm{NEW})} = \mathbf{E}$,

   $\mathbf{N}_{(\mathrm{NEW})} = \mathbf{B}_{(\mathrm{OLD})}$,

   $\mathbf{W}_{(\mathrm{NEW})} = \mathbf{N}_{(\mathrm{OLD})}$,

   where $\mathbf{E} = \mathbf{P} + \beta(\mathbf{P} - \mathbf{W})$, $\beta > 1.0$; usually 2.0.

2. If $\mathbf{F(B)} > \mathbf{F(R)} \geq \mathbf{F(N)}$ use:

   $\mathbf{B}_{(\mathrm{NEW})} = \mathbf{B}_{(\mathrm{OLD})}$,

   $\mathbf{N}_{(\mathrm{NEW})} = \mathbf{R}$,

   $\mathbf{W}_{(\mathrm{NEW})} = \mathbf{N}_{(\mathrm{OLD})}$.

3. If $\mathbf{F(N)} > \mathbf{F(R)} > \mathbf{F(W)}$ use:

   $\mathbf{B}_{(\mathrm{NEW})} = \mathbf{B}_{(\mathrm{OLD})}$,

   $\mathbf{N}_{(\mathrm{NEW})} = \mathbf{N}_{(\mathrm{OLD})}$,

   $\mathbf{W}_{(\mathrm{NEW})} = \mathbf{C}_{\mathbf{R}}$,

   where

   $\mathbf{C}_{\mathbf{R}} = \mathbf{P} + \beta(\mathbf{P} - \mathbf{W})$, $0.0 < \beta < 1.0$.

4. If $\mathbf{F(W)} > \mathbf{F(R)}$ use:

   $\mathbf{B}_{(\mathrm{NEW})} = \mathbf{B}_{(\mathrm{OLD})}$,

   $\mathbf{N}_{(\mathrm{NEW})} = \mathbf{N}_{(\mathrm{OLD})}$,

   $\mathbf{W}_{(\mathrm{NEW})} = \mathbf{C}_{\mathbf{W}}$,

   where

   $\mathbf{C}_{\mathbf{R}} = \mathbf{P} - \beta(\mathbf{P} - \mathbf{W})$, $0.0 < \beta < 1.0$.

5. If the step size get to the point where the input changes are insignificant, stop the iteration.

In the above, each new point tested is defined as **F**(**R**). Therefore, even if **E** is the next point of reflection, the tested value is referred to as **F**(**R**) in the algorithm. It can now be seen that the modified simplex has an ability to expand and contract in order for it to move rapidly at first and then shrink as it gets close to the optimum. In practice, it is best to start with a very large simplex with initial points being at the edges of the allowed input space. The operations then consist of contractions towards the optimum.

In most cases, the inputs are bounded in the algorithm to avoid problems. In the atomic emission example it is obvious that the flow rates must be constrained to be within ranges that will avoid flameouts or flashbacks. When the simplex moves beyond the input boundary constraints, it can simply be contracted to rest on the boundary. If it continually tries to move into the nonpermitted region, the process is halted and the best set of values on the boundary is used.

A word of caution is in order. The simplex will always find an optimum. However, it may only be a *local optimum* and not the sought-for *global optimum*. There is no general recipe for avoiding local optima. It is considered wise to start a new simplex in a new position if possible, in the hope that the same optimum will be found. Depending upon the cost of the experiments, the process can even be stated several times with randomly initiated starting simplexes.

Simplex optimization is beginning to be used extensively by chemists since its introduction to chemistry by Ernst (10), Long (11), and Deming and co-workers (12). The chemometrics section of Fundamental Reviews published in even numbered years in *Analytical Chemistry* provides an exhaustive summary of simplex applications.

In addition to the many simplex variations, the methods available for optimizing Figure 7.7-type systems are many and can be found under such headings as linear and nonlinear programming (13–15). These methods are open to exploration by the new chemometrician but are beyond the scope of this introductory text.

MULTIPLE INPUT/MULTIPLE OUTPUT SYSTEMS

The optimization and control of systems with both multiple inputs as well as multiple outputs is an extremely complex problem and many difficulties arise even when the functions describing system behavior are multivariate linear. Nevertheless, many chemical systems fall under this heading and control theory developed in various fields of engineering will probably see increased use in chemistry in the future.

For the case where a single input affects only a single output, the methods described early in this chapter suffice. In the more common case involving interacting or coupled inputs and outputs, a number of questions require answers before attempts at control can be made. These questions include the suitability of a selected model, the type of feedback or feedforward control required, and the level of stability needed. Additionally, what is meant by the ideas of poles, zeros, frequency response, root-loci, Nyquist diagrams, and the generalized notion of transfer functions in higher dimensional control problems are complex questions currently being addressed by control engineers.

The most common approach is to model these systems with multivariate linear or nonlinear models and use state variable models of process dynamics (16) coupled with parameter estimation (17).

The more sophisticated and relevant the model, the more intelligence can be built into the control strategy. Of the methods used to build more intelligence into control systems, the model algorithmic (18) and the internal model control (19) methods seem to be enjoying a considerable amount of attention in engineering at this time. The latter method is considered to be fairly robust in that it is tolerant of typical noise levels in process signals. These and other new methods have yet to be shown to work properly in complex industrial processes (20, 21) and are mentioned here only for completeness.

In spite of the complex nature of multiple input/multiple output systems, chemists are beginning to address the optimization and control of these systems by modifying existing methods as well as developing innovative new ones. An example of the former is the use of simplex optimization to optimize a function of multiple outputs instead of each individual output. While this approach can lead to useful selections of inputs, it is heavily dependent on selecting the proper function of the outputs.

An example of the development of an innovative solution to a multiple output optimization problem is the use of "window diagrams" in chromatography (22, 23). Using this ingenious method, the chemist can take full advantage of separation theory and find experimental conditions that maximize the separation of any pair of eluting peaks including the poorest separated of the lot.

REFERENCES

1. G. S. G. Beveridge and R. S. Schechter, *Optimization: Theory and Practice*, McGraw-Hill, New York, 1970, pp. 180–193.
2. T. F. Edgar, Ed., *AIChE Modular Instruction*, Series A, Volumes 1–3, American Institute of Chemical Engineers, New York, 1982.
3. C. L. Pomernacki, *Rev. Sci. Instrum.* **48**, 1420 (1977).
4. G. E. P. Box, *Appl. Statist.* **6**, 81 (1957).
5. G. E. P. Box and N. R. Draper, *Evolutionary Operation*, Wiley, New York, 1969.
6. W. Spendley, G. R. Hext, and F. R. Himsworth, *Technometrics* **4**, 441 (1962).
7. J. A. Nelder and R. Mead, *Comput. J.* **7**, 308 (1965).
8. S. L. S. Jacoby, J. S. Kowalik, and J. T. Pizzo, *Iterative Methods for Nonlinear Optimization Problems*, Prentice-Hall, Englewood Cliffs, NJ, 1972, p. 117.
9. S. L. S. Jacoby, J. S. Kawalik, and J. T. Pizzo, *Iterative Methods for Nonlinear Optimization Problems*, Prentice-Hall, Englewood Cliffs, NJ, 1972, p. 129.
10. R. R. Ernst, *Rev. Sci. Instrum.* **39**, 988 (1968).
11. D. E. Long, *Anal. Chim. Acta* **46**, 193 (1969).
12. S. N. Deming and S. L. Morgan, *Anal. Chem.* **45**, 278A (1973).
13. See ref. 1.
14. D. M. Himmelblau, *Applied Nonlinear Programming*, McGraw-Hill, New York, 1972.
15. F. A. Lootsma, *Numerical Methods for Nonlinear Optimization*, Academic Press, New York, 1972.

16. D. H. Owens, *Multivariable and Optimal Systems*, Academic Press, New York, 1981.
17. P. Eykhoff, *System Identification: Parameter and State Estimation*, Wiley-Interscience, New York, 1974.
18. R. Rouhani and R. K. Mehra, *Automatica* **18**, (4), 410 (1982).
19. C. E. Garcia and M. Morari, *Ind. Eng. Chem. Processes Des. Dev.* **21**, 308 (1982a).
20. C. J. Harris and S. A. Billings (Ed.), *Self-Tuning and Adaptive Control: Theory and Applications*, Peter Paregrinus Ltd., London, 1981.
21. K. J. Astrom (Ed.), *Optimal Control Appl. Methods* **3**, (4) (1982).
22. R. J. Laub and J. H. Purnell, *Anal. Chem.* **48**, 799 (1976).
23. S. N. Deming and M. L. H. Turoff, *Anal. Chem.* **50**, 547 (1978).

LIST OF TABLES AND THEIR ORIGINS

Table A.1 Normal Distribution (Single-sided)
Condensed and adapted from the *Biometrika Tables for Statisticians*, **1**, Table 1, with permission from the Trustees of *Biometrika*.

Table A.2 Probability Points of the *t*-Distribution (Single-sided)
Condensed and adapted from the *Biometrika Tables for Statisticians*, **1**, Table 12, with permission of the Trustees of *Biometrika*, and from *Statistical Tables for Biological, Agricultural and Medical Research*.

Table A.3 Probability Points of the χ^2 Distribution
Condensed and adapted from the *Biometrika Tables for Statisticians*, **1**, Table 8, with permission of the Trustees of *Biometrika*, and from *Statistical Tables for Biological, Agricultural and Medical Research*.

Table A.4 Probability Points of the Variance Ratio (*F*-Distribution)
Condensed and adapted from the *Biometrika Tables for Statisticians*, **1**, Table 18, with permission of the Trustees of *Biometrika*, and from *Statistical Tables for Biological, Agricultural and Medical Research*.

Table A.5 Wilcoxon (one-input) Test: Upper Tail Probabilities for the Null Distribution, adapted from *A Nonparametric Introduction to Statistics*, with permission.

Table A.1(a) Normal Distribution (Single-sided)

Proportion (P) of Whole Area Lying to Right of Ordinate Through $u = (x - \mu)/\sigma$

Deviate u	0.00	0.01	0.02	0.03	0.04	0.05	0.06	0.07	0.08	0.09
0.0	.5000	.4960	.4920	.4880	.4840	.4801	.4761	.4721	.4681	.4641
0.1	.4602	.4562	.4522	.4483	.4443	.4404	.4364	.4325	.4286	.4247
0.2	.4207	.4168	.4129	.4090	.4052	.4013	.3974	.3936	.3897	.3859
0.3	.3821	.3783	.3745	.3707	.3669	.3632	.3594	.3557	.3520	.3483
0.4	.3446	.3409	.3372	.3336	.3300	.3264	.3228	.3192	.3156	.3121
0.5	.3085	.3050	.3015	.2981	.2946	.2912	.2877	.2843	.2810	.2776
0.6	.2743	.2709	.2676	.2643	.2611	.2578	.2546	.2514	.2483	.2451
0.7	.2420	.2389	.2358	.2327	.2296	.2266	.2236	.2206	.2177	.2148
0.8	.2119	.2090	.2061	.2033	.2005	.1977	.1949	.1922	.1894	.1867
0.9	.1841	.1814	.1788	.1762	.1736	.1711	.1685	.1660	.1635	.1611
1.0	.1587	.1562	.1539	.1515	.1492	.1469	.1446	.1423	.1401	.1379
1.1	.1357	.1335	.1314	.1292	.1271	.1251	.1230	.1210	.1190	.1170
1.2	.1151	.1131	.1112	.1093	.1075	.1056	.1038	.1020	.1003	.0985
1.3	.0968	.0951	.0934	.0918	.0901	.0885	.0869	.0853	.0838	.0823
1.4	.0808	.0793	.0778	.0764	.0749	.0735	.0721	.0708	.0694	.0681
1.5	.0668	.0655	.0643	.0630	.0618	.0606	.0594	.0582	.0571	.0559
1.6	.0548	.0537	.0526	.0516	.0505	.0495	.0485	.0475	.0465	.0455
1.7	.0446	.0436	.0427	.0418	.0409	.0401	.0392	.0384	.0375	.0367
1.8	.0359	.0351	.0344	.0336	.0329	.0322	.0314	.0307	.0301	.0294
1.9	.0287	.0281	.0274	.0268	.0262	.0256	.0250	.0244	.0239	.0233
2.0	.0228	.0222	.0217	.0212	.0207	.0202	.0197	.0192	.0188	.0183
2.1	.0179	.0174	.0170	.0166	.0162	.0158	.0154	.0150	.0146	.0143
2.2	.0139	.0136	.0132	.0129	.0125	.0122	.0119	.0116	.0113	.0110
2.3	.0107	.0104	.0102		.00964		.00914		.00866	
2.4	.00820		.00776		.00734		.00695		.00657	
2.5	.00621		.00587		.00554		.00523		.00494	
2.6	.00466		.00440		.00415		.00391		.00368	
2.7	.00347		.00326		.00307		.00289		.00272	
2.8	.00256		.00240		.00226		.00212		.00199	
2.9	.00187		.00175		.00164		.00154		.00144	
	0.00		0.02		0.04		0.06		0.08	

Table A.1(b) Extension for Higher Values of the Deviate

Deviate u	Proportion of Whole Area P	Deviate u	Proportion of Whole Area P	Deviate u	Proportion of Whole Area P	Deviate u	Proportion of Whole Area P
3.0	.001 35	3.5	.000 233	4.0	$.0^4$ 317	4.5	$.0^5$ 340
3.1	.000 968	3.6	.000 159	4.1	$.0^4$ 207	4.6	$.0^5$ 211
3.2	.000 687	3.7	.000 108	4.2	$.0^4$ 133	4.7	$.0^5$ 130
3.3	.000 483	3.8	$.0^4$ 723	4.3	$.0^5$ 854	4.8	$.0^6$ 793
3.4	.000 337	3.9	$.0^4$ 481	4.4	$.0^5$ 541	4.9	$.0^6$ 479
						5.0	$.0^6$ 287

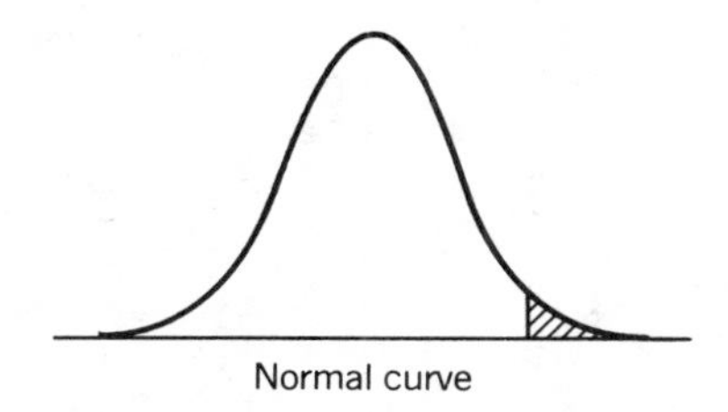

Normal curve

The illustration shows the *normal curve*. The shaded portion is the area P, which is given in the table.

The entries refer to positive values of the argument u. For negative values of u write down the complements $(1 - P)$ of the entries.

Examples

Let $u = +1.96$. Area to the right $= 0.0250$. Area to left $= 1 - 0.0250 = 0.9750$.

Let $u = -3.00$. The tabulated value $= 0.00135$. Since u is negative, this represents the area of the *left*. Area to right $= 1 - 0.00135 = 0.99865$.

Let $u = +4.50$. Tabulated value $= 0.00000340$. Area to left $= 0.99999660$.

To find the value of u corresponding to a given P we use the table in reverse, thus:

Let area to right (i.e., $P) = 0.10$. The two adjacent tabulated values are $P = 0.1003$ for $u = 1.28$, and $P = 0.0985$ for $u = 1.29$. We interpolate linearly to obtain the required value of u. Thus $u = 1.28 + (3)(0.01)/18 = 1.2817$.

Table A.2 Probability Points of the t-Distribution (Single-sided)

	P				
ϕ	0.1	0.05	0.025	0.01	0.005
1	3.08	6.31	12.70	31.80	63.70
2	1.89	2.92	4.30	6.96	9.92
3	1.64	2.35	3.18	4.54	5.84
4	1.53	2.13	2.78	3.75	4.60
5	1.48	2.01	2.57	3.36	4.03
6	1.44	1.94	2.45	3.14	3.71
7	1.42	1.89	2.36	3.00	3.50
8	1.40	1.86	2.31	2.90	3.36
9	1.38	1.83	2.26	2.82	3.25
10	1.37	1.81	2.23	2.76	3.17
11	1.36	1.80	2.20	2.72	3.11
12	1.36	1.78	2.18	2.68	3.05
13	1.35	1.77	2.16	2.65	3.01
14	1.34	1.76	2.14	2.62	2.98
15	1.34	1.75	2.13	2.60	2.95
16	1.34	1.75	2.12	2.58	2.92
17	1.33	1.74	2.11	2.57	2.90
18	1.33	1.73	2.10	2.55	2.88
19	1.33	1.73	2.09	2.54	2.86
20	1.32	1.72	2.09	2.53	2.85
21	1.32	1.72	2.08	2.52	2.83
22	1.32	1.72	2.07	2.51	2.82
23	1.32	1.71	2.07	2.50	2.81
24	1.32	1.71	2.06	2.49	2.80
25	1.32	1.71	2.06	2.48	2.79
26	1.32	1.71	2.06	2.48	2.78
27	1.31	1.70	2.05	2.47	2.77
28	1.31	1.70	2.05	2.47	2.76
29	1.31	1.70	2.05	2.46	2.76
30	1.31	1.70	2.04	2.46	2.75
40	1.30	1.68	2.02	2.42	2.70
60	1.30	1.67	2.00	2.39	2.66
120	1.29	1.66	1.98	2.36	2.62
∞	1.28	1.64	1.96	2.33	2.58

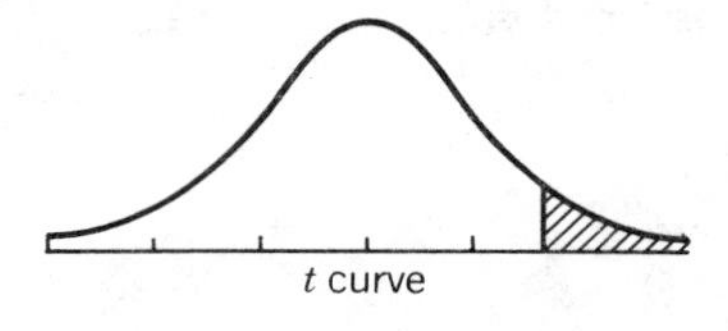

The illustration shows the t-curve for $\phi = 3$. The shaded area corresponds to the columnar headings of the table and the unshaded area to their complements.

It is of interest to note that t is related to the first column of the F-distribution (Table A.4), where $\phi_N = 1$. Setting $\phi_D = \phi$, F is in fact equal to t^2, providing the value of P in Table A.2 is doubled to make the comparison double-sided. Thus for $\phi = 8$ and $P = 2 \times 0.005$ the present table gives $t = 3.36$; in Table A.4 with $\phi_N = 1$, $\phi_D = 8$ and $P = 0.01$ we obtain $F = 11.3 = 3.36^2$.

Example

Single-Sided Test. For $\phi = 10$ the deviate of the t-curve which cuts off a single tail equivalent to $P = 0.05$ is given by $t = 1.81$. For the normal curve the corresponding value of u is 1.64.

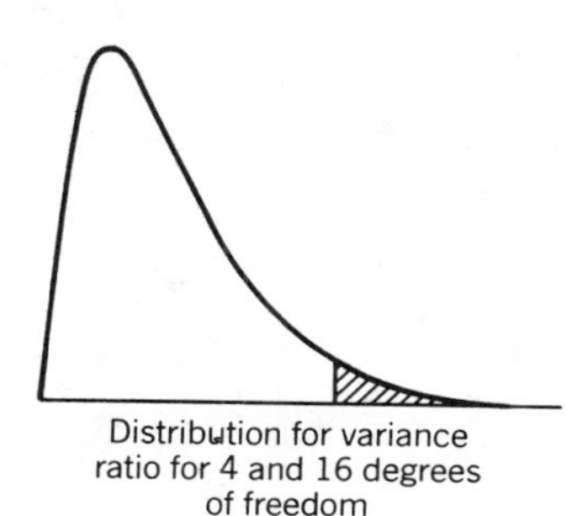

Distribution for variance ratio for 4 and 16 degrees of freedom

Table of the Variance Ratio (*F*-Distribution). The illustration shows the distribution of the variance ratio for 4 and 16 degrees of freedom. The shaded area, expressed as a proportion of the total area under the curve, is the argument in the first column of Table A.4.

The variance ratio is always calculated with the *larger* estimate of variance in the *numerator*, and ϕ_N and ϕ_D are the numbers of degrees of freedom in the numerator and denominator respectively.

Example

Let $F = 4.60$, $\phi_N = 5$, $\phi_D = 24$. The 5% and 1% points are 2.62 and 3.90, so the result is significant.

In calculating confidence limits for the variance ratio we require the upper and lower tail areas of the F-distribution. The levels actually tabled refer to the single upper tail area $F_a(\phi_N, \phi_D)$. However, the value

Table A.3 Probability Points of the χ^2 Distribution

ϕ	P: 0.995	0.99	0.975	0.95	0.90	0.75	0.50	0.25	0.10	0.05	0.025	0.01	0.005	0.001	ϕ [a]
1	—	—	—	—	.016	.102	.455	1.32	2.71	3.84	5.02	6.63	7.88	10.8	1
2	.010	.020	.051	.103	.211	.575	1.39	2.77	4.61	5.99	7.38	9.21	10.6	13.8	2
3	.072	.115	.216	.352	.584	1.21	2.37	4.11	6.25	7.81	9.35	11.3	12.8	16.3	3
4	.207	.297	.484	.711	1.06	1.92	3.36	5.39	7.78	9.49	11.1	13.3	14.9	18.5	4
5	.412	.554	.831	1.15	1.61	2.67	4.35	6.63	9.24	11.1	12.8	15.1	16.7	20.5	5
6	.676	.872	1.24	1.64	2.20	3.45	5.35	7.84	10.6	12.6	14.4	16.8	18.5	22.5	6
7	.989	1.24	1.69	2.17	2.83	4.25	6.35	9.04	12.0	14.1	16.0	18.5	20.3	24.3	7
8	1.34	1.65	2.18	2.73	3.49	5.07	7.34	10.2	13.4	15.5	17.5	20.1	22.0	26.1	8
9	1.73	2.09	2.70	3.33	4.17	5.90	8.34	11.4	14.7	16.9	19.0	21.7	23.6	27.9	9
10	2.16	2.56	3.25	3.94	4.87	6.74	9.34	12.5	16.0	18.3	20.5	23.2	25.2	29.6	10
11	2.60	3.05	3.82	4.57	5.58	7.58	10.3	23.7	17.3	19.7	21.9	24.7	26.8	31.3	11
12	3.07	3.57	4.40	5.23	6.30	8.44	11.3	14.8	18.5	21.0	23.3	26.2	28.3	32.9	12
13	3.57	4.11	5.01	5.89	7.04	9.30	12.3	16.0	19.8	22.4	24.7	27.7	29.8	34.5	13
14	4.07	4.66	5.63	6.57	7.79	10.2	13.3	17.1	21.1	23.7	26.1	29.1	31.3	36.1	14
15	4.60	5.23	6.26	7.26	8.55	11.0	14.3	18.2	22.3	25.0	27.5	30.6	32.8	37.7	15
16	5.14	5.81	6.91	7.96	9.31	11.9	15.3	19.4	23.5	26.3	28.8	32.0	34.3	39.3	16
17	5.70	6.41	7.56	8.67	10.1	12.8	16.3	20.5	24.8	27.6	30.2	33.4	35.7	40.8	17
18	6.26	7.01	8.23	9.39	10.9	13.7	17.3	21.6	26.0	28.9	31.5	34.8	37.2	42.3	18
19	6.84	7.63	8.91	10.1	11.7	14.6	18.3	22.7	27.2	30.1	32.9	36.2	38.6	43.8	19

20	7.43	8.26	9.59	10.9	12.4	15.5	19.3	23.8	28.4	31.4	34.2	37.6	40.0	45.3	20
21	8.03	8.90	10.3	11.6	13.2	16.3	20.3	24.9	29.6	32.7	35.5	38.9	41.4	46.8	21
22	8.64	9.54	11.0	12.3	14.0	17.2	21.3	26.0	30.8	33.9	36.8	40.3	42.8	48.3	22
23	9.26	10.2	11.7	13.1	14.8	18.1	22.3	27.1	32.0	35.2	38.1	41.6	44.2	49.7	23
24	9.89	10.9	12.4	13.8	15.7	19.0	23.3	28.2	33.2	36.4	39.4	43.0	45.6	51.2	24
25	10.5	11.5	13.1	14.6	16.5	19.9	24.3	29.3	34.4	37.7	40.6	44.3	46.9	52.6	25
26	11.2	12.2	13.8	15.4	17.3	20.8	25.3	30.4	35.6	38.9	41.9	45.6	48.3	54.1	26
27	11.8	12.9	14.6	16.2	18.1	21.7	26.3	31.5	36.7	40.1	43.2	47.0	49.6	55.5	27
28	12.5	13.6	15.3	16.9	18.9	22.7	27.3	32.6	37.9	41.3	44.5	48.3	51.0	56.9	28
29	13.1	14.3	16.0	17.7	19.8	23.6	28.3	33.7	39.1	42.6	45.7	49.6	52.3	58.3	29
30	13.8	15.0	16.8	18.5	20.6	24.5	29.3	34.8	40.3	43.8	47.0	50.9	53.7	59.7	30

[a] ϕ is the number of degrees of freedom.

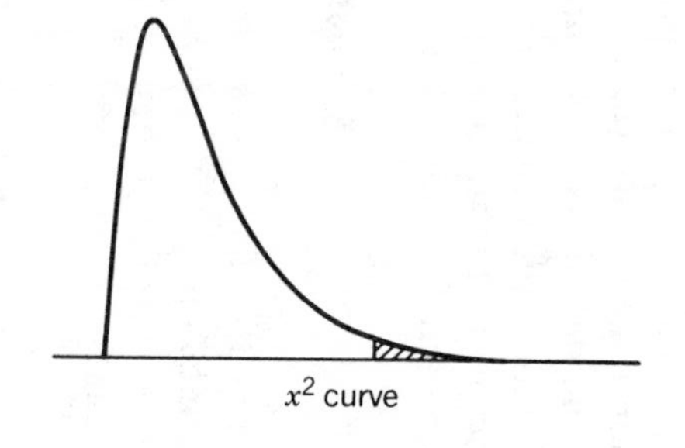

$F_{1-a}(\phi_N, \phi_D)$ (i.e., the value of F *below which* a proportion α of the whole curve lies) is given by $1/F_a(\phi_D, \phi_N)$.

Example

To obtain 90% confidence limits for the variance ratio we require the values of $F_{0.95}(\phi_N, \phi_D)$ and $F_{0.05}(\phi_N, \phi_D)$. If $\phi_N = 4$ and $\phi_D = 20$, then:

$$F_{0.95}(4,20) = \frac{1}{F_{0.05}(20,4)} = \frac{1}{5.80} = 0.172 \qquad \text{and} \qquad F_{0.05}(4,20) = 2.87.$$

The illustration shows the χ^2 curve for $\phi = 3$. The shaded portion, expressed as a proportion of the total area under the curve, is the columnar heading in the table.

It is of interest to note that χ^2 is related to the bottom row of the table of the F-distribution (Table A.4), where the number of degrees of freedom associated with the denominator is infinite. These values of F are in fact values of χ^2 divided by the number of degrees of freedom. Thus for $\phi = 8$ and $P = 0.01$ the present table gives $\chi^2 = 20.1$; in Table A.4 with $\phi_N = 8$ and $\phi_D = \infty$ we obtain $F = 2.51 = 20.1/8$.

Examples

Let $\chi^2 = 3.80$, $\phi = 3$. This is between the 0.50 and 0.25 points, and is therefore not significant.

Let $\chi^2 = 20.1$, $\phi = 9$. This is between the 0.025 and 0.01 points, and is therefore significant.

Values of χ^2 for $\phi > 30$. For values of $\phi > 30$ the expression $\sqrt{(2\chi^2)} - \sqrt{(2\phi - 1)}$ may be used as a normal deviate with unit variance, remembering that the probability for χ^2 corresponds to that of a single tail of the normal curve.

Example

Let $\chi^2 = 124.3$, $\phi = 100$. Then $u = \sqrt{248.6} - \sqrt{199} = 1.66$. For $u = 1.66$, the value of $P = 0.0485$. χ^2 is therefore just significant.

Table A.4 Probability Points on the Variance Ratio (*F*-Distribution)

Probability Point	ϕ_D	ϕ_N (corresponding to greater mean square) 1	2	3	4	5	6	7	8	9	10	12	15	20	24	30	40	60	120	∞	ϕ_D	Probability Point
0.100	1	39.9	49.5	53.6	55.8	57.2	58.2	58.9	59.4	59.9	60.2	60.7	61.2	61.7	62.0	62.3	62.5	62.8	63.1	63.3	1	0.100
0.050		161	199	216	225	230	234	237	239	241	242	244	246	248	249	250	251	252	253	254		0.050
0.025		648	800	864	900	922	937	948	957	963	969	977	985	993	997	1001	1006	1010	1014	1018		0.025
0.010		4052	4999	5403	5625	5764	5859	5928	5982	6022	6056	6106	6157	6209	6235	6261	6287	6313	6339	6366		0.010
0.100	2	8.53	9.00	9.16	9.24	9.29	9.33	9.35	9.37	9.38	9.39	9.41	9.42	9.44	9.45	9.46	9.47	9.47	9.48	9.49	2	0.100
0.050		18.5	19.0	19.2	19.2	19.3	19.3	19.4	19.4	19.4	19.4	19.4	19.4	19.4	19.5	19.5	19.5	19.5	19.5	19.5		0.050
0.025		38.5	39.0	39.2	39.2	39.3	39.3	39.4	39.4	39.4	39.4	39.4	39.4	39.4	39.5	39.5	39.5	39.5	39.5	39.5		0.025
0.010		98.5	99.0	99.2	99.2	99.3	99.3	99.4	99.4	99.4	99.4	99.4	99.4	99.4	99.5	99.5	99.5	99.5	99.5	99.5		0.010
0.100	3	5.54	5.46	5.39	5.34	5.31	5.28	5.27	5.25	5.24	5.23	5.22	5.20	5.18	5.18	5.17	5.16	5.15	5.14	5.13	3	0.100
0.050		10.1	9.55	9.28	9.12	9.01	8.94	8.89	8.85	8.81	8.79	8.74	8.70	8.66	8.64	8.62	8.59	8.57	8.55	8.53		0.050
0.025		17.4	16.0	15.4	15.1	14.9	14.7	14.6	14.5	14.5	14.4	14.3	14.3	14.2	14.1	14.1	14.0	14.0	13.9	13.9		0.025
0.010		34.1	30.8	29.5	28.7	28.2	27.9	27.7	27.5	27.3	27.2	27.1	26.9	26.7	26.6	26.5	26.4	26.3	26.2	26.1		0.010
0.100	4	4.54	4.32	4.19	4.11	4.05	4.01	3.98	3.95	3.94	3.92	3.90	3.87	3.84	3.83	3.82	3.80	3.79	3.78	3.76	4	0.100
0.050		7.71	6.94	6.59	6.39	6.26	6.16	6.09	6.04	6.00	5.96	5.91	5.86	5.80	5.77	5.75	5.72	5.69	5.66	5.63		0.050
0.025		12.2	10.6	10.0	9.60	9.36	9.20	9.07	8.98	8.90	8.84	8.75	8.66	8.56	8.51	8.46	8.41	8.36	8.31	8.26		0.025
0.010		21.2	18.0	16.7	16.0	15.5	15.2	15.0	14.8	14.7	14.5	14.4	14.2	14.0	13.9	13.8	13.7	13.7	13.6	13.5		0.010
0.100	5	4.06	3.78	3.62	3.52	3.45	3.40	3.37	3.34	3.32	3.30	3.27	3.24	3.21	3.19	3.17	3.16	3.14	3.12	3.10	5	0.100
0.050		6.61	5.79	5.41	5.19	5.05	4.95	4.88	4.82	4.77	4.74	4.68	4.62	4.56	4.53	4.50	4.46	4.43	4.40	4.36		0.050
0.025		10.0	8.43	7.76	7.39	7.15	6.98	6.85	6.76	6.68	6.62	6.52	6.43	6.33	6.28	6.23	6.18	6.12	6.07	6.02		0.025
0.010		16.3	13.3	12.1	11.4	11.0	10.7	10.5	10.3	10.2	10.1	9.89	9.72	9.55	9.47	9.38	9.29	9.20	9.11	9.02		0.010
0.100	6	3.78	3.46	3.29	3.18	3.11	3.05	3.01	2.98	2.96	2.94	2.90	2.87	2.84	2.82	2.80	2.78	2.76	2.74	2.72	6	0.100
0.050		5.99	5.14	4.76	4.53	4.39	4.28	4.21	4.15	4.10	4.06	4.00	3.94	3.87	3.84	3.81	3.77	3.74	3.70	3.67		0.050
0.025		8.81	7.26	6.60	6.23	5.99	5.82	5.70	5.60	5.52	5.46	5.37	5.27	5.17	5.12	5.07	5.01	4.96	4.90	4.85		0.025
0.010		13.7	10.9	9.78	9.15	8.75	8.47	8.26	8.10	7.98	7.87	7.72	7.56	7.40	7.31	7.23	7.14	7.06	6.97	6.88		0.010
0.100	7	3.59	3.26	3.07	2.96	2.88	2.83	2.78	2.75	2.72	2.70	2.67	2.63	2.59	2.58	2.56	2.54	2.51	2.49	2.47	7	0.100
0.050		5.59	4.74	4.35	4.12	3.97	3.87	3.79	3.73	3.68	3.64	3.57	3.51	3.44	3.41	3.38	3.34	3.30	3.27	3.23		0.050
0.025		8.07	6.54	5.89	5.52	5.29	5.12	4.99	4.90	4.82	4.76	4.67	4.57	4.47	4.42	4.36	4.31	4.25	4.20	4.14		0.025
0.010		12.2	9.55	8.45	7.85	7.46	7.19	6.99	6.84	6.72	6.62	6.47	6.31	6.16	6.07	5.99	5.91	5.82	5.74	5.65		0.010
0.100	8	3.46	3.11	2.92	2.81	2.73	2.67	2.62	2.59	2.56	2.54	2.50	2.46	2.42	2.40	2.38	2.36	2.34	2.32	2.29	8	0.100
0.050		5.32	4.46	4.07	3.84	3.69	3.58	3.50	3.44	3.39	3.35	3.28	3.22	3.15	3.12	3.08	3.04	3.01	2.97	2.93		0.050
0.025		7.57	6.06	5.42	5.05	4.82	4.65	4.53	4.43	4.36	4.30	4.20	4.10	4.00	3.95	3.89	3.84	3.78	3.73	3.67		0.025
0.010		11.3	8.65	7.59	7.01	6.63	6.37	6.18	6.03	5.91	5.81	5.67	5.52	5.36	5.28	5.20	5.12	5.03	4.95	4.86		0.010

Table A.4 (*Continued*)

Probability Point	ϕ_D	ϕ_N (corresponding to greater mean square) 1	2	3	4	5	6	7	8	9	10	12	15	20	24	30	40	60	120	∞	ϕ_D	Probability Point
0.100	9	3.36	3.01	2.81	2.69	2.61	2.55	2.51	2.47	2.44	2.42	2.38	2.34	2.30	2.28	2.25	2.23	2.21	2.18	2.16	9	0.100
0.050		5.12	4.26	3.86	3.63	3.48	3.37	3.29	3.23	3.18	3.14	3.07	3.01	2.94	2.90	2.86	2.83	2.79	2.75	2.71		0.050
0.025		7.21	5.71	5.08	4.72	4.48	4.32	4.20	4.10	4.03	3.96	3.87	3.77	3.67	3.61	3.56	3.51	3.45	3.39	3.33		0.025
0.010		10.6	8.02	6.99	6.42	6.06	5.80	5.61	5.47	5.35	5.26	5.11	4.96	4.81	4.73	4.65	4.57	4.48	4.40	4.31		0.010
0.100	10	3.28	2.92	2.73	2.61	2.52	2.46	2.41	2.38	2.35	2.32	2.28	2.24	2.20	2.18	2.16	2.13	2.11	2.08	2.06	10	0.100
0.050		4.96	4.10	3.71	3.48	3.33	3.22	3.14	3.07	3.02	2.98	2.91	2.84	2.77	2.74	2.70	2.66	2.62	2.58	2.54		0.050
0.025		6.94	5.46	4.83	4.47	4.24	4.07	3.95	3.85	3.78	3.72	3.62	3.52	3.42	3.37	3.31	3.26	3.20	3.14	3.08		0.025
0.010		10.0	7.56	6.55	5.99	5.64	5.39	5.20	5.06	4.94	4.85	4.71	4.56	4.41	4.33	4.25	4.17	4.08	4.00	3.91		0.010
0.100	12	3.18	2.81	2.61	2.48	2.39	2.33	2.28	2.24	2.21	2.19	2.15	2.10	2.06	2.04	2.01	1.99	1.96	1.93	1.90	12	0.100
0.050		4.75	3.89	3.49	3.26	3.11	3.00	2.91	2.85	2.80	2.75	2.69	2.62	2.54	2.51	2.47	2.43	2.38	2.34	2.30		0.050
0.025		6.55	5.10	4.47	4.12	3.89	3.73	3.61	3.51	3.44	3.37	3.28	3.18	3.07	3.02	2.96	2.91	2.85	2.79	2.72		0.025
0.010		9.33	6.93	5.95	5.41	5.06	4.82	4.64	4.50	4.39	4.30	4.16	4.01	3.86	3.78	3.70	3.62	3.54	3.45	3.36		0.010
0.100	15	3.07	2.70	2.49	2.36	2.27	2.21	2.16	2.12	2.09	2.06	2.02	1.97	1.92	1.90	1.87	1.85	1.82	1.79	1.76	15	0.100
0.050		4.54	3.68	3.29	3.06	2.90	2.79	2.71	2.64	2.59	2.54	2.48	2.40	2.33	2.29	2.25	2.20	2.16	2.11	2.07		0.050
0.025		6.20	4.77	4.15	3.80	3.58	3.41	3.29	3.20	3.12	3.06	2.96	2.86	2.76	2.70	2.64	2.59	2.52	2.46	2.40		0.025
0.010		8.68	6.36	5.42	4.89	4.56	4.32	4.14	4.00	3.89	3.80	3.67	3.52	3.37	3.29	3.21	3.13	3.05	2.96	2.87		0.010
0.100	20	2.97	2.59	2.38	2.25	2.16	2.09	2.04	2.00	1.96	1.94	1.89	1.84	1.79	1.77	1.74	1.71	1.68	1.64	1.61	20	0.100
0.050		4.35	3.49	3.10	2.87	2.71	2.60	2.51	2.45	2.39	2.35	2.28	2.20	2.12	2.08	2.04	1.99	1.95	1.90	1.84		0.050
0.025		5.87	4.46	3.86	3.51	3.29	3.13	3.01	2.91	2.84	2.77	2.68	2.57	2.46	2.41	2.35	2.29	2.22	2.16	2.09		0.025
0.010		8.10	5.85	4.94	4.43	4.10	3.87	3.70	3.56	3.46	3.37	3.23	3.09	2.94	2.86	2.78	2.69	2.61	2.52	2.42		0.010

0.100	24	2.93	2.54	2.33	2.19	2.10	2.04	1.98	1.94	1.91	1.88	1.83	1.78	1.73	1.70	1.67	1.64	1.61	1.57	1.53	24	0.100
0.050		4.26	3.40	3.01	2.78	2.62	2.51	2.42	2.36	2.30	2.25	2.18	2.11	2.03	1.98	1.94	1.89	1.84	1.79	1.73		0.050
0.025		5.72	4.32	3.72	3.38	3.15	2.99	2.87	2.78	2.70	2.64	2.54	2.44	2.33	2.27	2.21	2.15	2.08	2.01	1.94		0.025
0.010		7.82	5.61	4.72	4.22	3.90	3.67	3.50	3.36	3.26	3.17	3.03	2.89	2.74	2.66	2.58	2.49	2.40	2.31	2.21		0.010
0.100	30	2.88	2.49	2.28	2.14	2.05	1.98	1.93	1.88	1.85	1.82	1.77	1.72	1.67	1.64	1.61	1.57	1.54	1.50	1.46	30	0.100
0.050		4.17	3.32	2.92	2.69	2.53	2.42	2.33	2.27	2.21	2.16	2.09	2.01	1.93	1.89	1.84	1.79	1.74	1.68	1.62		0.050
0.025		5.57	4.18	3.59	3.25	3.03	2.87	2.75	2.65	2.57	2.51	2.41	2.31	2.20	2.14	2.07	2.01	1.94	1.87	1.79		0.025
0.010		7.56	5.39	4.51	4.02	3.70	3.47	3.30	3.17	3.07	2.98	2.84	2.70	2.55	2.47	2.39	2.30	2.21	2.11	2.01		0.010
0.100	40	2.84	2.44	2.23	2.09	2.00	1.93	1.87	1.83	1.79	1.76	1.71	1.66	1.61	1.57	1.54	1.51	1.47	1.42	1.38	40	0.100
0.050		4.08	3.23	2.84	2.61	2.45	2.34	2.25	2.18	2.12	2.08	2.00	1.92	1.84	1.79	1.74	1.69	1.64	1.58	1.51		0.050
0.025		5.42	4.05	3.46	3.13	2.90	2.74	2.62	2.53	2.45	2.39	2.29	2.18	2.07	2.01	1.94	1.88	1.80	1.72	1.64		0.025
0.010		7.31	5.18	4.31	3.83	3.51	3.29	3.12	2.99	2.89	2.80	2.66	2.52	2.37	2.29	2.20	2.11	2.02	1.92	1.80		0.010
0.100	60	2.79	2.39	2.18	2.04	1.95	1.87	1.82	1.77	1.74	1.71	1.66	1.60	1.54	1.51	1.48	1.44	1.40	1.35	1.29	60	0.100
0.050		4.00	3.15	2.76	2.53	2.37	2.25	2.17	2.10	2.04	1.99	1.92	1.84	1.75	1.70	1.65	1.59	1.53	1.47	1.39		0.050
0.025		5.29	3.93	3.34	3.01	2.79	2.63	2.51	2.41	2.33	2.27	2.17	2.06	1.94	1.88	1.82	1.74	1.67	1.58	1.48		0.025
0.010		7.08	4.98	4.13	3.65	3.34	3.12	2.95	2.82	2.72	2.63	2.50	2.35	2.20	2.12	2.03	1.94	1.84	1.73	1.60		0.010
0.100	120	2.75	2.35	2.13	1.99	1.90	1.28	1.77	1.72	1.68	1.65	1.60	1.54	1.48	1.45	1.41	1.37	1.32	1.26	1.19	120	0.100
0.050		3.92	3.07	2.68	2.45	2.29	2.18	2.09	2.02	1.96	1.91	1.83	1.75	1.66	1.61	1.55	1.50	1.43	1.35	1.25		0.050
0.025		5.15	3.80	3.23	2.89	2.67	2.52	2.39	2.30	2.22	2.16	2.05	1.94	1.82	1.76	1.69	1.61	1.53	1.43	1.31		0.025
0.010		6.85	4.79	3.95	3.48	3.17	2.96	2.79	2.66	2.56	2.47	2.34	2.19	2.03	1.95	1.86	1.76	1.66	1.53	1.38		0.010
0.100	∞	2.71	2.30	2.08	1.94	1.85	1.77	1.72	1.67	1.63	1.60	1.55	1.49	1.42	1.38	1.34	1.30	1.24	1.17	1.00	∞	0.100
0.050		3.84	3.00	2.60	2.37	2.21	2.10	2.01	1.94	1.88	1.83	1.75	1.67	1.57	1.52	1.46	1.39	1.32	1.22	1.00		0.050
0.025		5.02	3.69	3.12	2.79	2.57	2.41	2.29	2.19	2.11	2.05	1.94	1.83	1.71	1.64	1.57	1.48	1.39	1.27	1.00		0.025
0.010		6.63	4.61	3.78	3.32	3.02	2.80	2.64	2.51	2.41	2.32	2.18	2.04	1.88	1.79	1.70	1.59	1.47	1.32	1.00		0.010

Table A.5 Wilcoxon (one-input) Test: Upper Tail Probabilities for the Null Distribution

	n						
x	3	4	5	6	7	8	9
3	.625						
4	.375						
5	.250	.562					
6	.125	.438					
7		.312					
8		.188	.500				
9		.125	.406				
10		.062	.312				
11			.219	.500			
12			.156	.422			
13			.094	.344			
14			.062	.281	.531		
15			.031	.219	.469		
16				.156	.406		
17				.109	.344		
18				.078	.289	.527	
19				.047	.234	.473	
20				.031	.188	.422	
21				.016	.148	.371	
22					.109	.320	
23					.078	.273	.500
24					.055	.230	.455
25					.039	.191	.410
26					.023	.156	.367
27					.016	.125	.326
28					.008	.098	.285
29						.074	.248
30						.055	.213
31						.039	.180
32						.027	.150
33						.020	.125
34						.012	.102
35						.008	.082
36						.004	.064
37							.049
38							.037
39							.027
40							.020
41							.014
42							.010
43							.006
44							.004
45							.002

Table A.5 (*Continued*)

	n					
x	10	11	12	13	14	15
28	.500					
29	.461					
30	.423					
31	.385					
32	.348					
33	.312	.517				
34	.278	.483				
35	.246	.449				
36	.216	.416				
37	.188	.382				
38	.161	.350				
39	.138	.319	.515			
40	.116	.289	.485			
41	.097	.260	.455			
42	.080	.232	.425			
43	.065	.207	.396			
44	.053	.183	.367			
45	.042	.160	.339			
46	.032	.139	.311	.500		
47	.024	.120	.285	.473		
48	.019	.103	.259	.446		
49	.014	.087	.235	.420		
50	.010	.074	.212	.393		
51	.007	.062	.190	.368		
52	.005	.051	.170	.342		
53	.003	.042	.151	.318	.500	
54	.002	.034	.133	.294	.476	
55	.001	.027	.117	.271	.452	
56		.021	.102	.249	.428	
57		.016	.088	.227	.404	
58		.012	.076	.207	.380	
59		.009	.065	.188	.357	
60		.007	.055	.170	.335	.511
61		.005	.046	.153	.313	.489
62		.003	.039	.137	.292	.467
63		.002	.032	.122	.271	.445
64		.001	.026	.108	.251	.423
65		.001	.021	.095	.232	.402
66		.000	.017	.084	.213	.381
67			.013	.073	.196	.360
68			.010	.064	.179	.339
69			.008	.055	.163	.319
70			.006	.047	.148	.300
71			.005	.040	.134	.281
72			.003	.034	.121	.262
73			.002	.029	.108	.244
74			.002	.024	.097	.227
75			.001	.020	.086	.211
76			.001	.016	.077	.195
77			.000	.013	.068	.180

Table A.5 (*Continued*)

	n					
x	10	11	12	13	14	15
78			.000	.011	.059	.165
79				.009	.052	.151
80				.007	.045	.138
81				.005	.039	.126
82				.004	.034	.115
83				.003	.029	.104
84				.002	.025	.094
85				.002	.021	.084
86				.001	.018	.076
87				.001	.015	.068
88				.001	.012	.060
89				.000	.010	.053
90				.000	.008	.047
91				.000	.007	.042
92					.005	.036
93					.004	.032
94					.003	.028
95					.003	.024
96					.002	.021
97					.002	.018
98					.001	.015
99					.001	.013
100					.001	.011
101					.000	.009
102					.000	.008
103					.000	.006
104					.000	.005
105					.000	.004
106						.003
107						.003
108						.002
109						.002
110						.001
111						.001
112						.001
113						.001
114						.000
115						.000
116						.000
117						.000
118						.000
119						.000
120						.000

Source: Adapted from *A Nonparametric Introduction to Statistics*, by C. H. Kraft and C. van Eeden, Copyright © 1968, The Macmillan Company. Published by permission.

INDEX

Analysis of variance, 28, 32, 63
Anodic stripping voltammetry, 171
Anova, 28, 32, 43, 46, 167
  for linear models, 50
  one-way, 32
  two-way, 37, 40
ARTHUR, results for ARCH data interpreted, 257
Artificial intelligence, 183
Atomic absorption, 64
  flame, 30
  graphite furnace, 30
Atomic flame emission spectrometry, 307
Autocorrelation function, 237
Autoscaling, 258

Background, 66
Bayes:
  classification rule, 256
  decision rule, 235
Bias, 13
Biller-Biemann technique, 169
Binomial distribution, 12
Blank, 36, 66, 171
Block, randomized, 36
Box, George, 306
Boxcar:
  averaging, 102
  integration, 87

Calibration, 48, 71, 97, 119, 146
  confidence interval for, 132
  error propagation, 134
  estimating detection limits, 128
  with external standards, 122
  heteroscedastic data, 127
  linear, 122
  linear dynamic range in, 122
  linear segments in, 132
  model, 123
  non-linear, 132, 146
  regression band in, 124
  residuals, 125
  utilizing for chemical analysis, 126
Catalysts, promoters, 44
Category:
  continuous, 182
  discrete, 182
Category data, 182
χ^2-distribution, 313
Chromatography, 109, 114, 149
  Biller-Biemann method, 169
  gas, 54, 65
  liquid, 55
  resolution, 54
    in liquid, 161
  window diagrams for optimization, 311
Classification analysis, 228, 267
  binary, 246
  SIMCA, 246
Clinical chemistry, pattern recognition in, 203
Cluster analysis, 219, 258
  centroid method, 221
  complete link method, 221
  dendrogram, 220
  distance measures, 220
  minimal spanning tree, 227
  Q-mode, 220, 265
  R-mode, 227, 228, 263
  single link method, 221
Colinearity problem, 285
Communality, 209
Condition number, 141
Confidence interval, 130, 259
Continuous property, 182
  data, 182
Control:
  PID, 297, 301
  steady state offset in, 303
Control algorithms, 297
  Fibonacci search, 299
  internal model control, 310

Control algorithms (*Continued*)
model algorithmic, 310
model based control (MBC), 305
Convolution, 108
Correlation, interfeature, 259
Correlation matrix, 156, 199
Covariance, 181
Cross-validation, 242, 254
Curve fitting, 97, 113
area under peak in, 100
conjugate gradient, 116
estimating peak parameters, 98
Fletcher and Powell, 116
Marquardt, 116
multiple, 116
non-linear functions, 97
non-linear least squares, 116
Curve resolution, using factor analysis, 170
Cytochrome *c* protein, 223

Data reduction, 202
Deconvolution, 108
by curve fitting, 159
of overlapping signals, 159
Dendrogram, 220, 224
Detection, sequential, 112
Detection efficiency, sine and square waves, 113
Detection limit, 71, 112
in calibration, 128
Kaiser, 78
as minimum signal-to-noise, 81
point estimation, 76
by *t*-tests, 79
precision, 85
Detector, multichannel, 90
Differentiating function, 110
Dimensionality, 155
Display methods, 263
mappings, 217
non-linear, 217
projection, 216
variable *x* variable plots, 216
Distance, in SIMCA, 245
Distance measures:
city block, 220
Euclidian, 220
similarity, 219
Distribution, 21
binomial, 12
χ^2, 11, 313
F-, 313
Gaussian, 5
mean of, 5
of means, 8
normal, 5, 313
Poisson, 67
probability, 1
student's t, 10
t-, 313
variance of, 5
Wilcoxon test, 84

Eigenanalysis, of rainwater data, 206
Eigenvalue, 158, 199
Eigenvector, 158, 239, 259
plot, 217
projection, 216, 252
Emission spectroscopy, 132
Evaluation set, 228
EVOP, *see* Evolutionary operation (EVOP), 306
Experimental design, 23, 63, 135
condition in, 121
factors, 23
in GSAM, 142
level, 24
treatment, 23
Exploratory data analysis, 179, 258

Factors, 23
classification, 23
level of, 23
Factor analysis, 155, 158, 172, 214
non-linear, 214
Factorial design, 24, 44
two-level, 56
False detection, 73
Fast Fourier transform, 95
F-distribution, 313
Feature selection, 239, 280
Feature weighting, 195
Fibonacci search, 299
Filtering, 88, 285
Fisher weight, 195, 267
Flame spectrometry, 113
Flow injection analysis, 171
Fourier transform, 113
discrete, 95
fast, 95

Fourier transform spectroscopy, 92
F-test, 29, 30, 33, 63

Gaussian signals, 71
Generalized inverse, 54, 136
Generalized standard addition method (GSAM), 63, 139, 146
  error amplification in, 142
  incremental difference calculation, 144
  partition method, 145
  total difference calculation, 142
Global maximum, 59
Gravimetric analysis, 121
Grid search, 163

Hadamard transform spectroscopy, 91, 113
Hessian matrix, 165
Hierarchical cluster analysis, 219
Homoscedastic, 49

Inflection points, 160
Information theory, 112
Integration:
  numerical, 101, 114
  Simpson rule, 101
  trapezoidal rule, 101
Interference effect, 63, 141
Internal model control, 310
Inverse regression, 126
Ion selective electrode, 64

Kaiser, 78
Kaiser detection limit, 78
K-Matrix, 137
K-nearest neighbor method (KNN), 234, 237, 252
KNN, *see* K-nearest neighbor Method (KNN)

Latent variables, 281
Latin squares, 41
Lawton and Sylvestre, 170
Least squares, 49, 145
  multilinear, 137
  nonlinear, 99, 214
  non-negative, 168
  unweighted, 49, 123
  weighted, 49
Leave-one-out procedure, 228
Linear discriminant analysis, 237
Linear discriminant function, 229
Linear dynamic range, 122
Linear learning machine (LLM), 229
Loadings, 202, 259
Lock-in amplifiers, 113

Mann-Whitney test, 112
Mappings, 217
Mass spectrometry, 149
Matrix:
  algebraic operations, 63
  condition number of, 141
  covariance, 203
  diagonal, 199
  generalized inverse, 54
  Hessian, 165
  misclassification, 267
Matrix effect, 132, 141
Minimal spanning tree, 227
Missing data, 258
  methods to treat, 188
Model:
  errors in x and y, 131
  linear, 48
  linearizing the, 98
  principal component, 242
  quadratic, 52
  signal detection, 74
  SIMCA, 242
  test for curvature of linear, 58
  unweighted linear in 2 variables, 52
  unweighted nonlinear in 2 variables, 53
Model algorithmic control, 310
Model based control (MBC), 305
Modeling, 46, 64
Modeling power, in SIMCA, 242, 245
Modified simplex method, 306
Moving window averaging, 103
Multicomponent analysis, 63, 135, 139, 146
Multidimensional scaling, 214
Multiplex spectroscopy, 89

Neutron activation analysis, 249
Newton-Raphson method, 164
NLM, *see* Non-linear mapping (NLM)
NMR spectra, 235
Noise, 66
  flame spectrometry, 88
  flicker, 88

Noise (*Continued*)
  random, 67
  types, 89
Nonlinear mapping (NLM), 217, 225–226, 265
Nonparametric statistical procedures, 112
Normal distribution, 313
Normalization, in preprocessing, 190

Optimization, 23, 55, 297, 305
  to increase S/N, 86
  method of steepest ascent, 56
  window diagrams in chromatography, 311
Outliers, 242, 258

Paired observations, 34
Parametric mapping, 214
Partial least squares path modeling (PLS), 281
Pattern recognition:
  application:
    to archeological data set (ARCH), 257
    to construction of phylogenetic trees, 223
    to oil spill analysis, 248
    to rainwater study, 206
    to structure elucidation from NMR spectra, 235
  applications, 184
  approach, 183
  Bayles classification rule, 256
  in clinical chemistry, 203
  display, 187
  eigenvector rotation, 198
  feature selection, 239
  feature weighting, 195
  forensic application, 196
  intrinsic dimensionality, 202
  K-nearest neighbor (KNN), 234
  levels of, 246
  linear learning machine, 229
  missing data, 188
  outliers, 242
  paper analysis, 196
  parametric and nonparametric methods, 256
  preprocessing, 186
  scaling, 191
  SIMCA, 242
  software, 257
  structure-activity studies, 248
  supervised learning, 187, 228
  translation, 189
  unsupervised learning, 187, 219
  Varimax rotation, 206
Phylogenetic tree, 223
Poisson distribution, 67
Polarography, 47
Population, 7
  normally distributed, 28
  parameters, 7
Potentiometry, 121
Preprocessing, 188, 235
PRESS (predictive residual error sum of squares), 255
Principal Component Analysis, 63, 155, 259
Probability, 26, 256
  density function, 3, 68, 256
  distributions, 1
  objective, 1
  relative, 1
  signal distribution, 73
  subjective, 1
Projection:
  eigenvector, 216
  linear, 216
  nonlinear, 214
Propagation of error, 21, 134, 146

Rank annihilation, 168, 174
Regression:
  confidence interval in, 130, 146
  intersection of two lines, 129
  inverse, 126
  matrix solution, 53
  PLS, 285
Residuals, 125, 145
  PRESS (predictive residual error sum of squares), 255
  SIMCA, 244, 267
Resolution, 149, 172
  chromatography, 55
Response, 47, 65
  volume-corrected, 140
Response function, 121
  nonlinear, 121
Response surfaces, 54, 298
  known shapes, 55
  unknown shape, 56
Risk, in Bayes classification, 256

Rotation matrix, 216
Rotation of coordinate axes, 197

Samples, 7
  parameters, 7
Sampling, 15, 21
  error, 15
  variance, 17
Satellite pattern, 151
Savitzky and Golay, 108
Scaling, 191
  autoscaling, 193
  range, 193
Scores, 202
Select, 240, 280
Sensitivity, 134, 137, 139, 152
Sensitivity factor, 134
Sensors, 137, 142
Sequential detection, 73, 112
Signal, 65
  area under, 100
  background, 66
  differentiation, 109, 150
  derivative of, 110
  false detection, 74
  Gaussian, 71, 152
  Lorentzian, 152
  no-decision range, 72
  resolution of, 149
  resolution using regression, 166
  threshold, 74
Signal averaging, 86
  boxcar, 102
  moving window, 103
Signal detection, 65, 68, 112
  limit of guaranteed, 70
Signal domain, 71
Signal filtering, 88
Signal manipulation, 96
  curve fitting in, 97
  deconvolution, 97
  differentiation in, 109
  smoothing in, 102
Signal processing, *see* Signal manipulation
Signal-to-noise:
  enhancement, 139, 285
  methods to increase, 86
  ratio, 80, 83, 86, 113
Signed rank, 85
SIMCA, 242
  analysis of oil spills, 248
  ARCH data, 267
  class envelopes, 244
  classification, 246
  cross-validation in, 242
  distance between categories, 245
  discriminating power, 245
  levels of, 246
  misclassification matrix, 267
  modeling power, 245
  residuals in, 244
Simplex, 59, 307
Simplex method:
  modified, 306
  simple, 59
Simplex optimization, 64, 306
Simpson rule, 101
Simpson (or Newton) three-eights rule, 101
Smoothing, 102, 114
  formulas, 106
  Fourier transform, 108
  least-squares polynomial, 104
  multiple, 106
  Savitzky and Golay, 108
S/N, *see* Signal-to-noise ratio
Spectrometry, optimization in, 307
Spectroscopy:
  derivative, 114
  Fourier transform, 91
  Hadamard, 91
  Multiplex, 89
  pattern recognition in, 229
Standard addition, 140
Standard additional method (SAM), 133
Standard normal variate, 6
Steady state offset, 303
Steepest descent, method of, 164
Structure activity studies, 248
Student's *t*, 259
  distribution, 10, 35
Supervised learning, 228, 267
  leave-one-out procedures, 228
  misclassification matrix, 267

Taylor Series linearization, 160
t-Distribution, 313
Test set, 228

Training set, 228
Transforms:
  Fourier, 91
  Hadamard, 91
t-Test, 82

Univariate statistics, 259
Unsupervised learning, 219

Variance, pooled, 29
Variance weight, 195, 249, 267
Varimax rotation, 206, 259
Varivector, 210
Voigt profile, 159

Weight:
  in Bayes classification, 256
  Fisher, 267
  pair-wise, 267
  Variance, 267
Wilcoxon test, 84, 112, 313
  signed rank values, 85
Wold, Herman, 281

X-ray emission, 64
X-ray fluorescence, 132, 258